Statistical Physics for Cosmic Structures

A. Gabrielli   F. Sylos Labini
M. Joyce   L. Pietronero

Statistical Physics for Cosmic Structures

With 119 Figures

Dr. Andrea Gabrielli
Statistical Mechanics and Complexity Center - INFM
c/o Dipartimento di Fisica
Università "La Sapienza"
Piazzale A. Moro, 2
00185 Roma, Italy

Dr. Francesco Sylos Labini
Laboratoire de Physique Théorique
Bâtiment 210
Université de Paris XI
91405 Orsay Cedex, France

Professor Michael Joyce
Laboratoire de Physique Nucléaire et des Hautes Energies
Université de Paris VI
4, Place Jussieu, Tour 33
75252 Paris Cedex 05, France

Professor Luciano Pietronero
Dipartimento di Fisica
Università "La Sapienza"
Piazzale A. Moro, 2
00185 Roma, Italy

ISBN 3-540-40745-6 Springer Berlin Heidelberg New York

Library of Congress Control Number: 2004110247

This work is subject to copyright. All rights are reserved, whether the whole or part of the material is concerned, specifically the rights of translation, reprinting, reuse of illustrations, recitation, broadcasting, reproduction on microfilm or in any other way, and storage in data banks. Duplication of this publication or parts thereof is permitted only under the provisions of the German Copyright Law of September 9, 1965, in its current version, and permission for use must always be obtained from Springer. Violations are liable to prosecution under the German Copyright Law.

Springer is a part of Springer Science+Business Media

springeronline.com

© Springer-Verlag Berlin Heidelberg 2005
Printed in Germany

The use of general descriptive names, registered names, trademarks, etc. in this publication does not imply, even in the absence of a specific statement, that such names are exempt from the relevant protective laws and regulations and therefore free for general use.

Typesetting: Data prepared by the authors using a Springer TeX macro package
Cover design: *design & production* GmbH, Heidelberg

Printed on acid-free paper   57/3141/ts   5 4 3 2 1 0

Preface

This book has its roots in a series of collaborations in the last decade at the interface between statistical physics and cosmology. The specific problem which initiated this research was the study of the clustering properties of galaxies as revealed by large redshift surveys, a context in which concepts of modern statistical physics (e.g. scale-invariance, fractality..) find ready application. In recent years we have considerably broadened the range of problems in cosmology which we have addressed, treating in particular more theoretical issues about the statistical properties of standard cosmological models. What is common to all this research, however, is that it is informed by a perspective and methodology which is that of statistical physics. We can say that, beyond its specific scientific content, this book has an underlying thesis: such interdisciplinary research is an exciting playground for statistical physics, and one which can bring new and useful insights into cosmology. The book does not represent a final point, but in our view, a marker in the development of this kind of research, which we believe can go very much further in the future. Indeed as we complete this book, new developments - which unfortunately we have not been able to include here - have been made on some of the themes described here. Our focus in this book is on the problem of structure in cosmology. Our aim is to elucidate what the principal concepts of modern statistical physics are in this context, and then to illustrate their application to specific aspects of this problem.

At the outset we wish to acknowledge our debt to the numerous colleagues and friends - both statistical physicists and cosmologists - with whom parts of the work presented here have been done. We hope we have given the proper credit to each of them by citing the relevant papers at the appropriate points throughout the book. The responsibility for what is written here is, however, entirely ours.

We acknowledge in particular the contribution of Marco Montuori, with whom we have collaborated extensively on the analysis of galaxy clustering data. On this subject we have also benefitted from collaboration with Hélène di Nella-Courtois. Ruth Durrer has been involved in several projects which have provided the basis for the presentation given here: on the problems of biasing and of angular projection of fractals. She has also contributed to our work on the statistical properties of density fields in standard cosmological models through many very useful discussions. We thank Thierry Baertschiger for collaboration on the problem of gravitational clustering and related issues which has stimulated the work on discretisation and initial conditions which we describe in Chap. 7. We are especially indebted to him for the many simulations his has kindly provided us: most of the numerical results presented are based on his computer codes. Bruno Marcos has kindly provided us with some recent results on superhomogeneous distributions, a subject on which we have benefitted from collaboration with him.

We thank Yuri V. Baryshev and Pekka Teerikorpi for the the many studies performed together on galaxy structures, and for many stimulating discussions over the years. Georges Paturel has been for many years our advisor on problems related to the studies of galaxy samples.

We have had the pleasure to work on several issues presented here in collaboration with Philip W. Anderson, Jean-Pierre Eckmann, Bernard Jancovici and Joel L. Lebowitz.

We also thank for discussions, comments and suggestions: Luca Amendola, Daniel Amit, Tibor Antal, Robin Ball, Maurizio Bottaccio, Paul Coleman, Marc Davis, Paolo De Bernardis, George Djorgovski, Roberto Capuzzo-Dolcetta, Hector de Vega, Daniel Eiseinstein, Pedro G. Ferreira, Petar V. Grujic, Leo Kadanoff, David Hogg, Dominique Levesque, Benoit Mandelbrot, Adrian Melott, Miguel A. Muñoz, Giorgio Parisi, Jim P.E. Peebles, Daniel Pfenniger, Itamar Procaccia, Zoltan Rácz, Norma Sanchez, Bernard Sapoval, William C. Saslaw, Daniele Steer, Ann Pier Siebesma, Salvatore Torquato, Jean-Philippe Uzan, Filippo Vernizzi, Alessandro Vespignani and Pascal Viot. Finally we acknowledge suggestions and comments from Adolfo Paolo Masucci, Thierry Sousbi and Nickolay Vasilyev.

The institutions which have provided financial support for our research, and which we thank, are: the European Community (TMR network "Fractal structures and self-organization" ERB4061PL970910 and Marie-Curie Fellowship HPMF-CT-2001-01443), the "Istituto Nazionale di Fisica della Materia" (INFM, Sezione di Roma 1), the "Centro Studi e Ricerche *Enrico Fermi*" (Rome, Italy), the Swiss National Science Foundation (Bourse pour chercheurs avancés no. 8220-64668), the Physics Department of the University of Rome "La Sapienza", the Department of Theoretical Physics of the University of Geneva (Geneva, Switzerland), the Laboratoire de Physique Théorique at the University of Paris XI (Orsay, France), the Laboratoire de Physique Nucléaire et de Hautes Energies at the University of Paris VI (Paris, France) and the Observatory of Paris (Lerma) (Paris, France).

Rome (Italy) *Andrea Gabrielli*
Rome (Italy) *Francesco Sylos Labini*
September 2004 *Michael Joyce*
Luciano Pietronero

Contents

1 Introduction 1
   1.1 Motivations and Purpose of the Book 1
   1.2 Structures in Statistical Physics: A New Perspective 2
   1.3 Structures in Statistical Physics: The Methods 8
   1.4 Applications to Cosmology 11
   1.5 Perspectives for the Future 22

Part I Statistical Methods

2 Uniform and Correlated Mass Density Fields 27
   2.1 Introduction 27
   2.2 Basic Statistical Properties and Concepts.................. 31
        2.2.1 Spatial Averages and Ergodicity..................... 34
        2.2.2 Homogeneity and Homogeneity Scale 34
   2.3 Correlation Functions 35
        2.3.1 Characteristic Function and Cumulants Expansion.... 36
        2.3.2 Correlation Length 39
        2.3.3 Other Properties of the Reduced Two-Point
               Correlation Function............................. 40
        2.3.4 Mass Variance 41
   2.4 Poisson Point Process 44
   2.5 Stochastic Point Processes with Spatial Correlations 46
        2.5.1 Conditional Properties 48
        2.5.2 Integrated Conditional Properties 50
        2.5.3 Detection of the Homogeneity Scale of a Discrete SPP 50
   2.6 Nearest Neighbor Probability Density
        in Point Processes 52
        2.6.1 Poisson Case..................................... 52
        2.6.2 Particle Distributions with Spatial Correlations 54
   2.7 Gaussian Continuous Stochastic Fields 55
   2.8 Power-Laws and Self-Similarity........................... 58
   2.9 Mass Function and Probability Distribution................. 61
   2.10 The Random Walk and the Central Limit Theorem 64

	2.10.1 Probability Distribution of Mass Fluctuations in Large Volumes..................................	68

2.11 Gaussian Distribution as the Most Probable Probability Distribution 69
2.12 Summary and Discussion 71

3 The Power Spectrum and the Classification of Stationary Stochastic Fields........................... 73
3.1 Introduction 73
3.2 General Properties....................................... 73
    3.2.1 Mathematical Definitions.......................... 73
    3.2.2 Limit Conditions 76
3.3 The Power Spectrum for the Poisson Point Process and Other SPP 77
3.4 The Power Spectrum and the Mass Variance: A Complete Classification 78
    3.4.1 The Complete Classification of Mass Fluctuations versus Power Spectrum 83
3.5 Super-Homogeneous Mass Density Fields 84
    3.5.1 The Lattice Particle Distribution 85
    3.5.2 The One Component Plasma 88
3.6 Further Analysis of Gaussian Fields....................... 91
    3.6.1 Real Space Composition of Gaussian Fields, Correlation Length and Size of Structures 95
3.7 Summary and Discussion 96

4 Fractals... 101
4.1 Introduction 101
4.2 The Metric Dimension 102
4.3 Conditional Density 107
    4.3.1 Conditional Density and Smooth Radial Particle Distributions 109
    4.3.2 Statistically Homogeneous and Isotropic Distribution of Radial Density Profiles 113
    4.3.3 Nearest Neighbor Probability Density for Radial and Fractal Point-Particle Distributions 113
4.4 The Two-Point Conditional Density 116
4.5 The Conditional Variance in Spheres...................... 118
4.6 Corrections to Scaling.................................... 119
    4.6.1 Correction to Scaling: Deterministic Fractals 120
    4.6.2 Correction to Scaling: Random Fractals 124
4.7 Fractal with a Crossover to Homogeneity 127
4.8 Correlation, Fractals and Clustering 127
4.9 Probability Distribution of Mass Fluctuations in a Fractal 130

	4.10 Intersection of Fractals 132
	4.11 Morphology and Voids 134
	4.12 Angular and Orthogonal Projection of Fractal Sets 134
	4.12.1 On the Uniformity of the Angular Projection 137
	4.13 Summary and Discussion 141

5 Multifractals and Mass Distributions 143
   5.1 Introduction 143
   5.2 Basic Definitions 144
   5.3 Deterministic Multifractals 145
   5.4 The Multifractal Spectrum 149
   5.5 Random Multifractals 151
   5.6 Self-Similarity of Fluctuations and Multifractality
       in Temporal Multiplicative Processes....................... 154
   5.7 Spatial Correlation in Multifractals 158
   5.8 Multifractals and "Mass" Distributions..................... 159
   5.9 Summary and Discussion 161

Part II Applications to Cosmology

6 Fluctuations in Standard Cosmological Models: A Real Space View 167
   6.1 Introduction 167
   6.2 Basic Properties of Cosmological Density Fields 167
   6.3 The Cosmological Origin of the HZ Spectrum 171
   6.4 The Real Space Correlation Function
       of CDM/HDM Models 173
   6.5 $P(0) = 0$ and Constraints in a Finite Sample................ 177
   6.6 CMBR Anisotropies in Direct Space 179
       6.6.1 CMBR Anisotropies and the Matter Power Spectrum . 180
       6.6.2 The Origin of Oscillations in the Power Spectrum 183
       6.6.3 A Simple Example of k-Oscillations................. 184
       6.6.4 Oscillations in the CDM PS 185
       6.6.5 Oscillations in the CMBR Anisotropies.............. 187
   6.7 Summary and Discussion 189

7 Discrete Representation of Fluctuations in Cosmological Models...................................... 193
   7.1 Introduction 193
   7.2 Discrete versus Continuous Density Fields 194
   7.3 Super-Homogeneous Systems in Statistical Physics.......... 196
   7.4 HZ as Equilibrium of a Modified OCP 197
   7.5 A First Approximation to the Effect of Displacement Fields.. 199
   7.6 Displacement Fields: Formulation of the Problem 200

		7.7	Effects of Displacements on One and Two-Point Properties of the Particle Distribution 203
			7.7.1 Uncorrelated Displacements 206
			7.7.2 Asymptotic Behavior of $P(\mathbf{k})$ for Small k 208
			7.7.3 The Shuffled Lattice with Uncorrelated Displacements 209
		7.8	Correlated Displacements................................ 212
			7.8.1 Correlated Gaussian Displacement Field............. 214
		7.9	Summary and Discussion 217

8 Galaxy Surveys: An Introduction to Their Analysis 219
8.1 Introduction 219
8.2 Basic Assumptions and Definitions 220
8.3 Galaxy Catalogs and Redshift............................ 221
8.4 Volume Limited Samples 224
8.5 The Discovery of Large Scale Structure in Galaxy Catalogs .. 227
8.6 Standard Characterization of Galaxy Correlations and the Assumption of Homogeneity 228
8.7 Summary and Discussion 233

9 Characterizing the Observed Distribution of Visible Matter I: The Conditional Average Density in Galaxy Catalogs 235
9.1 Introduction 235
9.2 The Conditional Average Density in Finite Samples......... 236
9.3 Sample Size Smaller than the Homogeneity Scale 240
  9.3.1 The Reduced Correlation Function for a Particle Distribution with Fractal Behavior in the Sample..... 240
9.4 Sample Size Greater Than the Homogeneity Scale 242
  9.4.1 Critical Case...................................... 243
  9.4.2 Substantially Poisson Case 244
  9.4.3 Super-Homogeneous Case 245
  9.4.4 Some Remarks 245
9.5 Estimating the Average Conditional Density in a Finite Sample... 246
  9.5.1 Estimators of the Average Conditional Density 247
  9.5.2 Effective Depth of Samples 250
9.6 The Average Conditional Density (FS) in Real Galaxy Catalogs 250
  9.6.1 Normalization of the Average Conditional in Different VL Samples 257
  9.6.2 Estimation of the Conditional Average Luminosity Density 259
  9.6.3 Measuring the Average Mass Density Ω from Redshift Surveys 260
9.7 Summary and Discussion 263

10 Characterizing the Observed Distribution of Visible Matter II: Number Counts and Their Fluctuations 265
- 10.1 Introduction 265
- 10.2 Number Counts in Real Space 266
- 10.3 Number Counts as a Function of Apparent Magnitude 268
  - 10.3.1 Poisson Distribution 268
  - 10.3.2 Simple Fractal Distribution 271
  - 10.3.3 Effect of Long-Ranged Correlations in Homogeneous Distributions 273
- 10.4 Normalization of the Magnitude Counts to Real Space Properties in Euclidean Space 276
  - 10.4.1 Average Distance 276
  - 10.4.2 Normalization of Distance to Magnitude Counts 277
- 10.5 Galaxy Counts in Real Catalogs 278
  - 10.5.1 Real Space Counts 279
  - 10.5.2 Magnitude Space Counts 283
- 10.6 Summary and Discussion 288

11 Luminosity in Galaxy Correlations 291
- 11.1 Introduction 291
- 11.2 Standard Methods for the Estimation of the Luminosity Function 292
- 11.3 Multifractality, Luminosity and Space Distributions 293
- 11.4 Summary and Discussion 297

12 The Distribution of Galaxy Clusters 299
- 12.1 Introduction 299
- 12.2 Cluster Correlations and Multifractality 300
- 12.3 Galaxy Cluster Correlations 303
  - 12.3.1 The Average Conditional Density for Galaxy Clusters . 306
  - 12.3.2 Galaxy-Cluster Mismatch 306
- 12.4 Luminosity Bias and the Richness-Clustering Relation 308
- 12.5 Summary and Discussion 311

13 Biasing a Gaussian Random Field and the Problem of Galaxy Correlations 313
- 13.1 Introduction 313
- 13.2 Biasing of Gaussian Random Fields 314
- 13.3 Biasing and Real Space Correlation Properties 318
- 13.4 Biasing and the Power Spectrum 325
- 13.5 Summary and Discussion 330

14 The Gravitational Field in Stochastic Particle Distributions 335
14.1 Introduction 335
14.2 Nearest Neighbor Force Distribution 336
14.3 Gravitational Force PDF in a Poisson Particle Distribution 338
14.4 Gravitational Force in Weakly Correlated Particle Distributions: the Gauss-Poisson Case 342
14.5 Generalization of the Holtzmark Distribution to the Gauss-Poisson Case 343
    14.5.1 Large F Expansion 344
    14.5.2 Small F Expansion 347
    14.5.3 Comparison with Simulations 347
    14.5.4 Nearest-Neighbor Approximation for the Gauss-Poisson Case 348
14.6 Gravitational Force in Fractal Point Distributions 350
14.7 An Upper Limit in the Fractal Case 351
14.8 Average Quadratic Force in a Fractal 354
14.9 The General Importance of the Force-Force Correlation 358
14.10 Summary and Discussion 360

Part III Appendixes

A Scaling Behavior of the Characteristic Function for Asymptotically Small Values of k 365

B Fractal Algorithms 369
B.1 Cantor Set and Random Cantor Set 369
B.2 Levy Flight 372
B.3 Random Trema Dust 372

C Cosmological Models: Basic Relations 375
C.1 Cosmological Parameters 376
    C.1.1 Comoving (Radial) Distance 376
    C.1.2 Comoving (Transverse) Distance 377
    C.1.3 Luminosity Distance 377
    C.1.4 Magnitude 377
C.2 Cosmological Corrections in the Analysis of Redshift Surveys 378
    C.2.1 Flat Cosmologies: FMD and FLD 378
    C.2.2 Open Model: OBD 380

D Cosmological and k-Corrections to Number Counts ... 381
- D.1 k-Corrections ... 381
- D.2 k-Corrections and the Radial Number Counts ... 382
- D.3 Dependence on the Cosmological Model ... 383

E Fractal Matter in an Open FRW Universe ... 385
- E.1 Introduction ... 385
- E.2 Friedmann Solution in an Empty Universe ... 386
- E.3 Curvature Dominated Phase ... 387
- E.4 Radiation Dominated Era ... 390
- E.5 Fluctuations in the CMBR ... 391
- E.6 Other Remarks ... 392

F Errors in Full Shell Estimators ... 395
- F.1 Bias and Variance of Estimators ... 395
- F.2 Unconditional Average Density ... 396
- F.3 Conditional Number of Points in a Sphere ... 397
- F.4 Integrated Conditional Density ... 398
- F.5 Conditional Average Density in Shells ... 399
- F.6 Reduced Two-Point Correlation Function ... 402

G Non Full-Shell Estimation of Two Point Correlation Properties ... 405
- G.1 Estimators with Simple Weightings ... 406
- G.2 Other Pair Counting Estimators ... 407
- G.3 Estimation of the Conditional Density Beyond R_s ... 409

H Estimation of the Power Spectrum ... 411

References ... 413

Index ... 421

List of Acronyms

ACF	angular correlation function
CDM	cold dark matter
CMBR	cosmic microwave background radiation
DCR	deterministic Cantor set
DSI	discrete scale-invariance
GDF	Gaussian density fields
GP	Gauss-Poisson
FLD	flat Lambda dominated
FMD	flat matter dominated
FT	Fourier transform
FRW	Friedmann Robertson Walker
HZ	Harrison Zeldovich
HDM	hot dark matter
LF	luminosity function
MDCR	multiplicative deterministic Cantor set
MRCS	multiplicative random Cantor set
MF	multifractals
MLR	mass-length relation
Mpc	megaparsec
NBS	N-body simulation
nn	nearest neighbors
OBM	open baryon density
OCP	one component plasma
PDF	probability density functional
PS	power spectrum
RG	renormalization group
rms	root mean square
SBBN	standard big-bang nucleosynthesis
SSI	statistical stationarity and isotropy
SPP	stationary point process
SSP	stationary stochastic process
VL	volume limited

1 Introduction

"When inhomogeneities are considered (if at all) they are treated as unimportant fluctuations amenable to first order variational treatment. Mathematical complexity is certainly an understandable justification, and economy or simplicity of hypotheses is a valid principle of scientific methodology: but submission of all assumptions to the test of empirical evidence is an even more compelling law of science." Gerard de Vaucouleurs (1970)

1.1 Motivations and Purpose of the Book

Cosmology is the attempt to build a coherent physical theory to explain the many and diverse observations of the universe at the largest scales. Statistical physics is concerned with the understanding of systems with many degrees of freedom. Clearly the second should therefore have much to say about and to contribute to the former. And indeed it does. Cosmology has, in particular in the development of its formalism since the 1970s, borrowed tools from statistical physics. The mathematical language used to describe correlations in galaxy catalogs, for example, was imported from the theory of liquids. Direct exchange or even collaboration between people in the two communities has, however, been rare. Cosmology has thus tended to seek instruments when necessary from statistical physics, but has not been very much in contact with or influenced by the many developments which have taken place in statistical physics in the last decades. Statistical physics too – which has not hesitated to apply its methods to many other fields even outside physics (e.g. biology, geology, economics) – has been relatively shy with respect to cosmology, despite the fact that this has become one of the most vibrant and exciting fields in contemporary science. This book has come out of a series of collaborations between researchers from both communities, collaborating on a range of problems in cosmology approached with insights or methodologies coming from statistical physics.

The primary aim of this book is to present in a systematic manner some of the principal instruments used in modern statistical physics to describe and understand stochastic structure in different signals, and then to illustrate their use in their application to various problems in cosmology. Its intended audience is the large one of both cosmologists and statistical physicists interested in cosmology, at the graduate or research level. For cosmologists we hope that the first part of book can serve as a clear and accessible presentation of methods and concepts of statistical physics for the description of structures. The second part should then allow them to see the usefulness of these methods and concepts when applied to problems in cosmology. For statistical physicists the first part of the book should be a useful summary of material with which they will in large part be familiar, but treated and presented in a way which foreshadows the problems to which they are applied in the second part of the book. In this second part we have attempted to make the presentation of the cosmological problems treated such that they are for the most part accessible to statistical physicists. We emphasize, however, that the book is not intended to be or to substitute a textbook of cosmology. For readers without any background in cosmology, this second part would ideally be complemented with the reading of parts of the many texts available on structure formation in cosmology (e.g. [170, 179, 185, 208]).

In this introduction we give first a little bit of background on developments in statistical physics in the last decades relevant to the problem of structures in cosmology. This is intended primarily to help to give to cosmologists a first impression of why statistical physicists have a different perspective on such problems. We then give a brief summary of the organization and content of the book, in two sections paralleling its structure: we treat first the appropriate methods of statistical physics, and then some of their applications in cosmology.

1.2 Structures in Statistical Physics: A New Perspective

"More is different": This epochal paper of 1972 by Phil Anderson [4] has set the paradigm for what has now evolved into the science of complexity. The idea that "reality has a hierarchical structure in which at each stage entirely new laws, concepts, and generalizations are necessary, requiring inspiration and creativity to just as great a degree as in the previous one" has set a new perspective in our view of many natural phenomena. The reductionist view focuses on the elementary bricks of which matter is made, but then these bricks are put together in marvelous structures with highly elaborated architectures. Complexity is the study of those architectures which depend only in part on the nature of the bricks, but also have their fundamental laws and properties which cannot be deduced from the knowledge of the elementary bricks.

It has been realized that, in the physical sciences, the geometric complexity of structures very often corresponds to fractal or multifractal properties [150, 151]. It is not clear whether this is a unique and complete description of complexity, or simply the only framework which has been found for this purpose. In the study of dynamical processes which can give rise to complex geometrical structures, we have as basic concepts: chaos, fractals, avalanche dynamics, and 1/f noise (see [188] for an overview of the state-of-the-art in the field). Often complex structures arise from processes which are strongly out of equilibrium and dissipative. There is a broad field, however, which is in between equilibrium and non-equilibrium phenomena. This is the field of glasses and spin glasses which leads to highly complex energy landscapes and to the concept of *frustrated* configurations. Another important field in which this set of new concepts can also be applied is that of biological adaptation via evolution which is characterized by a degree of self-organization and a critical balance between periods of smooth evolution and dramatic changes resulting in bursts of extinctions.

A fundamental notion in this context is that of irregularity, one which is completely new to physics. Let us explain with a simple example why this notion is so fundamental and important: it requires, in particular, a whole new mathematical language which supersedes that usually employed in physics.

In Fig. 1.1 we show examples of regular and irregular structures. In the top left we have a distribution of points characterized by a small scale granularity which turns, at larger scales, into a well defined background density with a specific structure corresponding to an over-density around the center. One can also represent such a structure by defining a density profile along a line passing through the center as shown in the top right panel. This profile can be well approximated by a smooth (analytical) function, which in this case is a constant plus a Gaussian. If we consider the dynamical evolution of such a structure including the specific interactions between its constituent points, usually we can write differential equations for the smooth function of the density profile. The structure is thus essentially represented by the three elements: position, size and intensity (amplitude). The typical result of this study is to understand whether the structure moves, if it becomes more or less extended or more or less intense. This is the traditional approach to the study of structures based on the implicit assumption of regularity or analyticity which has been the one adopted in statistical physics before the advent of critical phenomena in the seventies [3, 148].

The bottom left panel of Fig. 1.1 shows instead a strongly irregular distribution. In this case it is not possible to recognize any positive background density, there are structures in many zones and at various scales, hierarchically organized in such a way that it is not possible to assign them a unique characteristic size or intensity. This situation is also illustrated by the density profile (bottom right) which is highly irregular at all scales. All the previous "regular fluid-like" concepts and theoretical methods loose their meaning and, in order to give a proper characterization of the properties of this system, one has to look at it from a new perspective. This highly irregular density field, which is in this case of a stochastic *multifractal*, has its "regularity" in

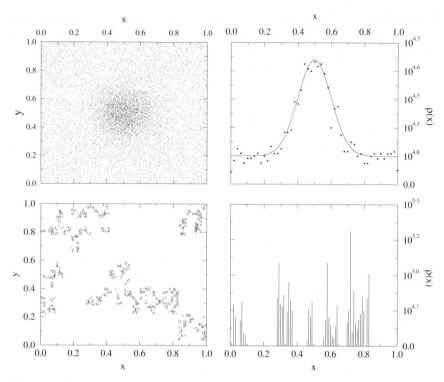

Fig. 1.1. Example of regular and irregular structures in mass distributions. *Top panels: (left)* A cluster in a homogeneous distribution. *Right:* Density profile. In this case the fluctuation corresponds to an enhancement of a factor 3 with respect to the average density. *Bottom panels: (left)* Multifractal distribution in the two dimensional Euclidean space. *Right:* Density profile. In this case the fluctuations are strongly irregular at all scales and there is no reference value, i.e. the average density. The average conditional density scales as a power law from any occupied point of the structure. (From [223])

scale transformations. This naturally leads to power-law density-density correlations mainly characterized by typical exponents: the fractal dimensions. Also from a theoretical point of view, the understanding of the origin of the irregular or fractal properties cannot arise from the traditional approach with differential equations but it requires new methods, e.g. scaling theories and the renormalization group (RG) [79, 80, 151].

Most of theoretical physics is based on analytical functions and differential equations for average values of the studied field. This implies that structures should essentially be smooth, and irregularities are treated as small fluctuations or isolated singularities. On the other hand, the study of critical phenomena and the development of the RG theory in the seventies was a major breakthrough [3, 245]. There one observes and describes phenomena in which *intrinsic self-similar irregularities develop at all scales* and fluctuations cannot be described in terms of sufficiently localized regular functions. The theoretical methods to describe this situation cannot be based on ordinary differential equations, in real or Fourier space, of the main physical fields as self-similar and strong fluctuations up to the largest scale of the field implies the absence of regularity everywhere and the usual "old" mathematical physics becomes redundant. The RG provides a new space – that of scale transformations – in which the problem again becomes analytical. These characteristics might seem to be specific only to thermodynamic critical phenomena, corresponding to a "critical" competition between order and disorder. However, in the past years, the development of fractal geometry has allowed us to realize that a larger variety of structures in nature are intrinsically irregular and self-similar [151]. Mathematically speaking these structures are described as singular in every point. This property can now be characterized in a quantitative mathematical way by using the concept of fractal dimension and other concepts developed in this field. However the construction of a theory to explain the physical origin of these structures is a very challenging task, which is still only in its infancy. This is actually the objective of much of the present activity in the field (see e.g. [79]). The main difference between "popular" fractals like coast-lines, mountains, clouds, lightnings etc., and the self-similarity of critical phenomena, is that criticality at phase transitions occurs only when there is an extremely accurate fine tuning of the relevant critical parameters (e.g. temperature, external fields, etc.). In the more familiar structures observed in nature, instead, the fractal properties are often self-organized, developing spontaneously out of some dynamical process. It is probably in view of this important difference that the two fields of critical phenomena and fractal geometry have proceeded somewhat independently.

The fact that we are traditionally accustomed to think in terms of smooth or analytical structures has a crucial effect on the type of questions we ask and on the methods we use to answer them. If one has never been exposed to the subtlety of strongly irregular and fractal-like structures, it is natural that the hypothesis of regularity of fluctuations is not even questioned. It is only after the above developments that we can realize that the property of regularity of structures can be tested experimentally and that it may, or may not, be present in a given physical system.

The physics of scale-invariant and complex systems is now a growing field which treats problems from several disciplines ranging from condensed matter physics to geology, biology, astrophysics and even economics [80, 240]. This broad interdisciplinarity has been possible because these new ideas allow us to look at many complex natural phenomena in a radically new and original way, with unifying concepts which are independent of the details of the system considered. The objective of this new approach is the study of complex, scale-invariant structures, appearing both in space and time in a wide variety of natural phenomena, and new types of collective behaviors, the comprehension of which is one of the most challenging problems in modern statistical physics. Research in this field involves a cooperative effort of numerical simulations, analytical and experimental work, which can be characterized by the following three levels:

- *(i) Mathematical or Geometrical Level*

This consists in applying the methods of fractal geometry, introduced and developed by Mandelbrot [150, 151], into new areas where strongly irregular structures appear at least up to a certain scale, in order to get new insights into important unresolved problems about the properties of these structures. It is an approach which brings into the realm of scientific investigation many phenomena characterized by intrinsic strong irregularities on sufficiently large spatial or temporal windows, which have been previously neglected because of the lack of an appropriate framework for their mathematical description.

- *(ii) Development of Physical Models: The Active Principles for the Generation of Fractal Structures*

Computer simulations represent an essential tool in the physics of complex and scale-invariant systems. A large number of models have been introduced to focus on specific mechanisms which can lead spontaneously to fractal structures. Here we list some of them which give, we believe, the active principles for certain processes which generate highly irregular structures characterized by scale-invariant properties based on physical dynamics[1]: diffusion limited aggregation (1983), dielectric breakdown model (1984). These models are the prototype of fractals in which an iteration process based on the Laplace equation leads spontaneously to very complex structures like that shown in Fig. 1.2. Further examples are cluster-cluster aggregation (1983), invasion percolation (1983) and the sandpile model (1987). The concept of self-organization is common to all these models, but it has been especially emphasized in relation to the sandpile model. Finally we mention the Kardar-Parisi-Zhang model of surface growth (1986) and the Bak-Sneppen model (1993) of biological evolution.

Going beyond these simplified models it is known, for example, that the vortexes and velocity fields generated in fluid turbulence (as described by Navier-Stokes equations) have multi-fractal properties (see e.g. [171]). Whether gravitational dynamics may generate fractal clustering from appropriate non-fractal initial conditions is still an open question.

- *(iii) Development of Theoretical Understanding*

In a phenomenological approach to these complex systems, scaling theory, following its application in critical phenomena, has been successfully used.

---

[1] In [80] one can find many papers on these models. In parentheses we show the year of publication of each model.

1.2 Structures in Statistical Physics: A New Perspective

Fig. 1.2. A dielectric breakdown model in cylindrical geometry, which is an example of a highly complex structure arising from a simple growth model in which the growth probability is proportional to the local electric field [79]. The zebra stripes around the structure represent equi-potential lines. (Elaboration and photo by B.B. Mandelbrot and C.J.G. Evertsz.)

This is essential for the rationalization of the results of computer simulations and experiments. This method allows us to identify the relations between different properties and to focus on the essential ones. From the point of view of the formulation of microscopic fundamental theories the situation is still in evolution. These systems are far from equilibrium and their dynamics is intrinsically irreversible. There does not seem to be a generalization of the notion of ergodicity in equilibrium statistical mechanics, and the temporal dynamics has to be explicitly considered in the theory. These, together with the concept of self-organization (in reference to criticality), represent the main new elements for the formulation of possible microscopic theories.

1.3 Structures in Statistical Physics: The Methods

In the first part of this book we aim to give a systematic overview of the principal tools used to describe structures in contemporary statistical physics. This subject is naturally divided into two parts: one which treats spatial mass distributions in which there are fluctuations about a well defined non-zero mean value, and another which treats distributions which are of the kind we have called *intrinsically irregular*. We devote two chapters to each of them.

Uniform Correlated Distributions

In Chap. 2 we describe systematically the framework which is used in statistical physics to describe systems with small amplitude fluctuations about a mean value e.g. density fluctuations in an ordinary fluid or charge fluctuations in a plasma. We refer to such systems as "uniform" because the fluctuating quantities possess a well-defined non-zero mean value. We will use this term interchangeably with "homogeneous", with the caveat that it should not be confused with the (weaker) property of statistical homogeneity. This framework is precisely the one which has been adopted widely in cosmology to describe structures. Its primary instruments are the reduced two-point correlation function $\xi(r)$ and the power spectrum $P(k)$. In our presentation we place emphasis on the physical meaning and mathematical definitions of two characteristic scales which are of primary importance in such distributions: the *homogeneity scale* and the *correlation length*. We focus on these concepts since there is often confusion concerning them in the cosmological literature. The origin of this confusion is that the term "correlation length" has been used sometimes in cosmology to refer to the scale which in fact corresponds to that characterizing homogeneity (or uniformity). We now briefly discuss this point.

In statistical physics the *correlation length* is defined as the distance up to which the spatial memory (i.e. spatial correlation) of local fluctuations persists, independently of their amplitude. In equilibrium statistical mechanics this is equivalent to the length scale up to which the effect of a small local perturbation is felt in the system considered. This is due to the *fluctuation-dissipation theorem* [148] which links the response of the system to a local perturbation and the large scale behavior of the two-point correlation function. There is no unique definition of the correlation length, but all definitions give a length which diverges for a correlation function which is non-integrable. For an exponentially decaying correlation function $\xi(r) = Ae^{-r/r_c}$, the correlation length is simply the scale r_c characteristic of the decay. For a sufficiently "slow" power-law i.e. $\xi(r) \sim r^{-\alpha}$, with α smaller than the spatial dimension, the correlation length is infinite. The former would be typical of a system of particles at equilibrium interacting through short-range forces (e.g. a gas at high temperature), while the latter is distinctive of the behavior of fluctuations in critical phenomena. In cosmology, on the other hand, a length scale, usually denoted r_0 and widely used, is very often referred to as the "correlation length". This scale is, however, defined by the condition $\xi(r_0) = 1$ i.e. through the *amplitude* of $\xi(r)$. This is not in line with the statistical physics use of the term which we have just described. For instance, following this definition, $\xi(r) = Ae^{-r/r_c}$ and $\xi(r) = (r_c \ln A)/r$ have both $r_0 = r_c \ln A$, which varies monotonically with the amplitude A. The "statistical physics" correlation length, however, as just described above, is equal to r_c in the former case, and diverges in the latter case (in any dimension). Therefore this "cosmological" correlation length has a completely different physical meaning, as the amplitude of $\xi(r)$ is related to the *local* relative amplitude of the fluctuations with respect to the average density, and cannot say anything about the large scale behavior of the *integrated* fluctuations. The length r_0 can be in fact simply a measure of the (completely distinct) *homogeneity scale*, i.e. a scale at which the fluctuations become of small amplitude with respect to the average density.

In Chap. 3 we discuss the power spectrum $P(k)$, which, for statistically homogeneous processes, is simply the Fourier transform of $\xi(r)$. We then give a simple classification (into three classes) of correlated stochastic fluctuations in uniform distributions, in terms of the behavior of $P(k)$ as k tends to zero. The classification divides systems according to the large scale behavior of integrated fluctuations. A first class, for which $0 < P(0) < \infty$, are those systems which have behavior at large scales like that in a Poisson system, i.e. with a variance of integrated fluctuations proportional to the volume of the region. We refer to them therefore as "Poisson" or "substantially Poisson". The two other classes are characterized by how their integrated fluctuations behave with respect to this case: For $P(0) = \infty$ they grow more rapidly as a function of scale and we refer to them as "super-Poisson". In this class are all systems with an infinite correlation length, as the non-integrability of $\xi(r)$ corresponds to the divergence of $P(k)$ as k tends to zero. Finally there is a third class, with $P(0) = 0$, which have integrated fluctuations growing less rapidly as a function of scale than in a Poisson distribution. For this reason – they have fluctuations which are suppressed compared to those in a completely uncorrelated distribution of points – we refer to them as "super-homogeneous".

We examine this last class in some detail as it is to it that the "primordial" (i.e. very early time) fluctuations in standard cosmological models belong[2].

---

[2] We note that in this context there is another term which is used in cosmology with a sense very different from its original usage in statistical physics: the canonical "primordial" spectrum is referred to usually as the "scale-invariant" spectrum. This has nothing to do with the notion as used in statistical physics, where this refers to the invariance under scale transformations of a given system or of its fluctuations. In cosmology, as discussed further in Chap. 6, the term "scale-invariant" used in this context refers to a property of this particular spectrum: the amplitude of the fluctuations it defines remains invariant as the "horizon scale" increases.

Their characteristic sub-Poisson integrated fluctuations represent a very non-trivial property. It requires that there be a fine-tuned balance between correlation and anti-correlation in the distribution (so that $\xi(r)$ integrates to zero). We note and discuss in some detail a relevant and interesting result about the behavior of the variance of the mass fluctuations integrated in a volume: there is in fact a lower limit to its growth, corresponding to a growth proportional to the surface of the volume (compared to the volume itself in the Poisson case). To illustrate these properties we consider various simple examples: a regular lattice, or a "shuffled" lattice (which is obtained by randomly displacing each particle of a lattice). A more physical example, which we describe is the so called "one-component plasma" (OCP), which is simply a set of identical point charges interacting by the Coulomb force, embedded in a uniform oppositely charged background giving overall charge neutrality. At thermal equilibrium the distribution of the discrete charges is super-homogeneous at scales much larger than the Debye length characterizing the screening of the electric field.

Intrinsically Irregular Structures (Fractals and Multifractals)

In the following two chapters we describe the framework of fractal geometry which has, as we have described above, become the instrument which has allowed statistical physics to extend its analysis to a whole new class of systems: intrinsically irregular structures, i.e. systems which are never uniform in the sense used above. A fractal mass distribution (Chap. 4) is, in a simple sense, the most correlated of correlated systems: its average density is zero in the infinite volume limit [223], but the *conditional* average density $\langle n(r) \rangle_p$ (i.e. the average mass density in a small shell at a distance r from an occupied point), decays to zero as a slow power-law $r^{-\gamma}$, with $0 < \gamma \le d$ (the fractal dimension of the mass distribution being defined as $D = d - \gamma$, where d is the spatial dimension). This means that the fluctuation field, which coincides now with the full density field, is highly irregular at all scales and is intrinsically scale-invariant (see Chaps. 4–5). For this reason a fractal is not a uniform distribution (i.e. is non-homogeneous) at any scale, and the concept of average density in a finite sample centered on an occupied point has no intrinsic meaning because it depends on the sample size. There is a close link between fractals and critical systems: if we take a liquid-gas co-existence phase at the critical point and define a new density field as the subset of over-densities we obtain a fractal mass distribution. Since the asymptotic average density of a fractal distribution is zero, the functions $\xi(r)$ and $P(k)$, which are the basic statistical quantities characterizing the fluctuations of stochastic uniform distributions, are not well defined in the infinite volume limit and are strongly sample dependent if estimated in a finite sample. One has necessarily to work in this case with conditional statistics (i.e. conditional on the presence of a point of the system at the origin of coordinates). For the two-point properties, for example, one uses usually the conditional density $\langle n(r) \rangle_p$ mentioned above. In Chap. 4 we introduce the principal concepts for simple fractal mass distributions defined by an infinite set of identical elements characterized, as described, by strong and scale invariant irregularities at all distances, with a conditional density decaying to zero as a slow power law. In particular we focus on those spatial distributions of identical particles characterized by these features, as this is the most interesting case for the study in cosmology of the spatial distribution of objects such as galaxies. A complete two-point description of fractal mass distributions is given by characterizing the statistical behavior of two-point correlations through the conditional density and the fractal dimension, both describing the behavior of the mass distribution under a scale transformation or a coarse-graining. For what concerns the three-point statistical properties we discuss the typical behavior of the conditional mass variance by introducing the concept of *lacunarity*, the void distribution, the correction to scaling and log − log fluctuations. Further we discuss other important aspects of fractals potentially useful for cosmological applications, such as the problems of intersection between fractals and the effects of an orthogonal or an angular projection of the system. All these aspects are illustrated with both deterministic and stochastic examples.

In Chap. 5 we consider a wider class of strongly irregular distributions by extending the previous concepts to measure theory. This permits a deep characterization of irregular distributions of objects (e.g. point-particles) characterized by a hierarchy of measures (e.g. masses). For these more complex irregular mass distributions, a continuous spectra of fractal dimensions is introduced in order to describe the scaling behavior under coarse graining and scale transformations of the different objects characterized by different measures. Also in this case we illustrate the complex concept of multifractality through the direct presentation of both deterministic and stochastic paradigmatic examples and models.

1.4 Applications to Cosmology

In the second part of the book we describe various aspects of the puzzle of cosmic structures, with the perspective of the mathematical and conceptual framework of the first part.

Cosmological Models: The Super-Homogeneous Universe

We start with a discussion of the statistical properties of the *theoretical* models now used by most cosmologists to describe the universe: so-called "standard cold dark matter" (CDM) and its variants. In these models fluctuations are described with the framework described in Chaps. 2 and 3, i.e. small fluctuations about a well defined mean density. We focus in particular on the initial conditions for such models (i.e. very early in the history of the universe), specified by the so-called Harrison-Zeldovich power spectrum, examining their properties in real space. It is here that the concept of "super-homogeneity" introduced in Chap. 3 is relevant, as these models describe fluctuations which are in fact of this type. We discuss the ways in which this is manifest in the real space correlation properties. Standard type models are thus characterized by *surface quadratic fluctuations* (of the mass in spheres) and, for the particular form of primordial cosmological spectra, by a negative power-law in the reduced correlation function at large separations ($\xi(r) \sim -1/r^4$). This real space view of these models is complementary to that usually given, which privileges the k-space description. We also briefly consider, in a similar spirit, the so-called "acoustic oscillations" observed in the fluctuations of the cosmic microwave background radiation (CMBR), identifying what they correspond to in real (i.e. angular) space correlations.

Primordial Fluctuations and Models in Statistical Mechanics

Chapter 7 follows on from the previous one: we consider here the generation of particle distributions with the very specific properties we have discussed. This is a problem which is of both heuristic and practical interest, because the numerical simulations of formation of structures by gravity (cosmological N-body simulations) require the generation of distributions of particles which represent the continuous density field at the initial time. We first define the sense in which a particle distribution may be considered to be a discretization of the continuous field, and then describe two approaches to this problem. The first involves finding a microscopic thermodynamic system which at thermal equilibrium gives rise to the desired correlation properties. We describe how such a system can be obtained through a modification of the OCP (Chap. 2). Secondly we analyze in a very general way the effect of superimposing a stochastic displacement field on a point particles distribution with given correlation properties. This includes the standard method used in cosmological N-body simulations to generate initial conditions, which involves imposing displacements on a simple lattice. We obtain the approximate relation used in this context as a special case, but we are able as well to specify the full correlation properties described by these initial conditions.

Analysis of Galaxy Clustering Without the Assumption of Homogeneity

We next turn to the question which first stimulated the interest of the statistical physicists among us in cosmology: the correlation properties of the observed distribution of galaxies and galaxy clusters. It is here that the perspective of a statistical physicist, exposed to the developments of the last decades in the description of intrinsically irregular structures, is radically different from that of a cosmologist for whom the study of fluctuations means

1.4 Applications to Cosmology

the study of small fluctuations *about a positive mean density*. And it is here therefore that the instruments used to describe strong irregularity, even if limited to a finite range of scales, described in Chaps. 4 and 5, offer a wider framework in which to approach the problem of how to characterize the correlations in galaxy distributions.

Analyzing galaxy distributions in this wider context means treating these systems without the a priori assumption of homogeneity, i.e. without the assumption that the finite sample considered gives, to a sufficiently good approximation, the true (non-zero) mean density of the underlying distribution of galaxies. While this is a simple and evident step for a statistical physicist, it can seem to be a radical one for a cosmologist. After all the whole theoretical framework of cosmology (i.e. the Friedmann-Robertson-Walker (FRW) solutions of general relativity) is built on the assumption of an homogeneous and isotropic distribution of matter. Our approach is an empirical one, which surely is appropriate when faced with the characterization of data. Further it is evidently important for the formulation of theoretical explanations to understand and characterize the data correctly. Let us, however, note briefly two common *theoretical* objections to this approach. One is that fractals are incompatible with what is called the "cosmological principle", by which it is meant that there is no privileged point or direction in the universe. This is a simple misconception about fractals. These irregular distributions are in keeping with this principle in exactly the same way as inhomogeneities treated in the standard framework of perturbed FRW models, i.e. they can be considered *statistically* stationary and isotropic distributions in space in the sense that their statistical properties are invariant under translation and rotation in space. Another common objection is that there is an inconsistency in using the Hubble law, which is used to convert redshift to physical distance, if one does not assume homogeneity. This objection forgets that the Hubble law is an established empirical relation, independent of theories explaining it. Further it can be noted that what we are concerned with is the distribution of *visible* matter: given that standard cosmological models describe a universe whose energy density is completely dominated by several non-visible components, reconciling the two is not in principle impossible (see also Appendix E)[3].

---

[3] The idea of modeling the universe as a hierarchical structure is an old concept. In the book by Baryshev and Teerikorpi [18] one may find a comprehensive and self-contained account of the development of cosmological ideas related to the concept of fractality. Among the first scientists who considered such ideas was Charlier [44, 45] who applied the idea of a hierarchical (fractal in modern terms) distribution in order to explain "Olbers paradox". More recently de Vaucouleurs (quoted at the beginning of this chapter) [63] proposed a hierarchical cosmology to explain galaxy counts: after him Wertz [243] and then Haggerty & Wertz [104, 105] developed a Newtonian hierarchical model. We refer the interested reader to [18] and to [17, 103].

In Chap. 8 we give a brief description of galaxy surveys, focusing in particular on the essential properties which must be considered if one wishes to infer information from them about the correlation properties of the underlying galaxy distribution. One must take into account the extent to which they are conditioned by being observed in a particular manner by an observer located in our galaxy. These are known as selection effects. We describe the construction of so-called "volume-limited" samples which correct as much as possible for the most important of these effects, and which are appropriate for use in the statistical analysis which we describe in subsequent chapters. We then give a brief account of the evolution of observations probing the distribution of galaxies, describing how structures – superclusters, walls, voids, filaments – were revealed in the first large three dimensional surveys which were published in the eighties. These maps revealed structures at scales much larger than had been suspected from the previous (fairly isotropic) angular data. The simple visual impression (see Figs. 1.3–1.4) of such three-dimensional data – apparently showing large fluctuations up to the sample sizes – gives a strong *prima facie* case for an analysis which does not assume that the underlying distribution is uniform (at the scales probed), but rather encompasses the possibility that it may be intrinsically irregular. The quantitative results obtained with the usual analysis performed on these samples, which works with the reduced two-point correlation function $\xi(r)$ (or equivalently the power spectrum $P(k)$) and thus builds in the assumption of homogeneity, give further evidence in this direction. While the correlation function shows consistently a simple power-law behavior characterized by the same exponent, there is very considerable variation between samples, with different depth and luminosity cuts, in the measured *amplitude* of $\xi(r)$ (and, correspondingly, in the normalization of the power spectrum $P(k)$). This variation in amplitude is usually ascribed *a posteriori* to an intrinsic difference in the correlation properties of galaxies of different luminosity ("luminosity bias", or "luminosity segregation"). It may, however, have a much simpler explanation in the context of irregular distributions. In a simple fractal, for example, the density in a finite sample decreases on average as a function of sample size; samples of increasingly bright galaxies are in fact generically of greater mean depth, which corresponds to an increasing amplitude of the correlation function $\xi(r)$ normalized to the "apparent" average density in each sample.

With this motivation we move on in the following two chapters to describe in detail the application of simple statistical methods which allow a characterization of galaxy clustering, irrespective of whether the underlying galaxy distribution is homogeneous (or uniform in the sense we used it above) or irregular at the sample size. In Chap. 9 we discuss the use of the average conditional density $\langle n(r) \rangle_p$ (which measures, as noted above, the mean density in a shell at distance r from a given point (e.g. a galaxy). We describe its theoretical (ensemble average) behavior for a wide class of distributions,

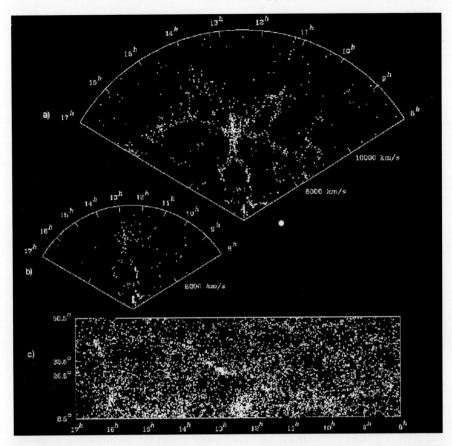

Fig. 1.3. Progress in redshift surveys in the eighties. (**a**) Maps of the CfA2 redshift survey (1986): Redshift distances cz are indicated (distances are simply given by $cz/100\,\mathrm{Mpc/h}$) and the thickness in declination is 6 degrees. The small circle has a diameter of 5 Mpc/h, the clustering length according to the standard interpretation of galaxy correlation. (**b**) Same region of the sky but in the CfA1 redshift survey (1983). (**c**) Angular region of the CfA2 survey from the Zwicky catalog (1970). (From [59])

simply characterized by the two scales we emphasized in Chap. 2: the homogeneity scale λ_0 and the correlation length r_c. The general form we consider encompasses both some irregular distributions (simple fractals) and uniform distributions, in particular those described by standard cosmological models. We consider the average behavior of an estimator of the conditional density as a function of the characteristic size of the sample R_s. If $R_s \gg \lambda_0$ one can detect using $\langle n(r)\rangle_p$ the existence and location of the homogeneity scale λ_0. If, on the other hand, $R_s \ll \lambda_0$ it allows only the determination of the fractal exponent characterizing the clustering on these scales.

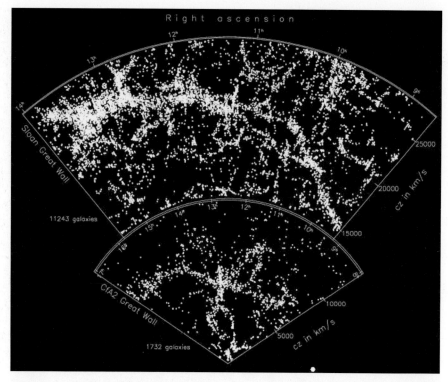

Fig. 1.4. Latest progress in redshift surveys. Sloan Great Wall (2003) compared to CfA2 (1986) Great Wall at the same scale. Redshift distances cz are indicated (distances are simply given by $cz/100\,\mathrm{Mpc/h}$). The *small circle* at the bottom has a diameter of $5\,\mathrm{Mpc/h}$, the clustering length according to the standard interpretation of galaxy correlation. The Sloan Digital Sky Survey slice is 4 degrees wide and the CfA2 slice is 12 degrees wide so that both slices have approximately the same physical width at the two walls. (Adapted from [101])

In Chap. 10 we discuss how the study of number counts from the origin (i.e. with respect to the observer in our galaxy) can be used to probe the nature of the underlying distribution, again considering the same wide class including both fractal-type distributions and the uniform distributions typical of cosmological models. These number counts are counts of objects (usually galaxies or galaxy clusters) in a given angular range on the sky, as a function either of real distance, or, more usually in the context of cosmology, as a function of the measured apparent luminosity of the objects. One can then consider these counts in different angular regions, and study their average behavior and variance. The slope of the average count (both for real space and magnitude counts) gives a direct measure of the fractal dimension (with homogeneity corresponding to the case that this dimension is three) on scales relevant for those magnitudes, while the fluctuations about the average behavior differs qualitatively between irregular and uniform distributions: for a simple fractal, it is a simple consequence of its scale-invariance at all scales that the fluctuations of the counts, normalized to the mean count, are approximately constant as a function of scale. This is also true in magnitude space, while for any uniform distribution with superimposed correlated stochastic fluctuations the normalized variance has a decaying behavior in both cases.

In both Chaps. 9 and 10 we consider, having given the methodological basis for our analysis, the application of these methods to some real data (surveys of galaxy and galaxy clusters). We do not attempt here to be comprehensive in our treatment of this issue, i.e. we do not attempt to present in a comprehensive manner all the data available to us at the time of writing, nor to determine very precisely what constraints they put, in particular, on the scale of homogeneity λ_0. Rather we take some representative samples which give both an illustration of the implementation of the methods we have outlined, and give an indication of the approximate constraints which can be obtained from current data. Using the average conditional density $\langle n(r) \rangle_p$ this leads us to conclude that the galaxy distribution is very well described as a simple fractal with dimension $D \approx 2$ over approximately two orders of magnitude in scale up to approximately 20 Mpc/h[4]. From the analysis of number counts we can constrain the distribution – but more weakly as these are un-averaged statistics affected by many corrections – from scales slightly larger than this up to about 100 Mpc/h. A single effective fractal dimension – albeit slightly larger with $D \approx 2.5$ – gives the best single power fit to the average count. Moreover fluctuations around this average behavior do not appear to decay rapidly as they would in the case of small scale uniformity of the distribution. We conclude therefore, that up to this larger scale, there is no robust evidence for homogeneity, and that future surveys (in particular, the Sloan Digital Sky Survey (SDSS) which is currently underway) will place much tighter constraints. We stress however the following important point. If the homogeneity scale is determined to be finite, the fractal-inspired analysis remains valid in two respects: (i) it is the unique way to detect the homogeneity scale itself without a priori assumptions, and (ii) it gives the right framework to obtain a full geometrical and statistical description of the strongly clustered region.

It is perhaps useful to clarify briefly two related questions: (i) what we mean here by "robust", and (ii) how these constraints compare with those commonly given in the cosmological literature, as there has been considerable controversy on this point[5]. What is widely agreed is that the galaxy distribution is fractal at small scales, at least up to 10 Mpc/h. So the first limit quoted above, based on the conditional density, is not particularly controversial.

---

[4] See Chap. 8 for a definition of this unit of length.
[5] We refer the reader to the relevant literature e.g. [223, 247], and references therein, rather than entering into all the details of this controversy here, where we wish to place the emphasis on methods and not results.

However there is not agreement on the value of the fractal dimension. It is usually quoted in the range $D = 1.2 - 1.5$ rather than the value we have determined through our analysis. Nor is there agreement about our second statement that there is no definite evidence for homogeneity at scales significantly larger than this. The discrepancy in the estimation of the fractal dimension has a simple explanation: it comes from a systematic effect which enters depending on whether the estimation is done with the statistical estimators of $\xi(r)$ or of $\langle n(r) \rangle_p$ (see Sect. 9.3.1 in Chap. 9 for details).

In order to explain the origin of the second (and principal) point of controversy, it is useful to clarify what we mean by the term "robust", as behind it lies our determination of the scale up to which one can obtain constraints from data sets which sample finite regions around us. In order to clarify it, one must analyze the role of finite size effects which enter when one estimates the conditional density in such a finite sample[6]. We consider our results to be robust up to the scale where we can determine an estimate of error for our estimator *without making assumptions beyond that of statistical isotropy and stationarity* (i.e. of translational and rotational invariance of the statistical properties) about the nature of the underlying distribution at the scales of the order of the sample size. In particular, of course, we do not make the a priori assumption of homogeneity. In practice we explain that this means that we use only estimates of the conditional average density which use shells (or spheres) fully enclosed within the sample. Further we obtain our error estimates from the variance between such estimates in non-overlapping regions. Thus to determine our "robust" results we only use our estimator up to scales of order of (but smaller than) the radius of the largest sphere that can be inscribed in the sample about an occupied point.

One can, however, estimate the conditional density using information also from shells partially enclosed in the volume. This is effectively what is done in the standard analysis (i.e. that prevalent in the literature on galaxy correlations), where pair-counting algorithms are used to estimate the reduced two-point correlation function $\xi(r)$, which allows results to be obtained up to the largest distance between points in the sample. Given that several important large redshift surveys (e.g. the Las Campanas survey and the 2dF survey) cover regions which are very narrow angular slices, this corresponds to a scale, of order of the sample depth, which is very much larger than the radius of the largest enclosed sphere, which gives an upper limit for estimation using our methods. One can convert estimates of $\xi(r)$ into an estimate for $\langle n(r) \rangle_p$, and thus extend our analysis apparently to these larger scales. The determination of a value of $r_0 \approx 5\,\mathrm{Mpc/h}$ then corresponds to a flattening of $\langle n(r) \rangle_p$ at a scale of order $10\,\mathrm{Mpc/h}$, and thus one concludes that there is a clear tendency to homogeneity at this scale. The problem, as we discuss in Chap. 9, is that, beyond the scale of the largest enclosed sphere, the estimates are dominated by partially enclosed shells. One can then only

---

[6] In Appendix G we describe these methods.

estimate an error (i.e. the typical deviation from the true ensemble density) by making a (completely untested) assumption about the real variance at these scales (e.g. by assuming that the variance is that estimated in some cosmological theoretical model). If one consistently found the same results for $\xi(r)$ within these error bars, one could make a stronger case for the validity of such models for the variance. However, as we have already mentioned, very considerable variation is actually observed across the literature of these estimates, in particular most strikingly in the amplitude of this function. In the standard analysis these variations are, as we mentioned above, ascribed to a new physical effect ("luminosity segregation"). Thus the question of homogeneity and this secondary hypothesis have become entangled.

Whatever the final outcome of this controversy – it is, fortunately, one which will be resolved by data in the next few years – it is evident that the only way to determine whether the distribution of visible matter is substantially uniform at the sample scale, and to characterize its transition to homogeneity, is to use a framework in which it is not assumed. It is precisely such a framework which is described and applied here.

The Problem of "Bias": Two Models

In the following three chapters the main focus is on the question of how clustering may depend on the kind of object considered. Firstly in Chap. 11 we explain how the analysis of the previous chapters, in which the simplifying assumption was made that space and luminosity correlation properties are independent, can be extended to the more general case where clustering depends on luminosity. This is important as such effects are known to exist in the real galaxy distribution (e.g. large elliptical galaxies are found preferentially in dense clusters). This difference in clustering properties according to luminosity is characterized by a set of multi-fractal exponents, using the formalism described in Chap. 5. Considering real galaxy data such luminosity dependence is found, with brighter galaxies showing a slightly steeper fractal exponent than that of the support (i.e. of the set of all galaxies). The qualitative statement that "bright galaxies are more clustered" thus corresponds quantitatively to the variation of the clustering exponent, rather than the variation of the amplitude of the correlation functions (as in the standard interpretation coming from analysis with $\xi(r)$).

In Chap. 12 we consider the correlation properties of galaxy clusters, of which quite extensive surveys have also been made. Because they are intrinsically brighter than galaxies, they allow one to probe much larger scales. We explain how such surveys can be understood as a coarse-graining of the underlying galaxy distribution, and consider how this allows for the determination of their correlation properties, in the case of a galaxy distribution which is fractal, with or without a cross-over to homogeneity at a finite scale. In particular we note that the fractal exponent of clusters should be the same as that of the galaxies. Using the standard $\xi(r)$ analysis an effect analogous to "luminosity bias" is found: while the exponent remains the same, the amplitude is much larger relative to that of galaxies (typically $r_0 \simeq 15 \div 25$ Mpc/h). In this case this is known as the "galaxy-cluster mismatch". We explain in more detail how both effects can be understood simply in the case that the cross-over from fractality to homogeneity in the galaxy distribution is actually at scales as large as ~ 50 Mpc/h. This is an interpretation of these effects which is also consistent with the analysis of galaxy number counts which we described above (with, however, the slight discrepancy in the best-fit dimension, which for clusters is $D \approx 2$).

In the standard interpretation of galaxy and cluster correlations resulting from the $\xi(r)$ analysis, these observations of the varying amplitude are ascribed to real physics, rather than being simply finite size effects as in the alternative explanation we present. We turn in Chap. 13 to the standard framework which has been developed to describe this effect. Generically this goes under the name of "bias", which is then simply taken to mean any difference between the correlation properties of any class of object (galaxy of different luminosity or morphology, galaxy cluster, quasar...) and that of the "cold" dark matter which, in standard models, dominates the gravitational clustering dynamics in the universe. The case that such a difference manifests itself as an overall normalization of $\xi(r)$ (or $P(k)$), as in the observations we have been discussing, is known as "linear bias".

We analyze in detail the original model, proposed by Kaiser in 1984 [123], to explain this phenomenon. While this is just a simple model, of which many variants have been proposed, it remains canonical in terms of the physical picture it proposes: different kinds of objects are hypothesized to correspond to selections from the underlying field above different thresholds for the density fluctuations, e.g. clusters are formed where there is over-density which is greater than that required to form a galaxy. We note that linear amplification of the "biased" correlation function with respect to the underlying one is found only in the region of weak correlation. In the regime of strong correlation (that which would be relevant to "luminosity bias" and the "galaxy-cluster" mismatch) the correlation functions are related in a very non-linear manner. Given that the amplification of the correlation function is linear at sufficiently large separations r (where the correlations are indeed weak) one might anticipate also a linear amplification for the power spectrum at small k. We show, however, that, for cosmological spectra for which $P(0) = 0$, this is not the case. The reason can be found in the property of super-homogeneity of such spectra: the sampling in the bias model necessarily destroys the surface nature of the fluctuations, as it introduces a volume (Poisson-like) term in the variance. The "primordial" form of the power spectrum is thus not detectable in the power spectrum determined from objects selected in this way. We explain that this conclusion, although shown formally only for the specific model, should hold for any generic model of bias. If a linear amplification is obtained in some regime of scales (as it can be in certain phenomenological models of bias) it is necessarily a result of a fine-tuning of the model parameters.

Some Instruments for the Study of Gravitational Clustering

In the last chapter of the book, we turn to one aspect of the *theoretical* problem of gravitational clustering. In cosmology the main instrument for treating this problem is numerical (in the form of N-body simulations), and the analytic understanding of this crucial problem is very limited. Other than in the regime of very small fluctuations where a linear analysis can be performed, the available models of clustering are essentially phenomenological models with numerous parameters which are fixed by numerical simulation[7]. In these simulations – which typically set out to follow over some range of scales the evolution of the "cold" dark matter – what is done in practice is a particle simulation, i.e. a simulation in which the smoothening length introduced in the gravitational force is much smaller than the initial inter-particle distance. While this evolution is usually (see e.g. [133]) followed essentially in k space, it is interesting to try also to study the evolution of clustering in real space. In this perspective it is very useful to know the statistical properties of the gravitational force on a particle due to all others in a stochastic point distribution. This is the subject of this chapter.

The starting point for such a study is the one-point statistics of this quantity, i.e. the probability distribution of the gravitational force felt by a particle in the system. In practice the problem has been exactly solved (by Chandrasekhar [43]) only in the case of uncorrelated Poisson point processes (when the probability distribution of this force is given by the so-called Holtzmark distribution). In this chapter we describe a simple extension of this result to the case of a so-called Gauss-Poisson process, which has non-trivial two-point correlation properties, but no non-zero higher order correlations. It can thus be considered the first step in increasing correlations beyond the Poisson case. We determine, within a well controlled approximation scheme, the one-point force distribution by identifying the way in which spatial correlations can significantly affect the behavior in the uncorrelated Poisson case. In both cases we consider the contribution to this force coming from the first nearest neighbor particle, and find it to be the dominant contribution. This implies that the gravitational force acting on the single particles is in general a strongly spatially fluctuating quantity. The same kind of study, restricted to the description of the first and second moments of the force and of their dependence on the lower and upper cut-offs, is also presented in the case of statistically stationary and isotropic pure fractal particle distributions. The scaling behavior of these quantities is found to change radically depending on whether the fractal dimension of the system is less than or greater than two (in three dimensions). Finally, we present a brief analysis of the spatial

---

[7] There are notable exceptions in the literature. See e.g. [208], and [64, 65, 66].

correlations between the gravitational forces experienced by different particles in a given distribution. It is observed that, even if the force acting on a particle is strongly fluctuating in space, it nevertheless displays in most cases long range correlation with the force felt by other particles.

1.5 Perspectives for the Future

The aim of this book is, as we said at the beginning, to give, firstly, a systematic introduction to methods of statistical physics used to describe structures in mass distributions with different degrees of geometrical complexity found in many natural phenomena, and then to show a series of applications of these methods to the problem of structure as encountered in cosmology. This second part does not aim to be complete, but is rather an opening onto these problems. Our emphasis has been on the observational questions related to galaxy distributions, and to some extent on how cosmology treats structure in its current models. We have underlined that the observational issues discussed here will, we believe, be resolved in the coming years by new data. In particular for the methods we have proposed to resolve the central questions raised - that of a satisfactory space characterization of clustering, that of homogeneity, and also the related question of luminosity properties of clustering – one needs the depth, uniformity of completeness and wide angular coverage which will be provided by the Sloan Digital Sky Survey within a few years.

These observational questions feed into many fascinating theoretical ones. We only address very tangentially in this book the primary question which arises from the successful application of the concepts of fractality to galaxy clustering: what is the dynamical origin of the observed fractal clustering? what is the origin of the weak multi-fractality when one considers also luminosity? Irrespective of the scale to which a simple fractal behavior is finally found to persist, this problem is posed. Current cosmological models invoke a complicated and fine-tuned chain of different physics (gravitational in the dark matter and non-gravitational in the visible) to explain phenomenologically the observation of power-law clustering over a limited range of scales, with amplitudes for $\xi(r)$ and $P(k)$ which depend on the kind of object. Basic questions remain unanswered: does gravitational dynamics give rise to fractal clustering? And from what initial conditions? N-body simulations of gravitational clustering in cosmology usually aim to reproduce in as much detail as possible the actual universe described by current cosmological models, rather than addressing basic physical questions of this type. We believe there is much unexplored space for a statistical physics approach to such problems, which should be able to shed light on the more general characteristics of gravitational clustering.

In a more strictly cosmological setting the question of the precise scale at which homogeneity in the distribution of visible matter is reached is of

1.5 Perspectives for the Future

considerable importance. Depending on the scale, the theoretical questions posed change. What theories can really predict is the clustering of dark matter, and its homogeneity scale is fixed (to be small, or order of a few Mpc/h) in most standard cosmological models, by the normalization to the fluctuations observed in the CMBR. What difference can be tolerated between this value and the observed homogeneity scale of the distribution of visible matter is, as we have discussed, linked to the theory of bias. By its nature a phenomenological approach, it is therefore difficult to say exactly what bound it places on the transition to homogeneity. As discussed, we highlight in Chap. 13 the difficulty these models have in accounting for the observations. Are there strong theoretical upper bounds on the scale of homogeneity? Many cosmologists would probably reply in the affirmative, citing two factors: incompatibility of inhomogeneity at large scale with FRW solutions of general relativity, and more specifically, with the isotropy of the CMBR. In Appendix E we discuss a model which illustrates that these objections are not well founded. It is a low density FRW model, containing only homogeneous radiation with a superimposed fractal distribution of baryons extending to arbitrarily large scales. It is shown that the latter can, because it is asymptotically empty, be treated as a perturbation to the former. Further it turns out that the CMBR fluctuations estimated are in fact too small to account for those actually observed. While such a model is not viable in detail as a cosmological model, it suggests to us that such unconventional avenues deserve further exploration.[8]

Beyond these problems there are certainly many others in cosmology on which an approach from statistical physics may shed light. An evident one, which we only touch on very superficially here in Chap. 6, is the CMBR. For this case the description of fluctuations in the framework of uniform distributions is valid and appropriate, as there is indeed a very well defined average for the temperature. While cosmologists describe and study the CMBR almost exclusively in terms of their theoretical models for the universe, it may prove fruitful to look at these rich data in a more phenomenological way with the many other instruments used by statistical physicists to study stochastic processes.

---

[8] We note, for example, that Ribeiro in a series of papers [202] has analyzed some numerical solutions of Tolman model which may be useful for the treatment of fractal objects in the framework of General Relativity. An interesting model describing an extremely inhomogeneous matter distribution in General Relativity has been recently presented in [1].

Part I

Statistical Methods

2 Uniform and Correlated Mass Density Fields

2.1 Introduction

In the analysis of finite samples of a stochastic field or a signal characterized by intrinsic fluctuations the simplest hypothesis is to represent it as a uniform average background with superimposed positive and negative stochastic fluctuations[1]. The typical examples for which this approach has been developed and extensively applied are those of an electromagnetic signal with noise in time, or the density field of a fluid at high temperature in space. In the context of astrophysics this approach has been widely applied to the temperature fluctuation field of the cosmic microwave background radiation (CMBR) for which it appears well justified. It has also been extensively applied to the study of the spatial distribution of galaxies in space, where its basic assumption appears to be less well founded. The basic reason behind this approach is of simplicity and convenience: In this perspective one can treat fluctuations as perturbations around a mean field, which represents a major simplification for the analysis and the theoretical description of the system. Then one can develop suitable perturbation theories, which are usually simpler than non-perturbative approaches, in order to capture the relevant aspects of the phenomenon. For all stochastic systems which can be represented in this way, specific statistical tools have been introduced in the past in order to permit a fine and complete statistical analysis. In particular this task is accomplished by the so-called *theory of stationary stochastic processes* (SSP). This chapter is devoted to the description of such a statistical framework.

Let us consider, first of all, which kinds of stochastic systems can be studied in this way. There are all the processes in which the fluctuating signal is continuous in space and such that, if averaged over sufficiently large spatial windows, it becomes practically constant in space. These systems are commonly called *continuous* SSP [68, 100]: The temperature field related to the CMBR belongs to this class. Many other examples can be found in physics and other disciplines: for instance the continuous mass density field of an

---

[1] In this book we are mainly interested in spatial systems (mass density or temperature stochastic fields), but most of statistical descriptions introduced in what follows can be simply translated to temporal processes. For a comprehensive review of stochastic processes see [53, 100].

homogeneous fluid at thermal equilibrium. In general the continuous stochastic field can be positive or negative at different points, while for stochastic mass densities the sign is strictly non-negative. Moreover fluctuations can be more or less spatially correlated.

A more complex case is when a stochastic system, usually a spatial mass (or measure) distribution, is representable as a discrete *point process* [53], that is the subclass of stochastic processes defining stochastic spatial distributions of separated massive point-particles. If it is statistically stationary in space it is called a *stationary point process* (SPP). A simple example is shown in Fig. 2.1 [53]. Note that in this case the stochastic field is discontinuous and strictly non-negative. However, if the average of this discrete mass density is well defined (i.e., sample size independent) and positive, we can apply also the previous representation (i.e., average value + positive and negative stochastic fluctuations) by smoothing the system on some *uniformity or homogeneity scale* and by analyzing small amplitude density fluctuations field around the average. As in the continuous case mass density fluctuations can be more or less spatially correlated (see Fig. 2.2). Thus, when such a point process is smoothed out on a scale much larger than the average inter-particle separation, it can be seen as a *continuous* stochastic process of the previous

Fig. 2.1. Spatial distribution of randomly placed point-particles without correlations (Poisson point process) in two-dimensional Euclidean space. The positive average density is in this case a well-defined concept, in the sense that its estimate from a single, sufficiently large, sample is independent of the sample size

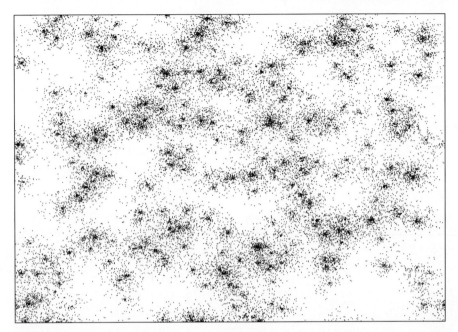

Fig. 2.2. Point process with small-scale correlations. On sufficiently large scales such a point-particle distribution is statistically homogeneous and isotropic as the one shown in Fig. 2.1

class (actually any mass distribution is discrete at sufficiently small scales, and the fluid-like continuous density field picture is permitted because of such a smoothing procedure on scale larger than the "atomic" length). Therefore most of statistical tools introduced for continuous SSP can be extended to this class of discrete distributions.

It is important to note, however, that there are mass (or measure) density fields for which some of the above mentioned statistical tools introduced in this chapter give spurious and invalid results. An important example of these mass densities is constituted by the so-called *fractal measures* [83]. Chap. 4 is devoted to this class of measures, and in particular to those point distributions presenting these features. A fractal mass density is strictly non-negative, but its average value in the infinite volume limit is zero i.e., it is *asymptotically empty* (see Fig. 2.3). However, if we take an arbitrary point belonging to the system, the mass density around it decays slowly (as a power-law) with distance. Therefore the estimator of the average mass density seen by any point of the system in a finite sample gives an infinite relative error with respect to the actual value (i.e., zero). This implies that only statistical averages are defined which are conditioned on the fact that the origin of coordinates is a point of the set, whilst free averages without this condition give the same results as for a completely empty set. The fact

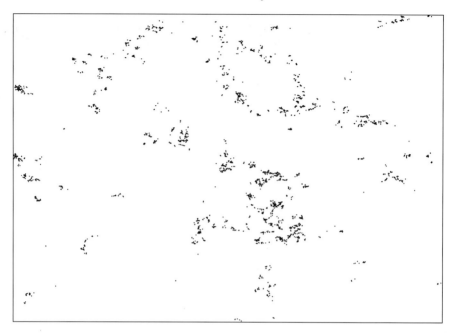

Fig. 2.3. Fractal point distribution with fractal dimension $D = 1.47$ in two dimensional Euclidean space. In this case the estimate of the average density from the single sample is strongly dependent on the sample size, and therefore gives intrinsically spurious estimates of the asymptotic average density: in fact this structure is asymptotically empty

that the asymptotic average density is zero while it decays slowly from any point of the set implies that: (i) it is an *extremely* irregular mass density which is "singular" at every distribution point; (ii) any point of the system sees at finite distance an average density mass fluctuation which is positive and infinite relative to the asymptotic average density.

Consequently, in studying fractals the interest is not in the amplitude of fluctuations, but rather in their *spatial scaling behavior*. All these features imply that many of the statistical tools and quantities introduced for the previous "more regular" stochastic systems are not well defined for fractals. The suitable statistical tools for fractals are presented in Chap. 4.

This brief excursion into the world of fractals has been made to underline that, before applying the statistical framework of ordinary (i.e., sufficiently uniform in space) SSP, presented in this chapter, to the analysis of a stochastic mass (or measure) density, one should verify that the schema "positive average value + small amplitude stochastic correlated (or uncorrelated) spatially decreasing fluctuations" is valid. This is done by verifying primarily the statistical stationarity of the process, that is the homogeneity in space of the statistical properties (see below), and the existence of a well defined positive (i.e., sample independent) average for the density. For this last step it is necessary to use the more general statistical approach suitable for the analysis of fractal densities discussed in Chaps. 4–5. Once the appropriate tests are passed, one may apply the statistical framework of SSP we now describe.

2.2 Basic Statistical Properties and Concepts

In this section we study the basic properties of those stochastic systems (mainly mass density fields) that can be represented as SSP, where *stationary* refers to the spatial statistical properties (it is sometimes called *statistical translational invariance* or *statistical homogeneity*). The single realization of such a SSP can be thought to be a particular mass density field. In the context of cosmology the requirement of statistical stationarity is justified by the fact that in the study of the mass distribution of the universe the cosmological principle is assumed: there are no preferential points or directions in the universe[2]. Clearly this principle must be intended in the statistical sense. It implies that the statistical properties of the considered SSP inside a sample volume should not depend on the location of the sample in the universe.

We now give a brief introduction to the theory of SSP, exposing the general features of correlation analysis, and presenting explicit examples of stochastic distributions, such as Poisson point processes and Gaussian continuous density fields (GDF), to illustrate the essential points.

Let us consider the random mass distribution represented by the microscopic density function or field $\rho(\bm{r})$. In what follows definitions and theorems are valid for density fields representing both cases of continuous fluid-like mass distributions, which are called in the literature *continuous* stochastic processes, and of a discrete particle distribution, called in the literature SPP [53]. In the former case $\rho(\bm{r})dV$ represents the mass (or more generally the measure) contained in the infinitesimal volume dV around the point $\bm{r}$. In the latter case, it gives the number of particles in the same small volume, assuming particles of identical unitary mass. Consequently, in this case $\rho(\bm{r})$ can be represented as

$$\rho(\bm{r}) = \sum_i \delta(\bm{r} - \bm{r}_i) , \qquad (2.1)$$

where $\bm{r}_i$ is the position vector of the particle i of the distribution and $\delta(\bm{x})$ is the Dirac delta function. More generally, if point-particles have different

---
[2] Note that the condition of statistical isotropy and homogeneity is also satisfied by stochastic processes with zero average density in the infinite volume limit (stochastic fractals – see Chap. 4). This condition is sometimes referred to as the "conditional cosmological principle" [47, 151, 223]. The requirement of exact (deterministic) rotational and traslational invariance, which is not satisfied in this case, is a stronger version of the "cosmological principle".

masses, we write:
$$\rho(r) = \sum_i m_i \delta(r - r_i) ,$$

where m_i is the mass of the particle at r_i. Unless explicitly specified, we will study mainly the case of unitary mass particles. Moreover we limit our analysis to *ordinary* or *regular* point processes, which are SPP such that, taking a small volume ΔV in an arbitrary point of the space, the probability of having more than one point-particle inside it is an infinitesimal of higher order than ΔV. This means basically that point-particles are separated from one another by a finite distance.

In both continuous and discrete cases the function $\rho(r)$ is to be thought of as a realization of a continuous SSP or of a SPP. In the case of a continuous field this means that to any point r is associated a positive random variable $\hat\rho(r)$ whose "extracted" value is $\rho(r)$. Therefore the random function $\hat\rho(r)$ can be seen as a continuous set of random variables which may be correlated or not. The stochastic process consists in extracting the value $\rho(r)$ at any point of the space, and it is completely characterized by its "probability density functional" (PDF) $\mathcal{P}[\rho(r)]$. This functional can be interpreted as the joint probability density function of the random variables $\hat\rho(r)$ at every point r. In order to clarify better its role, let us partition the space in a countable set of small cubic cells of volume ΔV centered around the lattice points $r_1, r_2, \ldots$. Let $\overline\rho(r_i)$ be the *local* average density in the cell around the generic point r_i:

$$\overline\rho(r_i) = \frac{1}{\Delta V} \int_{\Delta V(r_i)} d^3 r \rho(r) ,$$

where $\Delta V(r_i)$ is the cubic cell centered around r_i.

We call $p(\overline\rho(r_1), \overline\rho(r_2), \ldots)$ the joint probability density function of the local densities in all the cells. In the limit $\Delta V \to 0$ this function converges to the functional $\mathcal{P}[\rho(r)]$. The knowledge of $\mathcal{P}[\rho(r)]$ (or of $p(\overline\rho(r_1), \overline\rho(r_2), \ldots)$) permits formally the evaluation of the average of any function of the density field.

Once the average operation is defined, we can give a definition of continuity of the stochastic field using it. The usual definition of *continuity* of a stochastic process is the following:

$$\lim_{|\Delta| \to 0} \langle |\hat\rho(r + \Delta) - \hat\rho(r)|^2 \rangle = 0 . \tag{2.2}$$

where $\langle \ldots \rangle$ is the ensemble average over all the possible realizations of the stochastic process i.e., the average over $\mathcal{P}[\rho(r)]$.

For a discrete stochastic distribution of particles, on the other hand, the SPP generating the function $\rho(r)$, of the form of (2.1), can be considered as follows. Once the space is divided into sufficiently small cells of volume ΔV as in the previous case, the stochastic point process consists in occupying the cells with a point-particle with a certain probability (as clarified

2.2 Basic Statistical Properties and Concepts

above, we consider that the cells are so small that the probability of occupation of a single cell with two or more particles can be neglected). Let n_i be the random variable "number of particles" in the cell centered around $\boldsymbol{r}_i$ that takes the value 1 if the cell is occupied or 0 if the cell is unoccupied. The corresponding values of the *local average density* $\overline{\rho}(\boldsymbol{r}_i)$ in the ith cell are respectively $1/\Delta V$ and 0. In general the probabilities of occupation of different cells are not independent of each other. In terms of the variables n_i, a general configuration of such a discretized system is given by a sequence of 1 and 0 describing the state of occupation of all the cells, while using local densities the configuration will be given by a sequence of $1/\Delta V$ and 0. The joint probability of each possible occupation configuration of the whole set of cells determines completely the point process (and the average operation) in the limit $\Delta V \to 0$. Therefore in this case $\mathcal{P}[\rho(\boldsymbol{r})]$ can be seen as the limit $\Delta V \to 0$ of such a joint probability distribution.

Let us now clarify better the meaning of "spatial stationarity" for a stochastic process. By this it is meant that the functional $\mathcal{P}[\rho(\boldsymbol{r})]$ is invariant under spatial translations. This property is also called *statistical homogeneity* or translational invariance of the stochastic process. This definition implies the following more intuitive property for the case of a continuous mass density field: If we take an arbitrary set of N points $\boldsymbol{r}_1, \boldsymbol{r}_2, \ldots, \boldsymbol{r}_N$, the joint probability density $p[\rho(\boldsymbol{r}_1 + \boldsymbol{r}_0), \rho(\boldsymbol{r}_2 + \boldsymbol{r}_0), \ldots, \rho(\boldsymbol{r}_N + \boldsymbol{r}_0)]$ of the value of the random field at the translated points $\boldsymbol{r}_1 + \boldsymbol{r}_0, \boldsymbol{r}_2 + \boldsymbol{r}_0, \ldots, \boldsymbol{r}_N + \boldsymbol{r}_0$ does not depend on the translation vector $\boldsymbol{r}_0$. In other words it depends only on the relative vectorial separations among the different points. For the discrete case this property can be rephrased as follows: taken the arbitrary set of N points $\boldsymbol{r}_1, \boldsymbol{r}_2, \ldots, \boldsymbol{r}_N$ and a small volume ΔV, the joint probability that at the same time the cells of volume ΔV around the translated points $\boldsymbol{r}_1 + \boldsymbol{r}_0, \boldsymbol{r}_2 + \boldsymbol{r}_0, \ldots, \boldsymbol{r}_N + \boldsymbol{r}_0$ have an arbitrary state of occupation does not depend on the translation vector $\boldsymbol{r}_0$. In particular this implies that, if we take a finite volume of size V centered (in some well-defined way) around the point $\boldsymbol{r}_0$, then the probability of having N particles inside it does not depend on $\boldsymbol{r}_0$, but only on V and on the shape and orientation of the volume.

In what follows, we often suppose that the mass distribution is also *statistically isotropic*, that is $\mathcal{P}[\rho(\boldsymbol{r})]$ is also invariant under spatial rotation. Most of the definitions and properties introduced below are based only on spatial stationarity.

Let ρ_0 be the average value of the stochastic field:

$$\langle \hat{\rho}(\boldsymbol{r}) \rangle = \rho_0 \ . \tag{2.3}$$

As already mentioned if the analyzed stochastic field is a mass or measure density, then we can apply the ordinary SSP (continuous or SPP) framework if and only if $\rho_0 > 0$. It is this which we refer to as the condition of *uniformity*, or alternately *homogeneity*. In the case that the latter term is used, it should not be confused with the (much weaker) property of statistical homogeneity (i.e., stationarity) of the stochastic field. As we have discussed this latter refers only to the invariance of the statistical properties of the stochastic field under spatial translations.

2.2.1 Spatial Averages and Ergodicity

Before discussing the definition of correlation functions, it is important to introduce another property of the SSP or SPP we consider, which is needed to give a full statistical meaning to volume averages. A typical assumption in the statistical analysis of stochastic fields is the so-called *ergodicity* of the stochastic process which generates the mass field both in the continuous and discrete case. In order to clarify the meaning of *ergodicity*, let us take a generic observable $F = F(\rho(\mathbf{r}_1), \rho(\mathbf{r}_2), \ldots)$ of the mass distribution $\rho(\mathbf{r})$. Ergodicity means that $\langle F \rangle$ is equal to the spatial average $\overline{F}$ given by:

$$\overline{F} = \lim_{V \to +\infty} \frac{1}{V} \int_V d^3 r_0 F(\rho(\mathbf{r}_1 + \mathbf{r}_0), \rho(\mathbf{r}_2 + \mathbf{r}_0), \ldots) , \qquad (2.4)$$

where V is the integration volume and $\lim_{V \to +\infty}$ means that the limit of the integration is taken over all space Ω (with volume $\|\Omega\| = +\infty$). Finally, $\rho(\mathbf{r})$ is almost any realization of the mass distribution "extracted" from the probability functional $\mathcal{P}[\rho(\mathbf{r})]$. This property is also referred to as the *self-averaging* property of the distribution. Note that if the average in (2.4) is extended only to a finite sub-sample V of the whole space Ω, then (2.4) is only an *estimator* of $\langle F \rangle$ in the given sub-sample. In cosmology one typically has only such finite volume estimators. Therefore the assumption of ergodicity is necessary if we want to use these statistical estimators of some specific quantities to build or verify hypotheses and theories. An analysis of real space estimators is presented in Chap. 9 and in the Appendixes F-G.

The assumption of *ergodicity* is based on a theorem of continuous stochastic processes: the ergodic theorem of Birkhoff-Khinchin which states that if $\rho(\mathbf{r})$ has a well-defined average value ρ_0, then the volume average, in the infinite volume limit, converges with probability one to a well-defined limit [100].

2.2.2 Homogeneity and Homogeneity Scale

Let us now consider the meaning of *homogeneity* (or uniformity) given by (2.3) in terms of the spatial average in a single realization of a stochastic mass distribution. Due to the assumption of ergodicity of the stochastic process, for a single realization of the mass distribution (i.e., a strictly non-negative field) the existence of a well-defined average positive density implies that [90]

$$\lim_{R \to \infty} \frac{1}{\|C(R; \mathbf{x}_0)\|} \int_{C(R, \mathbf{x}_0)} \rho(\mathbf{r}) d^3 r = \rho_0 > 0 \ \forall \mathbf{x}_0 . \qquad (2.5)$$

where $\|C(R, \boldsymbol{x_0})\| \equiv 4\pi R^3/3$ is the volume of the sphere $C(R, \boldsymbol{x_0})$ of radius R, centered on an arbitrary point $\boldsymbol{x_0}$. When (2.5) is valid i.e., a well-defined positive average density exists for the mass distribution, the characteristic *homogeneity scale* λ_0 (or uniformity scale) can be defined as the scale such that

$$\left| \frac{1}{C(R; \boldsymbol{x_0})} \int_{C(R;\boldsymbol{x_0})} d^3 r\, \rho(\boldsymbol{r}) - \rho_0 \right| < \rho_0 \quad \forall R > \lambda_0, \forall \boldsymbol{x_0}\,. \tag{2.6}$$

In the next section we will see that this scale substantially depends on the nature of the correlations between density fluctuations at different points i.e., on the two-point correlation function. This scale gives basically the distance above which fluctuations can be considered small with respect to the mean density and a perturbative approach can be appropriate to describe the physics of the system. At smaller scales, however, fluctuations can be large and irregular, as in a fractal with a crossover to homogeneity, and must be treated in a different statistical and physical framework (see Chap. 4).

2.3 Correlation Functions

In this section we introduce the definitions and the principal properties of the auto-correlation functions for a spatially SSP. The quantity

$$\langle \hat{\rho}(\boldsymbol{r_1}) \hat{\rho}(\boldsymbol{r_2}) \ldots \hat{\rho}(\boldsymbol{r_l}) \rangle$$

is called the *complete* l-point correlation function of the density field. For more general correlation functions characterizing different morphological and statistical aspects of stochastic mass distributions, see e.g., [230]. Statistical stationarity imply that the l-point correlation functions $\langle \hat{\rho}(\boldsymbol{r_1}) \ldots \hat{\rho}(\boldsymbol{r_l}) \rangle$, for any l, depend only on the vectorial relative distances among the l points [90] $\boldsymbol{r_i} - \boldsymbol{r_j}$ with $i, j = 1, \ldots, l$.

In the discrete case of a SPP the quantity

$$\langle \hat{\rho}(\boldsymbol{r_1}) \hat{\rho}(\boldsymbol{r_2}) \ldots \hat{\rho}(\boldsymbol{r_l}) \rangle\, dV_1, dV_2, \ldots, dV_l$$

gives the a priori probability of finding l particles simultaneously, in a single realization, placed one to one in the infinitesimal volumes $dV_1, dV_2, \ldots, dV_l$ respectively around $\boldsymbol{r_1}, \boldsymbol{r_2}, \ldots, \boldsymbol{r_l}$. Indeed, in this case, because of (2.1), $\rho(\boldsymbol{r})dV$ is equal to unity if there is one particle in the volume element dV around $\boldsymbol{r}$, and zero if not.

Let us analyze in more detail the two and three-point correlation functions introducing the concept of *reduced* correlation functions. As mentioned above, due to the hypothesis of statistical stationarity and isotropy (SSI) $\langle \hat{\rho}(\boldsymbol{r_1}) \hat{\rho}(\boldsymbol{r_2}) \rangle$ depends only on $r_{12} = |\boldsymbol{r_1} - \boldsymbol{r_2}|$, and $\langle \hat{\rho}(\boldsymbol{r_1}) \hat{\rho}(\boldsymbol{r_2}) \hat{\rho}(\boldsymbol{r_3}) \rangle$ is a function only of $r_{12} = |\boldsymbol{r_1} - \boldsymbol{r_2}|$, $r_{23} = |\boldsymbol{r_2} - \boldsymbol{r_3}|$ and $r_{13} = |\boldsymbol{r_1} - \boldsymbol{r_3}|$. The reduced two and three-point correlation functions are defined respectively as [25, 182]

$$C_2(r_{12}) = \langle (\hat{\rho}(\mathbf{r}_1) - \rho_0)(\hat{\rho}(\mathbf{r}_2) - \rho_0) \rangle \qquad (2.7)$$
$$C_3(r_{12}, r_{23}, r_{13}) = \langle (\hat{\rho}(\mathbf{r}_1) - \rho_0)(\hat{\rho}(\mathbf{r}_2) - \rho_0)(\hat{\rho}(\mathbf{r}_3) - \rho_0) \rangle .$$

The quantity $C_2(r_{12})$ is also called *covariance* function: This is the principal function used to study and characterize spatial correlations between fluctuations from the average value. For a generic stochastic process ρ_0 can be positive, negative or zero. However, in the case of a discrete or continuous mass distribution with a well defined positive average density $\rho_0 > 0$, (2.7) are usually defined in a dimensionless form as

$$\tilde{\xi}(r_{12}) = \frac{C_2(r_{12})}{\rho_0^2} \qquad (2.8)$$

$$\tilde{\zeta}(r_{12}, r_{23}, r_{13}) = \frac{C_3(r_{12}, r_{23}, r_{13})}{\rho_0^3} . \qquad (2.9)$$

Note that in the case of a mass field $\hat{\rho}(\mathbf{r})$ is a non-negative quantity; therefore $\tilde{\xi}(r) \geq -1$ at any r. We can write in general the relations between complete and reduced two and three-point correlation functions as:

$$\langle \hat{\rho}(\mathbf{r_1})\hat{\rho}(\mathbf{r_2}) \rangle \equiv \rho_0^2 + C_2(r_{12}) \qquad (2.10)$$
$$\langle \hat{\rho}(\mathbf{r_1})\hat{\rho}(\mathbf{r_2})\hat{\rho}(\mathbf{r_3}) \rangle \equiv \rho_0^3 + C_2(r_{12}) + C_2(r_{23}) +$$
$$C_2(r_{13}) + C_3(r_{12}, r_{23}, r_{13}) . \qquad (2.11)$$

The reduced two and three-point correlation functions just introduced are also called *connected* because in field theory they are given by the sum only of the connected Feynman graphs defining the complete correlation functions [3]. Unfortunately for the connected n-points correlation functions with $n \geq 4$ we cannot extend (2.7) simply but need to introduce the cumulant expansion which is the usual way to isolate the connected parts of the complete correlation functions. This is done explicitly in the next subsection.

2.3.1 Characteristic Function and Cumulants Expansion

We now give the simple mathematical recipe to find the connected n-point correlation functions for any value of n [98]. To simplify the presentation we limit the discussion to a vector $\mathbf{X} \equiv \{X_1, X_2, \ldots, X_N\}$ of N random variables X_i whose value is x_i, instead of treating the more general case of a stochastic field $\hat{\rho}(\mathbf{r})$ depending on the continuous parameter $\mathbf{r}$. The vector $\mathbf{X}$ can be thought of as the field $\hat{\rho}$ in the discretized space.

Let $p(\mathbf{x})$ be the joint probability density function of all the components of $\mathbf{X}$. The characteristic function of the stochastic vector $\mathbf{X}$ is defined as [100]:

$$\tilde{p}(\mathbf{k}) = \langle \exp(-i\mathbf{k} \cdot \mathbf{X}) \rangle = \int d^N x \, p(\mathbf{x}) \exp(-i\mathbf{k} \cdot \mathbf{x}) \,, \tag{2.12}$$

where $\mathbf{k} \equiv \{k_1, k_2, \ldots, k_N\}$. The fact that $p(\mathbf{x})$ is a probability density function, and the general theorems of the Fourier transform (FT) imply, among others, the following simple properties of the characteristic function $\tilde{p}(\mathbf{k})$:

1) $\tilde{p}(\mathbf{0}) = 1$.

2) $|\tilde{p}(\mathbf{k})| \leq 1$.

3) $\tilde{p}(\mathbf{k})$ is a uniformly continuous function of its argument for all real $\mathbf{k}$.

4) If the moment $\left\langle \prod_{i=1}^{N} X_i^{l_i} \right\rangle$ exists, it is given by:

$$\left\langle \prod_{i=1}^{n} X_i^{l_i} \right\rangle = \left[\prod_{i=1}^{n} \left(i \frac{\partial}{\partial k_i} \right)^{l_i} \tilde{p}(\mathbf{k}) \right]_{\mathbf{k}=\mathbf{0}}. \tag{2.13}$$

The moments given in (2.13) are also called the *complete* correlation coefficients of the stochastic vector $\mathbf{X}$, and they are the discrete analogues of the correlation functions $\langle \hat{\rho}(\mathbf{r_1}) \hat{\rho}(\mathbf{r_2}) \ldots \hat{\rho}(\mathbf{r_l}) \rangle$ for the stochastic field $\hat{\rho}(\mathbf{r})$ depending on the continuous parameter $\mathbf{r}$. Equation 2.13 implies that the Taylor series of $\tilde{p}(\mathbf{k})$, if it exists, can be written as:

$$\tilde{p}(\mathbf{k}) = \sum_{j=0}^{\infty} (-i)^j \sum_{l_1,\ldots,l_N}^{0,\infty} \frac{\left\langle X_1^{l_1} \ldots X_N^{l_N} \right\rangle}{l_1! \ldots l_N!} \delta\left(j, \sum_{m=1}^{N} l_m\right) k_1^{l_1} \ldots k_N^{l_N} \,, \tag{2.14}$$

where $\delta(a, b)$ is the Kronecker delta function which is 1 if $a = b$ and 0 otherwise.

5) Knowing $\tilde{p}(\mathbf{k})$ one can obtain $p(\mathbf{x})$ by simply inverting the FT:

$$p(\mathbf{x}) = (2\pi)^{-N} \int d^N k \, \tilde{p}(\mathbf{k}) \exp(i\mathbf{k} \cdot \mathbf{x}) \,. \tag{2.15}$$

6) If the variables $\{X_i\}$ are independent i.e., $p(\mathbf{x}) = \prod_i p_i(x_i)$, then

$$\tilde{p}(\mathbf{k}) = \prod_{i=1}^{N} \tilde{p}_i(k_i) \,, \tag{2.16}$$

where $\tilde{p}_i(k_i)$ is the FT of $p_i(x_i)$. 6-bis) If $Y = \sum_{i=1}^{N} X_i$, where the variables X_i are mutually independent, then the characteristic function $\phi_y(k)$ of the stochastic variable Y is simply given by

$$\phi_y(k) = \prod_{i=1}^{N} \tilde{p}_i(k) \,. \tag{2.17}$$

We can now introduce the *cumulant generating function* $\psi(\mathbf{k})$ which is defined by:
$$\psi(\mathbf{k}) = \log \tilde{p}(\mathbf{k}) . \tag{2.18}$$

Let us assume that all the moments of $p(\mathbf{x})$ exist and are finite, so that $\tilde{p}(\mathbf{k})$, and therefore $\psi(\mathbf{k})$, can be expanded in Taylor series. We then have

$$\psi(\mathbf{k}) = \sum_{j=0}^{\infty} (-i)^j \sum_{l_1,\ldots,l_N}^{0,\infty} \frac{\mathcal{C}(X_1^{l_1},\ldots,X_N^{l_N})}{l_1!\ldots l_N!} \delta\left(j, \sum_{m=1}^{N} l_m\right) k_1^{l_1}\ldots k_N^{l_N} . \tag{2.19}$$

The quantities $\mathcal{C}(X_1^{l_1},\ldots,X_N^{l_N})$ are called *cumulants* of the stochastic vector **X**. They are also called *connected* correlation coefficients. In order to find the relationship between cumulants and complete correlation coefficients one has to use (2.14) and (2.19) linked by (2.18) (note that for small $|\mathbf{k}|$ the characteristic function takes the from $\tilde{p}(\mathbf{k}) = 1 + \delta\tilde{p}(\mathbf{k})$ where $\delta\tilde{p}(\mathbf{k})$ is small). In general the relations found between $\mathcal{C}(X_1^{l_1},\ldots,X_N^{l_N})$ and the moments are not simple. The first three cumulants are given by:

$$\mathcal{C}(X_i) = \langle X_i \rangle \tag{2.20}$$
$$\mathcal{C}(X_i, X_j) = \langle X_i X_j \rangle - \langle X_i \rangle \langle X_j \rangle$$
$$\mathcal{C}(X_i, X_j, X_k) = \langle X_i X_j X_k \rangle - \langle X_i X_j \rangle \langle X_k \rangle - \langle X_i \rangle \langle X_j X_k \rangle -$$
$$\langle X_i X_k \rangle \langle X_j \rangle + 2 \langle X_i \rangle \langle X_j \rangle \langle X_k \rangle .$$

We see that if $\langle X_i \rangle = x_0$ for any $i = 1, \ldots, N$ then the second and the third equation of (2.20) are the discrete equivalent of (2.7).

As further clarified in what follows, the importance of the cumulants or connected correlation functions is due to their behavior in the context of Gaussian distributions. In fact if

$$p(\mathbf{x}) = B \exp\left[-\sum_{i,j}^{1,N} (x_i - \langle X_i \rangle) A_{i,j} (x_j - \langle X_j \rangle)\right] ,$$

where $\hat{A} \equiv ((A_{i,j}))$ is a positive definite matrix and B a normalization constant, then it is simple to show that all cumulants, except $\mathcal{C}(X_i)$ and $\mathcal{C}(X_i, X_j)$ with $i, j = 1, \ldots, N$, are zero. Therefore cumulants of order larger than two are a measure of deviations from Gaussianity for multivariate distributions. These results can be easily extended to the case of stochastic fields $\hat{\rho}(\mathbf{r})$ depending on a continuous parameter.

Finally, the origin of the term *connected* comes from quantum field theory (in fact a quantum field can be seen as a continuous stochastic process). More precisely, it comes from the Feynman graphs expansion of complete correlation functions in field theory [3]. In this context one can see that the cumulants are given by only the sum of *connected* graphs.

2.3.2 Correlation Length

Let us comment on the definition given in (2.8). First of all, we see that $C_2(r)$ (or equivalently $\tilde{\xi}(r)$ in the case of a mass distribution with $\rho_0 > 0$) measures the spatial *memory* of mass density fluctuations on the scale r. In fact, if $C_2(r)$ is different from zero, it means that the knowledge of the mass density at a generic point of the space influences the conditional probability density of the mass density at another point at distance r from the first one. In statistical physics, in order to characterize through a single number the persistence of correlations in the fluctuation field, the concept of *correlation length* r_c has been introduced. Different and practically equivalent definitions of r_c can be found in literature, all sharing the same characteristic of distinguishing systems with slow decay of correlations from systems where the decay is fast. A definition is [148]:

$$r_c^2 = \frac{\int_\Omega d^d r \, r^2 |C_2(r)|}{\int_\Omega d^d r \, |C_2(r)|} \ . \tag{2.21}$$

The concept of correlation length has been introduced basically to distinguish two cases (in a d-dimensional space): (i) $C_2(r) \sim r^{-\gamma}$ with $0 < \gamma < d$, and (ii) $C_2(r) \sim \exp(-r/r_*)$ at sufficiently large r. The case (i) is typical of the correlation function of the order parameter of a thermodynamic system at the critical point of a second-order phase transition, while (ii) is the ordinary behavior of it far from criticality. The main difference between (i) and (ii) is that in the first case $\int d^d r \, C_2(r) \to \infty$, while in the second one the same integral is finite. Every definition found in the literature will give $r_c \to +\infty$ for the former case, and $r_c \sim r_*$ for the latter. The reason for this kind of definition of r_c for statistical mechanical systems is that for such systems one can prove *fluctuation-dissipation* theorems [116] which state that the response of the system to a localized small perturbation is proportional to the integral of the reduced correlation function. Therefore a small localized perturbation is felt by the system in a region of size r_c, while in the case of a critical structure it is felt up to the system size [148]. Moreover, as shown in Sect. 2.10, the correlation length is the characteristic scale beyond which integrated fluctuations in a finite volume e.g., the mass fluctuations in a finite volume V in a stochastic mass density field, can be considered a Gaussian variable. In Fig. 2.4 we show the typical behaviors of $C_2(r)$ for a finite and infinite correlation length.

Note that the definition of r_c given by (2.21), as any other used in statistical physics, does not depend on the *amplitude* of $C_2(r)$ but only on the rate of its decay. This is an important point in the context of cosmology as the term *correlation length* is often used for the spatial scale marking the cross-over from large to small fluctuations. This can lead to confusion in particular when the system is characterized by power-law correlations. Instead, the scale λ_0 given by (2.6), as discussed in what follows, is a more appropriate definition of this cross-over scale which should be more appropriately called the *homogeneity* scale.

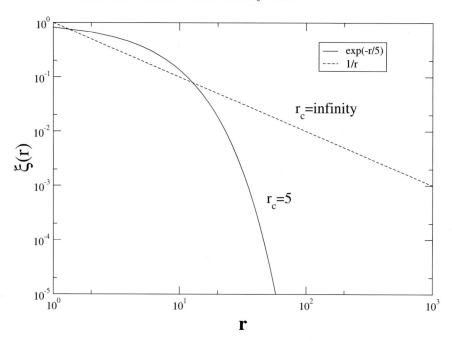

Fig. 2.4. Behavior of $C_2(r)$ for a stochastic field with a finite correlation length $r_c = 5$ (*solid line*), which is manifested by an exponential cut-off of the two-point correlation function. Instead, in the case of an infinite correlation length (*dashed line*) $C_2(r)$ has a power-law decay (as r^{-1} in this case). The amplitude of the correlation function is not related to the value of the correlation length, which instead describes the rate of decrease of $C_2(r)$ i.e., its spatial *persistence* at large scales

2.3.3 Other Properties of the Reduced Two-Point Correlation Function

Let us now analyze some general properties of $C_2(r)$ and $\tilde{\xi}(r)$ in d-dimensions.

- In the case of a continuous SSP it is possible to show from (2.2) that $C_2(r)$ is a continuous function of r.
- We have seen that, in the case of a continuous SSP (2.2) must hold. It is simple to show that this implies that $C_2(0) = s^2$, where s^2 is the variance (usually finite) of the density field in a single point i.e., $s^2 = (\langle \hat{\rho}(r)^2 \rangle - \rho_0^2)$. In general for a homogeneous *mass* density field ($\rho_0 > 0$) $s^2/\rho_0^2 < 1$ in order to have the mass density positive at any point. As shown below, for a point process (i.e., a discrete particle distribution) $s^2 \to +\infty$. Accordingly we have that $\tilde{\xi}(0) = +\infty$.
- It can be shown that, for a stationary ergodic process, the auto-correlation function has its maximum at $r = 0$:

$$C_2(0) \geq |C_2(r)|.\qquad(2.22)$$

- It can be demonstrated that the correlation function of any well defined SSP (here we impose only spatial stationarity and not isotropy) satisfies the property

$$\langle f|f\rangle = \int\int d^d r_1 d^d r_2 f(\bm{r}_1) C_2(\bm{r}_1 - \bm{r}_2) f(\bm{r}_2) \geq 0 \ . \tag{2.23}$$

for any real $f(\bm{r})$ in the appropriate class of functions (e.g., with a well defined FT). That is, $C_2(\bm{r}_1 - \bm{r}_2)$ is a positive definite *correlation kernel*. Therefore it is not necessarily possible to find a SSP for which a given arbitrary function represents the reduced two-point correlation function.

For a *continuous* SSP one can formulate such a property in a more strict way as a *necessary and sufficient* condition for the existence of a continuous SSP with a given $C_2(\bm{r})$. This is the so-called Wiener-Khinchin (or only Khinchin) theorem (which can be seen as a special case of the Bochner theorem) [100, 230]: Given a function $C_2(\bm{r})$ a necessary and sufficient condition for the existence of a SSP with $C_2(\bm{r})$ as reduced two-point correlation function is that $C_2(\bm{r})$ can be written as

$$C_2(\bm{r}) = \frac{1}{(2\pi)^d} \int P(\bm{k}) \exp(i\bm{k}\cdot\bm{r}) d^d k \tag{2.24}$$

where $P(\bm{k}) \geq 0 \ \forall \bm{k}$, and integrable in k-space. We address the properties of the FT (2.24) of the correlation function in Chap. 3.

For a discrete SPP a proposition similar to the Wiener-Khinchin theorem can be formulated. However, as shown more clearly in Chap. 3, it gives only a necessary condition for the existence of a SPP with a given correlation function $C_2(\bm{r})$. At present a necessary and sufficient condition for this class of stochastic process is not yet known.

Going back to continuous SSPs, note that, for instance, a box function

$$C_2(r) = \begin{cases} A & \text{if } r \leq r_c \\ 0 & \text{otherwise} \end{cases} \tag{2.25}$$

cannot be an acceptable two-point correlation function. One can see this because (i) it is not continuous (ii) its FT (in $d = 1$) is $2A\sin(kr_c)/k$ which is not positive definite [182].

2.3.4 Mass Variance

Let us now limit the discussion to (non-negative) mass fields (with well defined $\rho_0 > 0$) in $d = 3$ (extensions to other dimensions is straightforward), and analyze the mass fluctuations in finite volumes relating them to the density-density correlation functions. In particular in this section we consider the mass fluctuation in a generic sphere of radius R with respect to the average mass. Let $M(R) = \int_{C(R)} \hat{\rho}(\bm{r}) d^3 r$ be the stochastic variable giving the mass included inside the sphere $C(R)$ of radius R. In Sect. 2.10, by using the central limit theorem of probability theory, we will see that, if the mass density field has a finite correlation length r_c, the quantity $M(R)$ for $R \gg r_c$ can be considered a Gaussian variable, while this is not the case if $r_c \to \infty$.

Fluctuations of this mass in different spheres with the same radius can be estimated by the normalized mass variance $\sigma^2(R)$ defined as

$$\sigma^2(R) = \frac{\langle M(R)^2 \rangle - \langle M(R) \rangle^2}{\langle M(R) \rangle^2} , \qquad (2.26)$$

where

$$\langle M(R) \rangle = \int_{C(R)} d^3 r \langle \hat{\rho}(\mathbf{r}) \rangle = \rho_0 \|C(R)\| = \frac{4\pi}{3} \rho_0 r^3 , \qquad (2.27)$$

and

$$\langle M(R)^2 \rangle = \int_{C(R)} d^3 r_1 \int_{C(R)} d^3 r_2 \langle \hat{\rho}(\mathbf{r}_1) \hat{\rho}(\mathbf{r}_2) \rangle . \qquad (2.28)$$

Note that, because of the assumed translational invariance of $\mathcal{P}[\rho(\mathbf{r})]$, (2.28) does not depend on the location of the center of the sphere. Therefore we can assume the origin of coordinates as the center of the sphere. The extension of the definition of the mass variance to a generic volume V is straightforward from (2.26)–(2.28) by substituting $C(R)$ with V. In general, for a statistically stationary mass density field with well defined $\rho_0 > 0$, substituting (2.10) in (2.26), we obtain

$$\sigma^2(R) = \frac{1}{\|C(R)\|^2} \int_{C(R)} d^3 r_1 \int_{C(R)} d^3 r_2 \tilde{\xi}(|\mathbf{r}_1 - \mathbf{r}_2|) . \qquad (2.29)$$

Equation 2.29 makes explicit the relation between fluctuations in one-point properties (as in this case the mass or the number of points in a sphere) and two-point correlations. In general similar links can be found between fluctuations in n-point properties and $(n+1)$-point correlations.

One can show that for a SSP with a well defined positive average ρ_0, the statistical counterpart of the uniformity or homogeneity condition (2.5) is

$$\lim_{R \to \infty} \sigma^2(R) = 0 . \qquad (2.30)$$

This implies directly that $\tilde{\xi}(r) \to 0$ for $r \to +\infty$, which is thus a necessary condition for there to be a well defined average value of the density field: This is satisfied both by continuous stochastic mass densities and discrete point processes with $\rho_0 > 0$.

An alternative and slightly different definition to (2.6) for the scale characterizing homogeneity is given by the scale at which $\sigma^2(R)$ is equal to unity (or some other appropriate fiducial value). In the literature about the distribution of matter in the universe (galaxies, clusters etc.) there is no global convention about how this scale is defined; in fact it is a scale which is almost

2.3 Properties of the Two-Point Correlation Function

never discussed in precise terms. The two most commonly used quantities in the characterization of the two-point properties are (i) the scale r_0 defined by $\xi(r_0) = 1$, and (ii) the amplitude of the normalized mass variance at a fiducial physical scale, taken to be $8h^{-1}$ Mpc (see Chap. 8 for the definition of this unit of length). Given (or having determined) the dependence on scale of the correlation function or normalized mass variance, these can be easily related to one another.

The scale r_0 has, unfortunately, been widely referred to in the cosmological literature as the "correlation length". As already mentioned, it has no relation to the statistical physics use of the same term, which is a scale characterizing the rate of decay of fluctuations, and not of their amplitude. (See [97] for a clear discussion of this point. A practical working definition of the homogeneity scale applicable in the analysis of galaxy surveys, and a discussion of the current status of this scale, is given in [140] and in Chap. 8).

Let us return to a further analysis of (2.29) and (2.30). It is very important to note that the condition given by (2.30), which holds for any mass distribution generated by a SSP with a positive ρ_0, is very different from the requirement

$$\int d^3 r \tilde{\xi}(r) = 0 \, , \tag{2.31}$$

where the integral is extended to all the space. Equation (2.31) is a much stronger special condition which holds for certain mass density fields – the ones to which we will ascribe the name "super-homogeneous" [91] (in Chap. 3). Finally, we note that in cosmology the following approximation is often used (e.g., [184, 208])

$$\int_{C(R)} d^3 r_1 \int_{C(R)} d^3 r_2 \tilde{\xi}(|\mathbf{r_1} - \mathbf{r_2}|) \approx \|C(R)\| \int_{C(R)} d^3 r \tilde{\xi}(\mathbf{r}) \, , \tag{2.32}$$

in particular, in evaluating the variance through (2.29). Such an approximation is not always valid even for large R, and the convergence properties of the double integral need to be examined carefully to establish it. In particular it does not hold when the condition (2.31) is satisfied by the system. This will be evident following the analysis we give in Chap. 3, where we will show that, for any mass distribution (continuous or discrete), $\sigma^2(R) = R^{-a}$ where $0 < a \le d+1$ (where d is the space dimension) at large R. Using the approximation given by (2.32), one could apparently obtain through (2.29) arbitrarily rapidly decaying behaviors with an appropriate power-law behavior in the correlation function. A detailed analysis of mass fields satisfying (2.31), and a complete classification of all the possible mass distributions coming from a SSP or a SPP with $\rho_0 > 0$, based on an analysis of the mass variance, will be given in Chap. 3.

2.4 Poisson Point Process

In this section we present the simplest example of a SPP (i.e., spatially stationary random point-particle distribution): the three-dimensional Poisson point process. The generalization to other dimensions is straightforward. The stochastic process generating such a distribution can be defined as follows. We partition all of space into small cells of volume dV, and we consider the following Poisson occupation process of each cell independently of all other ones: Let us chose a positive number $\rho_0 > 0$ (dV must be chosen in such a way $dV \ll 1/\rho_0$) and either occupy the cell with a particle of unitary mass with probability $\rho_0 dV$ or leave it empty with the complementary probability $1 - \rho_0 dV$. This process defines the stochastic density field in each cell as follows

$$\hat{\rho}(r) = \begin{cases} \frac{1}{dV} & \text{with probability} \quad \rho_0 dV \\ 0 & \text{with probability} \quad 1 - \rho_0 dV \end{cases} \quad (2.33)$$

where r is the center of the cell of volume dV. Note that we have assumed no correlation between different cells i.e., the knowledge of the occupation status of a cell does not influence the probability of occupation of any other cell. Therefore $\rho_0 > 0$ is the only parameter characterizing this SPP. We have also excluded the possibility of a double occupation of a cell. This is not a limitation in the limit $dV \to 0$, as the probability of double occupation of a single cell is an infinitesimal of higher order in dV. Equation (2.33) means that in each cell the microscopic density follows a binomial distribution. Given this definition of the SSP it is simple now to calculate any average over the process (i.e., over $\mathcal{P}[\rho(r)]$). First of all one obtains

$$\langle \hat{\rho}(r) \rangle = \rho_0 \,, \quad (2.34)$$

and the homogeneity scale (using (2.6)) is given by

$$\lambda_0 \simeq \frac{1}{\rho_0^{1/3}} \quad (2.35)$$

which means that the average size of a void in this kind of particle distribution is of the order of the mean separation between two nearest neighbor particles (see Fig. 2.1 and Fig. 2.5).

The lack of correlations implies

$$\langle \hat{\rho}(r_1)\hat{\rho}(r_2) \rangle = \begin{cases} \rho_0^2 & \text{if } r_1 \neq r_2 \\ \frac{\rho_0}{dV} & \text{if } r_1 = r_2 \end{cases} \,. \quad (2.36)$$

In the limit $dV \to 0$ one obtains

$$\langle \hat{\rho}(r_1)\hat{\rho}(r_2) \rangle - \rho_0^2 = \rho_0 \delta(r_1 - r_2) \,. \quad (2.37)$$

Hence the reduced two-point correlation function (see (2.10)) can be written as

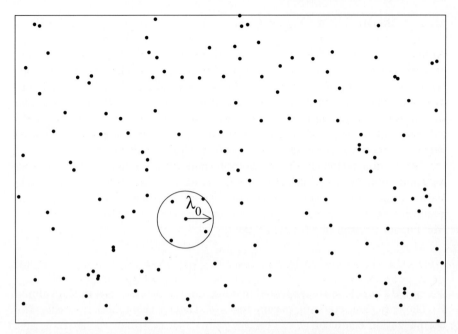

Fig. 2.5. Poisson distribution in the two-dimensional Euclidean space. The homogeneity scale λ_0 is shown. It is of the same order of the mean inter-particle separation

$$\tilde{\xi}(r) = \frac{\delta(r)}{\rho_0} . \qquad (2.38)$$

Note that, as discussed in Sect. 2.3.3 above, for any SPP $\tilde{\xi}(r=0) = +\infty$. Further one can obtain the reduced three-point correlation function (2.8):

$$\tilde{\zeta}(r_{12}, r_{23}, r_{13}) = \frac{\delta(r_1 - r_2)\delta(r_2 - r_3)}{\rho_0^2} . \qquad (2.39)$$

The two previous relations make evident that there is no correlation between different points. That is, the reduced correlation functions $\tilde{\xi}$ and $\tilde{\zeta}$ have only the so-called "diagonal" part. As explained below, this diagonal part is present in the reduced correlation functions of any other discrete SPP. At this point we can evaluate the normalized mass variance in a sphere of radius R by applying (2.29):

$$\sigma^2(R) = \frac{1}{\rho_0 \|C(R)\|} \equiv \frac{1}{\langle M(R) \rangle} \sim R^{-3} , \qquad (2.40)$$

which can be expressed in the following form valid in any dimension

$$\langle M^2(R) \rangle - \langle M(R) \rangle^2 = \langle M(R) \rangle , \qquad (2.41)$$

with $\langle M(R)\rangle \sim R^d$. In Chap. 3 we show that both for a continuous SSP or a discrete SPP, if $C_2(r)$ is mainly positive and rapidly decreasing to zero at large r (see (2.29)), we have always a relation of proportionality $\sigma^2(R) \sim R^{-d}$ at sufficiently large scales in d-dimensions. For this reason we will call this type of mass distribution "substantially Poisson". As we will discuss in Chap. 3, we can classify all mass distributions in three different classes depending on the large scale behavior of the mass fluctuations: (i) "substantially Poisson", (ii) "super-homogeneous" and (iii) "critical", if the normalized mass variance decays respectively (i) as in (2.40), (ii) faster, or (iii) slower.

As an application of the previous formalism we consider the following question: what is the probability of finding exactly M particles in a volume V? By a direct integration, it is simple to show that for a particle distribution generated by a homogeneous Poisson point process (2.33) this is given by the well-known Poisson law (see [142, 208]):

$$P(M,V) = \frac{\langle M\rangle^M \exp(-\langle M\rangle)}{M!} \qquad (2.42)$$

where $\langle M\rangle = \rho_0 V$ is the expected average number of particles in the volume V. The only characterizing parameter is ρ_0. Clearly this function satisfies the normalization condition

$$\sum_{M=0}^{\infty} P(M,V) = 1 \ . \qquad (2.43)$$

From (2.42) we can calculate the number variance obtaining $\langle M^2\rangle - \langle M\rangle^2 = \langle M\rangle$, which is the most famous relation of Poisson statistics and the generalization of (2.40) to an arbitrary volume V. Note that sometimes one would like to generate (for example with a computer simulation) a Poisson distribution by fixing a-priori the total number of points N, in the system volume V i.e., by throwing N particles in random positions in V. In this case one cannot know the ensemble value of the average density ρ_0 but only the sample estimate N/V and the properties of its probability distribution are given again by (2.42). For further discussions of the properties of this probability distribution see [208, 142].

2.5 Stochastic Point Processes with Spatial Correlations

In this section we analyze in detail some features of correlation functions for the case of a generic statistically stationary and isotropic stochastic distribution of point-particles with unitary mass (i.e., SPP). We show that for any SPP with $\rho_0 > 0$ we can write

$$\tilde{\xi}(r) = \frac{\delta(\mathbf{r})}{\rho_0} + \xi(r) \ ,$$

$$\tilde{\zeta}(r_1, r_2, r_{12}) = \frac{\delta(\mathbf{r_1})\delta(\mathbf{r_2})}{\rho_0^2} + \zeta(r_1, r_2, r_{12}) \ , \qquad (2.44)$$

2.5 Stochastic Point Processes with Spatial Correlations

where the functions ξ and ζ are the non-diagonal parts, which are defined only for $r > 0$ and $r_1, r_2 > 0$ respectively. In general, $\xi(r)$ (usually denoted $h(r)$ in statistical physics) is a smooth function of r [90]. The contribution at $r = 0$ to $\tilde{\xi}(r)$ and at $r_1 = r_2 = 0$ to $\tilde{\zeta}(r_1, r_2, r_{12})$, are called the *diagonal part* of the correlation functions[3]. In the simple Poisson case $\xi(r) \equiv \zeta(r_1, r_2, r_{12}) \equiv 0$ (an example of this kind of distribution is shown in Fig. 2.2). In order to show the first relation in (2.44) it is sufficient to observe that:

$$\langle \hat{\rho}(\mathbf{r})\hat{\rho}(\mathbf{r}')\rangle = \left\langle \sum_{i,j} \delta(\mathbf{r} - \mathbf{r}_i)\delta(\mathbf{r}' - \mathbf{r}_j) \right\rangle =$$

$$\rho_0 \delta(\mathbf{r} - \mathbf{r}') + \left\langle \sum_{i \neq j} \delta(\mathbf{r} - \mathbf{r}_i)\delta(\mathbf{r}' - \mathbf{r}_j) \right\rangle . \quad (2.45)$$

By calling

$$\left\langle \sum_{i \neq j} \delta(\mathbf{r} - \mathbf{r}_i)\delta(\mathbf{r}' - \mathbf{r}_j) \right\rangle = \rho_0^2 [1 + \xi(|\mathbf{r} - \mathbf{r}'|)] , \quad (2.46)$$

we obtain the first of (2.44). An analogous treatment can be applied to $\zeta(r_1, r_2, r_{12})$.

Let us now rewrite the expression linking the normalized mass variance with the two-point correlation function using (2.44):

$$\sigma^2(R) = \frac{1}{\rho_0 \|C(R)\|} + \frac{1}{\|C(R)\|^2} \int_{C(R)} d^3 r_1 \int_{C(R)} d^3 r_2 \xi(|\mathbf{r_1} - \mathbf{r_2}|) . \quad (2.47)$$

Note that the sign of the second term of (2.47) is not uniquely determined. In fact we can have, as we have mentioned above, particle distributions whose normalized mass variance decays faster than $\frac{1}{\|C(R)\|}$ for large R (see Chap. 3). In the discrete case, in order to measure $\sigma^2(R)$, one has to take into account both terms of (2.47), and not only the second one. Therefore, from (2.29), the variance (2.47) can, in general, be written as the sum of two contributions:

$$\sigma^2(R) = \sigma_{PN}^2(R) + \Xi(R) , \quad (2.48)$$

where the first term σ_{PN}^2 represents the intrinsic Poisson noise present in any discrete SPP, which is usually called *shot noise*, and the second term $\Xi(r)$ (which, as remarked above, does not have a determined sign) is the additional contribution due to correlations (i.e., due to $\xi(r) \neq 0$) between different particles.

---
[3] Note that in cosmology only the non-diagonal part $\xi(r)$ is usually called the two-point correlation function [184].

2.5.1 Conditional Properties

For the case of a discrete SPP it is often very important to consider observations from a point occupied by a particle. In order to characterize statistically these observations it is necessary to define a new kind of average for the observable F: the *conditional* average $\langle F \rangle_p$. This is defined as an ensemble average (i.e., over $\mathcal{P}[\rho(\boldsymbol{r})]$) with the condition that the origin of the coordinates is occupied by a particle of the distribution. When only one realization $\rho(\boldsymbol{r})$ extracted from $\mathcal{P}[\rho(\boldsymbol{r})]$ is available, $\langle F \rangle_p$ can be substituted by the spatial average in the following way. By calling $N(V)$ the number of particles in the volume V we can write:

$$\overline{F(\rho(\boldsymbol{r}_1), \rho(\boldsymbol{r}_2), \ldots)}_p = \lim_{V \to \infty} \frac{1}{N(V)} \sum_{i=1}^{N(V)} F(\rho(\boldsymbol{r'}_i + \boldsymbol{r}_1), \rho(\boldsymbol{r'}_i + \boldsymbol{r}_2), \ldots) , \tag{2.49}$$

where the sum is restricted to all the points $\boldsymbol{r'}_i$ occupied by point-particles. Again in the case in which only a finite sample of the system is available, the sum in (2.49) is restricted to the particles $\boldsymbol{r'}_i$ belonging to the sample, and (2.49) must be considered only as an *estimator* of $\langle F \rangle_p$.

Let us now introduce the *conditional* correlation functions. For a SPP the quantity

$$\langle \rho(\boldsymbol{r}_1) \rho(\boldsymbol{r}_2) \ldots \rho(\boldsymbol{r}_l) \rangle_p \, dV_1 dV_2 \ldots dV_l \tag{2.50}$$

gives the a-priori probability of finding l particles placed in the infinitesimal volumes $dV_1, dV_2, \ldots, dV_l$ respectively around $\boldsymbol{r}_1, \boldsymbol{r}_2, \ldots, \boldsymbol{r}_l$ with the condition that the origin of coordinates is *occupied* by a particle. We call $\langle \rho(\boldsymbol{r}_1) \rho(\boldsymbol{r}_2) \ldots \rho(\boldsymbol{r}_l) \rangle_p$ the conditional l-point average density. In particular for $l=1$ the function $\langle \rho(\boldsymbol{r}) \rangle_p$ represents the average density of particles seen by a fixed particle at a distance r from it.

It is possible to relate conditional correlation functions to the ordinary ones by applying the rules of conditional probability. Indeed, using the basic rule of conditional probability[4], one can write:

$$\langle \hat{\rho}(\boldsymbol{r}) \rangle_p = \frac{\langle \hat{\rho}(\boldsymbol{0}) \hat{\rho}(\boldsymbol{r}) \rangle}{\rho_0}$$

$$\langle \hat{\rho}(\boldsymbol{r}_1) \hat{\rho}(\boldsymbol{r}_2) \rangle_p = \frac{\langle \hat{\rho}(\boldsymbol{0}) \hat{\rho}(\boldsymbol{r}_1) \hat{\rho}(\boldsymbol{r}_2) \rangle}{\rho_0} . \tag{2.51}$$

However, in general, the following convention is assumed in the definition of the conditional densities: the particle at the origin does not count itself.

---

[4] Given two events A and B, the conditional probability $P(A|B)$ that the event A occurs if the event B has occurred, is given by $P(A|B) = P(A \cap B)/P(B)$, where $P(A \cap B)$ is the unconditional joint probability for the occurrence of both A, B and $P(B)$ is the unconditional probability for the occurrence of the event B without any condition [100].

Therefore $\langle \rho(r) \rangle_p$ is defined only for $r > 0$, and $\langle \rho(\boldsymbol{r_1})\rho(\boldsymbol{r_2}) \rangle_p$ for $r_1, r_2 > 0$. Consequently, we write

$$\langle \hat{\rho}(\boldsymbol{r}) \rangle_p = \rho_0 [1 + \xi(r)] \tag{2.52}$$
$$\langle \hat{\rho}(\boldsymbol{r_1})\hat{\rho}(\boldsymbol{r_2}) \rangle_p = \rho_0^2 \left[1 + \xi(r_1) + \xi(r_2) + \tilde{\xi}(r_{12}) + \zeta(r_1, r_2, r_{12}) \right] .$$

Note that, while $\langle \hat{\rho}(\boldsymbol{r}) \rangle$ is a one-point statistical quantity $\langle \hat{\rho}(\boldsymbol{r}) \rangle_p$ is a two-point quantity as we are conditioning the average to the fact that the origin of coordinates is occupied by a particle. Similarly $\langle \hat{\rho}(\boldsymbol{r_1})\hat{\rho}(\boldsymbol{r_2}) \rangle_p$ is a three-point statistical quantity.

Before passing to the application to the Poisson case, we note that $\langle \hat{\rho}(\boldsymbol{r_1})\hat{\rho}(\boldsymbol{r_2}) \rangle_p$, even though defined already for $r_1, r_2 > 0$, can be separated (in analogy with the unconditional two-point density) into a diagonal part giving the singular contribution for $\boldsymbol{r_1} = \boldsymbol{r_2}$ and a non-diagonal part giving the regular contribution for $\boldsymbol{r_1} \neq \boldsymbol{r_2}$. In order to obtain this result, we start directly from the definition

$$\langle \hat{\rho}(\boldsymbol{r_1})\hat{\rho}(\boldsymbol{r_2}) \rangle_p = \left\langle \sum_{i,j} \delta(\boldsymbol{r_1} - \boldsymbol{r_i})\delta(\boldsymbol{r_2} - \boldsymbol{r_j}) \right\rangle_p . \tag{2.53}$$

Equation (2.53) can be rewritten as

$$\langle \hat{\rho}(\boldsymbol{r_1})\hat{\rho}(\boldsymbol{r_2}) \rangle_p = \left\langle \sum_{i,j} \delta(\boldsymbol{r_1} - \boldsymbol{r_i})\delta(\boldsymbol{r_2} - \boldsymbol{r_j}) \right\rangle_p =$$
$$\left\langle \sum_{i} \delta(\boldsymbol{r_1} - \boldsymbol{r_i})\delta(\boldsymbol{r_2} - \boldsymbol{r_i}) \right\rangle_p + \left\langle \sum_{i,j}^{i \neq j} \delta(\boldsymbol{r_1} - \boldsymbol{r_i})\delta(\boldsymbol{r_2} - \boldsymbol{r_j}) \right\rangle_p =$$
$$\delta(\boldsymbol{r_1} - \boldsymbol{r_2}) \langle \hat{\rho}(\boldsymbol{r_1}) \rangle_p + \Gamma^{(2)}(\boldsymbol{r_1}, \boldsymbol{r_2}) , \tag{2.54}$$

where we have called

$$\Gamma^{(2)}(\boldsymbol{r_1}, \boldsymbol{r_2}) = \left\langle \sum_{i,j}^{i \neq j} \delta(\boldsymbol{r_1} - \boldsymbol{r_i})\delta(\boldsymbol{r_2} - \boldsymbol{r_j}) \right\rangle_p . \tag{2.55}$$

By applying (2.52) to the case of a Poisson distribution, we obtain:

$$\langle \hat{\rho}(r) \rangle_p = \rho_0$$
$$\langle \hat{\rho}(\boldsymbol{r_1})\hat{\rho}(\boldsymbol{r_2}) \rangle_p = \rho_0^2 \left[1 + \frac{\delta(\boldsymbol{r_1} - \boldsymbol{r_2})}{\rho_0} \right] . \tag{2.56}$$

The fact that $\langle \hat{\rho}(\boldsymbol{r}) \rangle_p = \langle \hat{\rho}(\boldsymbol{r}) \rangle$ correspond to the fact that in a Poisson distribution different points are spatially uncorrelated. In general, if a SPP has a short ranged two-point correlation (i.e., a finite correlation length), then conditional and unconditional properties at large distances are practically the same. Note that conditional correlations, appropriately normalized, have the important advantage of being well defined also for fractal structures as will be discussed Chap. 4.

2.5.2 Integrated Conditional Properties

We now study fluctuations of the number of point-particles seen by another particle of the distribution within a distance R from it. Let us call $N(R)$ the random variable representing the number of points seen by a given particle in a sphere of radius R around it. By integrating the first of (2.52) in the sphere $C(r)$ in $d=3$, we can write directly

$$\langle N(R)\rangle_p = \int_{C(R)} \langle \hat{\rho}(\boldsymbol{r})\rangle_p d^3r = \rho_0 \|C(R)\| + \rho_0 \int_{C(R)} \xi(r) d^3r \ . \qquad (2.57)$$

Therefore, with respect to unconditional observations, we have an extra-term, the second one, which take into account the *clustering* properties of the distribution compared with a Poisson distribution which has no clustering at all. Note that since the origin of the coordinates is occupied, (2.57) is a two-point quantity.

Let us now define the conditional normalized variance $\sigma_p^2(r)$, which is the variance of the variable $N(R)$:

$$\sigma_p^2(r) \equiv \frac{\langle N(R)^2\rangle_p - \langle N(R)\rangle_p^2}{\langle N(R)\rangle_p^2} \ , \qquad (2.58)$$

where

$$\langle N(R)^2\rangle_p = \int_{V(R)} \int_{V(R)} \langle \hat{\rho}(\boldsymbol{r}_1)\hat{\rho}(\boldsymbol{r}_2)\rangle_p d^3r_1 d^3r_2 \ . \qquad (2.59)$$

2.5.3 Detection of the Homogeneity Scale of a Discrete SPP

In this section we proceed to a further discussion of the concept of homogeneity (or uniformity) scale λ_0 for a discrete point process, classifying the different but substantially equivalent definitions given in different sections above, and explaining the principal method to detect it through the analysis of a finite sample of the distribution. Only in the case where one has successfully measured the homogeneity scale to be smaller than the size of a given sample, can the estimator of the reduced two-point correlation function in the sample be considered to approximate its real value (defined in the infinite volume limit or through an ensemble average). This is a crucial point that must be considered every time one considers the analysis of finite samples (e.g., *volume limited* samples of galaxy redshift surveys as discussed in Chap. 8) of spatial point-particle distributions.

2.5 Stochastic Point Processes with Spatial Correlations

The homogeneity scale marks the cross-over from large to small correlations or fluctuations. We have already defined this scale in (2.5). An alternative definition of λ_0 is the length-scale at which the normalized mass variance in spheres is of order unity

$$\sigma^2(\lambda_0) \simeq 1 \qquad (2.60)$$

and $\sigma^2(R) < 1$ for $\forall R > \lambda_0$. We stress that the definition of the homogeneity scale via (2.60) can be misleading when the average density is not a well-defined property in the given sample i.e., if the estimator of the average density measured in the sample is sample-dependent, as in fractal particle distributions (see Chaps. 4–5).

Clearly λ_0 can be equivalently defined as the length scale beyond which the average density becomes well-defined, and therefore we can write

$$|\langle \hat{\rho}(\boldsymbol{r}) \rangle_p - \rho_0| < \rho_0 \qquad (2.61)$$

for all $r > \lambda_0$. This is equivalent to

$$\xi(\lambda_0) \simeq 1 \, , \qquad (2.62)$$

and $\xi(r) < 1$ for all $r > \lambda_0$. Note that for both these definitions, we need to have a sufficiently precise estimator of the asymptotic average density ρ_0 in order to find λ_0. For a single sample, this requirement is satisfied if its size $R_s \gg \lambda_0$. It may seem that there is a problem here for analysis in a finite sample: to evaluate λ_0 we need to know ρ_0 accurately enough, but to know ρ_0 in turn we need to know λ_0 in such a way that the ratio between the number of points in the sample and its volume is a good estimator of the asymptotic average density. In order to break this loop and to make it possible that the function $\xi(r)$ can be accurately enough estimated in a single finite sample, we need to have a method for determining λ_0 independently of the knowledge of ρ_0. The operative way to detect the homogeneity scale is by looking for a flattening of the conditional density $\langle \hat{\rho}(r) \rangle_p$ with scale [223]. In fact, if the system reaches to a good approximation the condition of homogeneity well within the sample size, then the conditional density $\langle \hat{\rho}(r) \rangle_p$ must flatten toward ρ_0 for some $r < R_s$. The point where the flattening is observed can be taken to represent λ_0. Only if $\lambda_0 < R_s$ is detected in this way is it possible to estimate $\xi(r)$ from the finite sample. The latter then gives a more refined instrument than the conditional density to study correlations and their spatial persistence and to address the more difficult problem of detecting the correlation length r_c from a single sample (see also Chap. 9). Note that this method of detection of λ_0 is free from any a-priori assumption about the homogeneity in the sample. As will be discussed in Chaps. 8–9 this is a very important point in the context of the analysis of galaxy redshift surveys.

Let us finally emphasize again that, whichever of the above definitions is used for λ_0, this length scale is a completely distinct one from the correlation length r_c which measures a different property of the system. To clarify this point it suffices to consider a case in which we have a critical system with a well defined average density (and hence not a fractal) for which $\xi(r) = (r/r_0)^{-\gamma}$ with $0 < \gamma < d$. In this case $\lambda_0 = r_0$, while as discussed before $r_c \to +\infty$. This is due to the fact that λ_0 is related to the amplitude of fluctuations (in this case given by r_0), while r_c describe the persistence of the correlation in the fluctuations regardless of their amplitude.

2.6 Nearest Neighbor Probability Density in Point Processes

The probability density of the distance between nearest neighbor (nn) particles in a stochastic particle distribution (i.e., point process) is important for many physical applications. The simple Poisson case was first considered in [111]. It is possible to extend, with some approximations, this approach to the case of a particle distribution with correlations (for a complete analysis, see [230]).

Let us denote by $w(r)dr$ the probability that the nn particle to a given particle lies at a distance in the range $[r, r+dr]$. This probability is equal by definition to the joint probability that there are no particles in the range $[0, r]$, and a particle in the spherical shell between r and $r+dr$. Let us assume first that these two events can be considered independent of one another. Using the definitions of the conditional density $\langle \hat{\rho}(\mathbf{r}) \rangle_p$ and of the PDF $w(r)$, the function $w(r)$ itself must then satisfy the equation in $d=3$

$$w(r) = \left(1 - \int_0^r w(r')dr'\right) 4\pi r^2 \langle \hat{\rho}(\mathbf{r}) \rangle_p . \tag{2.63}$$

Clearly the probability of having no particles in $[0, r)$ and that of having one particle in $[r, r+dr)$ are really independent only in absence of density-density correlations between different spatial points (i.e., only in the case of a Poisson distribution). However, (2.63) can be used as an approximate equation also for the case that correlations are non-zero but sufficiently weak.

2.6.1 Poisson Case

In a Poisson distribution (2.63) is exact because of the absence of any spatial correlation between different points. Using (2.56), we obtain

$$w(r) = 4\pi \rho_0 r^2 \exp\left(-\frac{4\pi \rho_0 r^3}{3}\right), \tag{2.64}$$

by solving explicitly the integral (2.63) (see Fig. 2.6). This function satisfies the necessary normalization condition

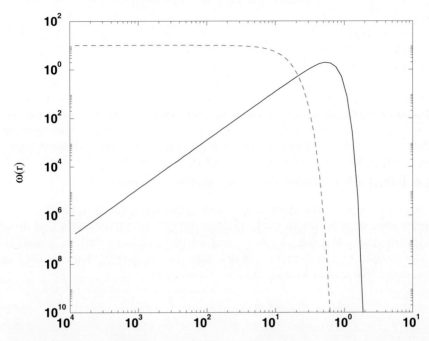

Fig. 2.6. Nearest-neighbor distribution for a Poisson point process (*solid line*) with average density $\rho_0 = 1$ and $\langle \Lambda \rangle \approx 0.5$ and for a particle distribution with power-law correlation of the type $\xi(r) = (r/10)^{-1}$ (*dashed line*)

$$\int_0^\infty \omega(r)dr = 1 \ . \tag{2.65}$$

A useful distance scale is the average distance between nearest neighbor particles:

$$\langle \Lambda \rangle = \int_0^\infty \omega(r)r\,dr \ . \tag{2.66}$$

We readily obtain

$$\langle \Lambda \rangle = \left(\frac{3}{4\pi\rho_0}\right)^{\frac{1}{3}} \Gamma_e\left(1 + \frac{1}{3}\right) , \tag{2.67}$$

where $\Gamma_e(x)$ is the Euler gamma-function. Therefore the average nn separation is of the same order as the average distance per particle $\rho_0^{-1/3}$, which in this case is also the homogeneity scale λ_0 (i.e., the only intrinsic length scale of the distribution). It is possible also to compute the typical fluctuation of the nn separation by considering that

$$\langle \Lambda^2 \rangle = \int_0^\infty \omega(r)r^2 dr = \Gamma_e(5/3)\left(\frac{4}{3}\rho_0\right)^{-\frac{2}{3}} , \tag{2.68}$$

from which we obtain that

$$\Sigma_\Lambda^2 = \langle (\Lambda - \langle \Lambda \rangle)^2 \rangle = \left(\Gamma_e(5/3) - \Gamma_e^2(4/3)\right) \left(\frac{4}{3}\rho_0\right)^{-\frac{2}{3}} \quad (2.69)$$

so that

$$\frac{\Sigma_\Lambda}{\langle \Lambda \rangle} \simeq \frac{1}{3} . \quad (2.70)$$

The probability of finding a point in the range $[0, r]$ is thus very small if $r \ll \langle \Lambda \rangle$. In Chap. 4 we will extend this result to the case of an inhomogeneous Poisson SPP, in which the average density of particles $\langle \hat{\rho}(\mathbf{r}) \rangle$ depends on r.

2.6.2 Particle Distributions with Spatial Correlations

Let us now extend the analysis in $d = 3$ of the nn distance to homogeneous SPPs with weak correlations. In this case (2.63) can be seen as a first order approximation. We denote $\langle \hat{\rho}(\mathbf{r}) \rangle_p = \Gamma(r)$ to make clear that for a statistically stationary and isotropic particle distribution $\langle \hat{\rho}(\mathbf{r}) \rangle_p$ depends only on the scalar distance r. From (2.63), we then obtain

$$w(r) = A(r_*) r^2 \Gamma(r) \exp\left(-4\pi \int_{r_*}^{r} r'^2 \Gamma(r') dr'\right) , \quad (2.71)$$

where r_* is an arbitrary normalization distance, and $A(r_*)$ is the normalization constant depending on r_* for a given $\Gamma(r)$. Let us consider two different cases corresponding to different functional behaviors of $\Gamma(r) = \rho_0(1 + \xi(r))$.

- Let us consider the case where the reduced correlation function has an exponential decay of the type

$$\xi(r) = A \exp(-r/r_c) , \quad (2.72)$$

with $A < 1$. In this case, from (2.52), we can further approximate

$$\Gamma(r) \approx \begin{cases} \rho_0(1 + A) & \text{for } r \ll r_c \\ \rho_0 & \text{for } r \gg r_c . \end{cases} \quad (2.73)$$

Hence for $r \gg r_c$ the behavior converges rapidly to that found for the purely Poisson case, but with the replacement $\rho_0 \to \rho_0' = \rho_0(1 + A)$, as the small distance behavior clearly gives the dominant contribution.

- A very different situation occurs when the reduced correlation function has a power-law behavior with distance:

$$\xi(r) \simeq \left(\frac{r}{r_0}\right)^{-\gamma} \quad \text{for } r \gg r_0 , \quad (2.74)$$

with $0 < \gamma < 3$. From (2.52) and (2.71), one has

$$w(r) = A\rho_0 r^2 \left[1 + \left(\frac{r_0}{r}\right)^{-\gamma}\right] \exp\left(-\frac{4\pi}{3}\rho_0 r^3 - \frac{4\pi \rho_0}{3-\gamma} r_0^\gamma r^{3-\gamma}\right) . \quad (2.75)$$

From (2.75) we see that for $r \gg r_0$ the behavior converges *slowly* to the Poisson one. Instead for $r \ll r_0$, where $\xi(r) \gg 1$, it is very different and, as discussed in Chap. 4, is very similar to the behavior typical of a fractal particle distribution.

2.7 Gaussian Continuous Stochastic Fields

In this section we present the basic properties of the most important case of a continuous SSP: the Gaussian field. Its relevance relies on two mathematical facts:

- The central limit theorem: as we will discuss in Sect. 2.10, if about a random variable Y can be seen as the appropriately normalized sum of many independent (or almost independent) random variables, then one can demonstrate, making appropriate hypotheses that Y is a Gaussian variable;
- Information theory [174]: as we will discuss in Sect. 2.11 if a random variable we know only the average value and the finite variance, then its most probable probability distribution is the Gaussian one. This can be seen by maximizing the Shannon entropy with the constraint given by the knowledge of the average value and the variance.

Therefore, when one studies stochastic signals, in many contexts *Gaussianity* arises as a natural working hypothesis at least as a zero order approximation. This is the case in cosmology, for example, of the temperature fluctuations in the CMBR.

Here we first introduce the generic Gaussian field without the requirement of spatial stationarity and subsequently we specialize the discussion to the stationary case.

In order to introduce the Gaussian continuous mass density field in a simple way, let us consider a partitioning of space into small cubic cells of volume ΔV whose center is denoted $\boldsymbol{r}_i$. The random variable representing the *local average density* in the ith cell is

$$\rho(\boldsymbol{r}_i; \Delta V) = \frac{1}{\Delta V} \int_{\Delta V(\boldsymbol{r}_i)} d^d r \hat{\rho}(\boldsymbol{r}) \,,$$

where $\hat{\rho}(\boldsymbol{r})$ is the microscopic stochastic field and $\Delta V(\boldsymbol{r}_i)$ is the small cell around the point $\boldsymbol{r}_i$. The probability of a realization $\{\rho(\boldsymbol{r}_i; \Delta V)\}$ of the Gaussian mass density field is defined by the joint probability density function $p(\{\rho(\boldsymbol{r}_i; \Delta V)\})$ given by:

$$p(\{\rho(\boldsymbol{r}_i; \Delta V)\}) = B \exp\left[-\frac{1}{2}\sum_{i,j}(\rho(\boldsymbol{r}_i; \Delta V) - m_i) A_{ij} (\rho(\boldsymbol{r}_j; \Delta V) - m_j)\right] \,,$$
(2.76)

where B is the normalization constant, $\hat{A} = ((A_{ij}))$ a positive definite and symmetric matrix and m_i are real constants. In general A_{ij} will depend on $\boldsymbol{r}_i, \boldsymbol{r}_j$ and on the discretization volume ΔV. If the continuous limit $\Delta V \to 0$ is appropriately taken, $p(\{\rho(\boldsymbol{r}_i; \Delta V)\})$ converges to the Gaussian functional $\mathcal{P}[\rho(\boldsymbol{r})]$ for the microscopic field $\hat{\rho}(\boldsymbol{r})$:

$$\mathcal{P}[\rho(\boldsymbol{r})] \sim \exp\left[-\frac{1}{2}\int_V \int_V d^d r\, d^d r'(\rho(\boldsymbol{r}) - m(\boldsymbol{r}))K(\boldsymbol{r},\boldsymbol{r}')(\rho(\boldsymbol{r}') - m(\boldsymbol{r}'))\right], \quad (2.77)$$

where V is the volume of the space of definition of the field,

$$K(\boldsymbol{r}_i, \boldsymbol{r}_j) = \lim_{\Delta V \to 0} \frac{A_{ij}}{\Delta V^2},$$

is the so-called *correlation kernel*, and $m(\boldsymbol{r})$ is an arbitrary function that, as we show below, defines the average of the field.

In order to see why the matrix $\hat{A}$ must be positive definite, it is sufficient to use the fact that $p(\{\rho(\boldsymbol{r}_i; \Delta V)\})$ is a probability density function. If $\hat{A}$ is not positive definite it is simple to show that $p(\{\rho(\boldsymbol{r}_i; \Delta V)\})$ cannot be properly normalized by integrating over all the possible discretized mass density functions $\rho(\boldsymbol{r}; \Delta V)$. Indeed let us call $\underline{\delta\rho}$ the vector whose ith component is $[\rho(\boldsymbol{r}_i; \Delta V) - m_i]$, therefore we can rewrite (2.76) in the following vectorial form

$$p(\{\rho(\boldsymbol{r}_i; \Delta V)\}) \equiv p(\underline{\delta\rho}) = B \exp\left[-\frac{1}{2}\underline{\delta\rho} \cdot \hat{A}\underline{\delta\rho}\right]. \quad (2.78)$$

Since $\hat{A}$ is symmetric, it is diagonalized by an orthonormal transformation $\hat{R}$ (orthogonality implies $\hat{R}^{-1} = \hat{R}^t$, where $\hat{R}^{-1}$ is the inverse matrix and $\hat{R}^t$ is the transposed).

Let us call $\underline{Q} \equiv \hat{R}\underline{\delta\rho}$. The components of $\underline{Q}$, being linear combinations of those of $\underline{\delta\rho}$, are stochastic variables. By calling $\hat{S} \equiv \hat{R}\hat{A}\hat{R}^t$, which is a diagonal matrix ($S_{ij} = S_{ii}\delta_{ij}$), we can write the probability density function $f(\underline{Q})$ of the "rotated" stochastic field $\underline{Q}$ as

$$f(\underline{Q}) \equiv p(\underline{\delta\rho} = R^t \underline{Q}) = B \exp\left[-\frac{1}{2}\sum_i S_{ii}Q_i^2\right] \equiv B \prod_i \exp\left[-\frac{1}{2}S_{ii}Q_i^2\right], \quad (2.79)$$

which is the product of one-dimensional Gaussian functions. The S_{ii} are the eigenvalues of $\hat{A}$. The normalization constant is unchanged because the change of variables determined by an orthonormal transformation has an unitary Jacobian determinant. From (2.79), in order to have $f(\underline{Q})$ (and hence $p(\{\rho(\boldsymbol{r}_i; \Delta V)\})$) integrable over all the variables Q_i (i.e., normalizable) we need $S_{ii} > 0\ \forall i$. This is equivalent to requiring $\hat{A}$ to be positive definite. We will see in Chap. 3 that this requirement, in the case in which A_{ij} depends only on $\boldsymbol{r}_i - \boldsymbol{r}_j$, is another formulation of the Khinchin or Bochner theorem for Gaussian fields.

It is simple to show that

$$\langle \rho(\boldsymbol{r}_i; \Delta V) \rangle = m_i \quad \forall i$$
$$\langle (\rho(\boldsymbol{r}_i; \Delta V) - m_i)(\rho(\boldsymbol{r}_j; \Delta V) - m_j) \rangle = C_{ij} ,$$
(2.80)

where $\hat{C} \equiv ((C_{ij})) = \hat{A}^{-1}$. Because of (2.80), $\hat{C}$ is called the *correlation matrix*. The diagonal element C_{ii} is the variance of the stochastic field $\hat{\rho}$ in the ith cell (i.e., in the cell around $\boldsymbol{r}_i$). The main feature of a Gaussian field is that the knowledge of the set of average values $\{m_i\}$ and of the correlation matrix $\hat{C}$ determines completely any other l-point correlation coefficient (i.e l-point correlation function in the continuous limit) for any $l > 2$. In field theory this property is called *Wick's theorem* and can be formulated as follows:

$$\langle (\rho(\boldsymbol{r}_{i_1}) - m_{i_1}) \ldots (\rho(\boldsymbol{r}_{i_l}) - m_{i_l}) \rangle =$$
$$\begin{cases} 0 & \text{if } l \text{ is odd} \\ \frac{l!}{(l/2)! 2^{l/2}} \{C_{i_1 i_2} \ldots C_{i_{l-1} i_l}\}_{symm} & \text{if } l \text{ is even} \end{cases}$$
(2.81)

where $\{C_{i_1 i_2} \ldots C_{i_{l-1} i_l}\}_{symm}$ represents the symmetric product of $C_{i_1 i_2} \ldots C_{i_{l-1} i_l}$. For instance for $l = 4$ with four different indices, we have

$$\{C_{i_1 i_2} C_{i_3 i_4}\}_{symm} = \frac{1}{3} [C_{i_1 i_2} C_{i_3 i_4} + C_{i_4 i_1} C_{i_2 i_3} + C_{i_1 i_3} C_{i_2 i_4}] .$$

A trivial case is

$$\{C_{ii}^2\}_{symm} = C_{ii}^2 .$$

It is simple to show that (2.81) is equivalent of saying that all the cumulants of order greater than two of the Gaussian distribution are zero.

The most important case is when the Gaussian field is spatially stationary (or homogeneous). In this case $m_i = \rho_0 \ \forall i$, $A_{ij} = A(\boldsymbol{r}_i - \boldsymbol{r}_j)$, and, as a consequence, $C_{ij} = C(\boldsymbol{r}_i - \boldsymbol{r}_j)$ which is the discretized two-point connected correlation function. If the field is also statistically isotropic A_{ij} and C_{ij} depend only on $|\boldsymbol{r}_i - \boldsymbol{r}_j|$. Moreover, if $\rho_0 \neq 0$, we can write the dimensionless, but discretized, connected two-point correlation function $\tilde{\xi}(\boldsymbol{r}_i, \boldsymbol{r}_j)$ as:

$$\tilde{\xi}(\boldsymbol{r}_i, \boldsymbol{r}_j) = \frac{C_{ij}}{\rho_0^2} .$$
(2.82)

The main property of a spatially stationary Gaussian field is that the diagonalizing operator for the correlation matrix $\hat{C}$ (and also its inverse $\hat{A}$) is the Fourier Trasform operator. This will be shown in Chap. 3, where the analysis of stationary Gaussian fields will be completed through the discussion of the properties in Fourier space, the problem of the continuum limit, and the presentation of a simple recipe to construct correlated Gaussian fields. For the moment, concerning this last point, we only specify that in the continuous limit $\Delta V \to 0$ the double sum in (2.76) becomes a double integral and $p(\{\rho(\boldsymbol{r}_i; \Delta V)\})$ converges to the functional $\mathcal{P}[\rho(\boldsymbol{r})]$ as shown by (2.79). Therefore the ensemble averages become path integrals where any functional form $\rho(\boldsymbol{r})$ of the microscopic field $\hat{\rho}(\boldsymbol{r})$ represents a single path.

2.8 Power-Laws and Self-Similarity

A power-law correlation function usually has deep conceptual implications. In particular it implies self-similarity of the fluctuation field. The term *self-similarity* was introduced in connection to fractal objects and second order phase transitions in order to characterize their shared property of giving spatial patterns whose high degree of complexity is invariant under a scale transformation [116]. The same concept is also referred to as *scale-invariance*.

In the context of stochastic processes self-similarity can be defined as follows: applying to space a re-scaling of the length by a factor $b > 0$

$$\boldsymbol{r} \to \boldsymbol{r}' = b\boldsymbol{r} \tag{2.83}$$

leaves the correlation function between fluctuations unchanged apart from a re-scaling of the global amplitude that depends on b but not on the variable r. This leads to the functional relation (we use the notation $C_2(r)$ in order to unify the analysis for the continuous and the discrete case):

$$C_2(r') = C_2(br) = A(b)C_2(r) \quad \text{for } r > 0 \tag{2.84}$$

which is clearly satisfied by a power-law with *any* exponent ($A(b)$ is a pre-factor depending on b only). Indeed, for

$$C_2(r) = \xi_0 r^{-\gamma} \tag{2.85}$$

we have

$$C_2(r') = \xi_0 [br]^{-\gamma} = b^{-\gamma} C_2(r) \,. \tag{2.86}$$

Note that (2.84) does not hold, for example, for an exponential function

$$C_2(r) = \xi_0 e^{-r/r_0} \,. \tag{2.87}$$

We have seen in Sect. 2.3.2 that r_0 is the correlation length r_c of this stochastic field, while fields satisfying (2.85) have $r_c \to \infty$.

Equation 2.86 reflects the fact that for a power-law $C_2(r)$ a re-scaling of distances does not change the *persistence* of the spatial correlation of the fluctuation field. That is the re-scaling of distances results only in a re-scaling of the amplitude of fluctuations, but not in a change of the rate of decrease of their spatial "memory". As the amplitude changes, the homogeneity scale (whatever its precise definition) will change. Instead for a $\xi(r)$ as in (2.87), a spatial re-scaling changes both properties. In both cases the correlation length actually changes as $r'_c = br_c$, but in the first case, if $0 < \gamma < d$, where d is the spatial dimension $r_c \to +\infty$. The case $\gamma > d$, even though we have a sort of scale-invariance, is usually assimilated in statistical physics to the exponential case because, from the point of view of spatially integrated fluctuations (see Sect. 2.3.4), we have a behavior of the same kind as that of

2.8 Power-Laws and Self-Similarity

an exponential correlation function. On the contrary, as explicitly shown in Chap. 3, for $0 < \gamma < d$ the spatially integrated fluctuations have a stronger persistence at any scale. This is the reason why for the case $\gamma > d$ one usually speaks of "non-dangerous" power-laws.

In Figs. 2.7–2.9 we show a visual example of "self-similarity" and of the real meaning of the correlation length. The system represented is the two-dimensional Ising model, in absence of an external magnetic field. It is the most widely studied model to investigate the critical behavior of a system at a second order phase transition (e.g., magnetic or liquid-gas systems at the critical point) [116]. This model shows scale-invariant properties when the temperature is set to the critical value T_c. In all the three figures the concentric boxes give the concentric reproduction of a finite portion of the system under three different re-scalings of the distances. This re-scaling procedure is also called *coarse graining*. In all the figures the two different colors represent the two local states of magnetization (positive and negative). Note that the color does not refer to the *amplitude* of fluctuations, but only to their sign. They could be equally 10^{-5} or 10^5 times the average value of the studied stochastic field: the self-similarity has nothing to do with their amplitude. Therefore the size or characteristic length of a single color area gives a measure of the persistence of fluctuations, and thus of the correlation length of the stochastic system. A similar construction can be made, for example, also for a stochastic mass density field representing areas where the density fluctuation from the average is positive with one color and areas of negative fluctuations with another color.

When $T > T_c$ (Fig. 2.7) both the states of magnetization are present in the system, and the regions with constant magnetization (i.e., of a single color) have a typical size which is a decreasing function of $(T - T_c)$ (i.e., $\sim (T - T_c)^{-\nu}$ with $\nu > 0$). This size, as stated above, is strictly related to the finite correlation length $r_c(T)$ at the temperature T. Therefore, under repeated re-scaling of distances, as shown by the figure, the average size of the regions magnetized in one direction decreases until they disappear completely. That is the coarse-graining does not act only on the amplitude of correlations, but also on their spatial scale. Consequently the fluctuation field is not self-similar. Instead when $T < T_c$ (Fig. 2.9) (spontaneously broken symmetry) one state dominates over the other. However the typical size of the residual state (the dark one in the figure), i.e. the size of the holes in the dominating state, is again finite. This is a measure of the correlation length which is therefore finite, and the re-scaling procedure has a similar effect as in the $T > T_c$ phase, marking the lack of self-similarity or scale-invariance. Finally at $T = T_c$ (see Fig. 2.8) again as at $T > T_c$, the two magnetization states are equally present in the system, but the size of regions of a given color only has no characteristic finite value, i.e., we have regions of a fixed state of magnetization of any size, and the correlation length $r_c(T_c)$ is infinite. A direct consequence of this situation is that under spatial re-scaling we obtain

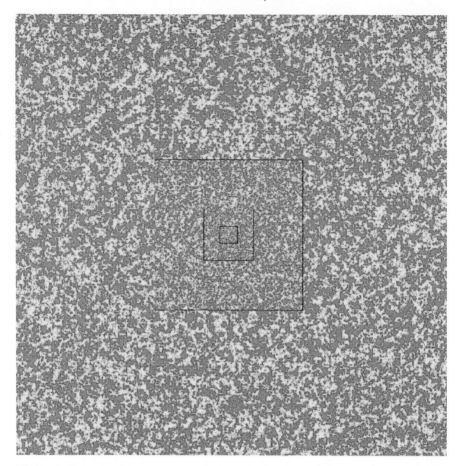

Fig. 2.7. Ising model at temperature $T > T_c$. Compact regions of both states of magnetization are present, but they have a small typical size which is a measure of the finite correlation length $r_c(T)$. The finiteness of the correlation length implies that under a *coarse graining* (i.e., rescaling of distances) the picture does not look the same and the typical size of compact regions changes. This means that the fluctuation field is not spatially *self-similar*.

images practically indistinguishable from the original one. Therefore we see that the concept of self-similarity and scale-invariance are strictly related to that of the correlation length in statistical physics. On the other hand, the concept of homogeneity scale is not of relevance in this context.

This discussion is closely related to the Gaussianity of fluctuations in the system. As shown explicitly in Sect. 2.10, the integrated fluctuations in a finite volume e.g., the mass fluctuations $\Delta M(V)$ in a volume V of a stochastic mass density field, can be considered Gaussian only if the size of the volume $V^{1/d}$ (where d is the space dimension) is much larger than the correlation

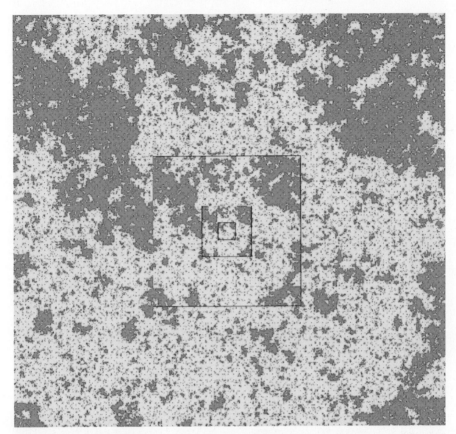

Fig. 2.8. Ising model at the critical temperature ($T = T_c$): There are clusters of fluctuations of a fixed sign of all sizes and the correlation length diverges $r_c(T_c) \to \infty$. The absence of an intrinsic characteristic scale is shown by the fact that at different "zooms" the system looks the same i.e., the fluctuation field is self-similar and has *fractal* features

length r_c. On the other hand, for a stochastic density field with an infinite correlation length the mass in a finite volume V of is not Gaussian distributed at any spatial scale.

2.9 Mass Function and Probability Distribution

Let us now consider the statistical properties of a stochastic distribution of point-particles with different masses. The ith particle has mass m_i. The microscopic density is then

$$\rho(\boldsymbol{x}) = \sum_i m_i \delta(\boldsymbol{r} - \boldsymbol{r}_i) \,. \tag{2.88}$$

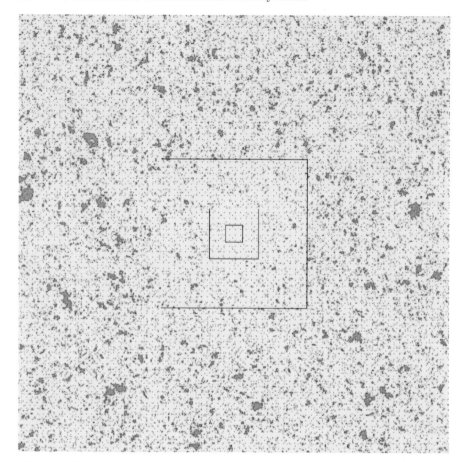

Fig. 2.9. Ising model below the critical temperature $(T < T_c)$. In this case the symmetry of the system is *spontaneously broken*, and one of the two states of magnetization dominates. The clusters of the opposite state of magnetization (i.e., the holes in the dominating magnetization) have a typical finite size which is a measure of the finite correlation length $r_c(T)$. Consequently, the system no longer appears *self-similar* under coarse graining

As discussed at the beginning of the chapter, point processes can be generalized to include point-particle distributions generated not only through a stochastic choice of the positions of the particles, but also of their masses. In general these two quantities are correlated. We limit now the discussion to the simplest case in which there is no correlation between mass and position. The opposite case of strong correlations between positions and masses will be presented in Chap. 5 in the context of *multifractal* measures. Thus the joint probability density function of having N particles of masses $m_1, m_2, \ldots, m_N$ placed respectively at the points $\boldsymbol{r}_1, \boldsymbol{r}_1 \ldots, \boldsymbol{r}_N$ is then of the form:

2.9 Mass Function and Probability Distribution

$$f(\boldsymbol{r}_1, \ldots, \boldsymbol{r}_N; m_1, \ldots, m_N) = p(\boldsymbol{r}_1, \ldots, \boldsymbol{r}_N) \prod_{i=1}^{N} \phi(m_i) ,$$

where $\phi(m)$ is the probability density function (or mass function) for the mass of a particle with $m > 0$. In particular, with this hypothesis, the joint probability of finding a particle of mass between m and $m+dm$ in the volume element dV at $\boldsymbol{r}$ can be written as

$$\nu(\boldsymbol{r}, m) dV \, dm = \langle \hat{\rho}(\boldsymbol{r}) \rangle \, \phi(m) dV \, dm . \tag{2.89}$$

For the conditional probability (i.e., conditioned on observing particles from another particle) we simply obtain

$$\langle \nu(\boldsymbol{r}, m) \rangle_p dV \, dm = \langle \hat{\rho}(\boldsymbol{r}) \rangle_p \phi(m) dV \, dm . \tag{2.90}$$

Let us now analyze the statistical properties of the macroscopic mass $M(V)$ in a finite volume V. We can write

$$M(V) = \int_V d^d r \sum_i m_i \delta(\boldsymbol{r} - \boldsymbol{r}_i) . \tag{2.91}$$

It is clear that

$$\langle M(V) \rangle = m_0 n_0 V , \tag{2.92}$$

where

$$m_0 = \int_0^\infty dm \, m \, \phi(m)$$

and $n_0 > 0$ is the average number density of particles, i.e., $\langle \sum_i \delta(\boldsymbol{r} - \boldsymbol{r}_i) \rangle$. On the other hand

$$\langle M^2(V) \rangle = m_1^2 n_0 V + m_0^2 n_0^2 \left(V^2 + \int_V \int_V d^d r \, d^d r' \xi(\boldsymbol{r}, \boldsymbol{r}') \right) , \tag{2.93}$$

where

$$m_1^2 = \int_0^\infty dm \, m^2 \, \phi(m)$$

and

$$\xi(\boldsymbol{r}, \boldsymbol{r}') = \frac{1}{n_0^2} \left\langle \sum_{i \neq j} \delta(\boldsymbol{r} - \boldsymbol{r}_i) \delta(\boldsymbol{r} - \boldsymbol{r}_j) \right\rangle - 1$$

is the usual non-diagonal part of the connected two-point correlation function of the number density (as usual we are particularly interested in the case in which the microscopic number density is statistically stationary, which gives $\xi(\boldsymbol{r}, \boldsymbol{r}') = \xi(\boldsymbol{r} - \boldsymbol{r}')$). Therefore the quadratic mass fluctuation is:

$$\langle M^2(V) \rangle - \langle M(V) \rangle^2 = m_1^2 n_0 V + m_0^2 n_0^2 \int_V \int_V d^d r \, d^d r' \xi(\boldsymbol{r}, \boldsymbol{r}') , \tag{2.94}$$

which is the extension to the case of distributions of point-particles with different masses of (2.29). Similarly we can extend (2.59) by using also the third moment of the probability density function $\phi(m)$ and the three-point correlation function of the number density.

An important question, in the present context, is whether there is a scale above which $M(V)$ (or its fluctuations appropriately normalized) can be considered a Gaussian variable. The arguments given in Sect. 2.10 below about the validity of the central limit theorem can be applied directly to the case of point-particles of equal mass. They can also be extended to include the case of different masses. In particular if m_1^2 is finite and the correlation length r_c of the microscopic number density is finite we can say that $M(V)$ is Gaussian if $V^{1/d} \gg r_c$. If, instead, m_1^2 and/or r_c are infinite we cannot draw such a conclusion.

Note that in general there are more difficult cases of spatial distributions of point-particles with masses. The first case is when $n_0 > 0$ is well defined, but there are non-trivial correlations between masses and positions. The second case is when the point-particle distribution is fractal ($n_0 = 0$ but the conditional average number density is slowly decreasing as a power-law), and masses and positions are not correlated. Finally, the last and more difficult case is when the point-particle distribution is fractal and positions and masses are correlated. All these possibilities will be discussed in Chap. 5.

2.10 The Random Walk and the Central Limit Theorem

In order to consider an example of a stochastic process leading to a Gaussian behavior of certain quantities, let us consider an ordinary one-dimensional *random walk* without memory [43]. To discuss this process is very important because it permits, through the introduction of the central limit theorem, to show the extent of the "Gaussian class" of processes. We will then consider how memory (i.e correlations) introduces mathematical complications which may lead to deviations from the central limit theorem [186].

Let us firstly consider the uncorrelated walk in one dimension. The walk consists of random independent steps of length x_n with $n = 1, 2, \ldots$ whose common probability density function is $p(x)$. For simplicity we assume that the distribution has zero average:

$$\langle x \rangle = \int_{-\infty}^{\infty} p(x)x\,dx = 0 \ . \tag{2.95}$$

Moreover let us consider the case in which $\langle x^2 \rangle > 0$ is finite. We are interested in the probability density function of the end-to-end distance X_N after N steps i.e., of the variable

$$X_N = \sum_{n=1}^{N} x_n \ . \tag{2.96}$$

2.10 The Random Walk and the Central Limit Theorem

The problem is easily generalized to the case in which $\langle x \rangle \neq 0$ by considering that on average it introduces a constant drift $N \langle x \rangle$ to the sum variable X_N.

It is convenient to introduce the re-scaled variable

$$U_N = \frac{X_N}{\sqrt{N}}. \tag{2.97}$$

Let us call $W(U_N)$ the PDF of U_N. The problem of determining it reduces therefore to the evaluation of the integral

$$W(U_N) = \int \cdots \int_{-\infty}^{+\infty} \left[\prod_{n=1}^{N} dx_n p(x_n) \right] \delta\left(\frac{\sum_{n=1}^{N} x_n}{\sqrt{N}} - U_N \right) \tag{2.98}$$

where $\delta(x)$ is the Dirac delta-function. The joint probability density of the path leading to X_N can be factorized due to the independence of different steps. By using the integral representation of the delta-function

$$\delta(y) = \frac{1}{2\pi} \int_{-\infty}^{\infty} \exp(iky) dk \tag{2.99}$$

we obtain

$$W(U_N) = \frac{1}{2\pi} \int_{-\infty}^{\infty} dk \exp(-ikU_N) \left[\prod_{n=1}^{N} \int_{-\infty}^{+\infty} dx_n p(x_n) \exp(ikx_n/\sqrt{N}) \right]. \tag{2.100}$$

Since the function $p(x_n)$ is the same at each step, all the terms of the product in (2.100) are identical. This gives

$$\left[\prod_{n=1}^{N} \int_{-\infty}^{+\infty} dx_n p(x_n) \exp(ikx_n/\sqrt{N}) \right] = \left[\int_{-\infty}^{\infty} dx p(x) \exp(ikx/\sqrt{N}) \right]^N. \tag{2.101}$$

By expanding the exponential term in the right side integral of (2.101), and considering that $\langle x^2 \rangle < +\infty$, we obtain

$$\left[\int_{-\infty}^{\infty} dx p(x) \exp(ikx/\sqrt{N}) \right]^N \simeq \left[1 + \frac{ik}{\sqrt{N}} \langle x \rangle - \frac{k^2}{2N} \langle x^2 \rangle + o(k^2/N) \right]^N, \tag{2.102}$$

where $o(a)$ means an infinitesimal of higher order than a. The second term in the sum of the right-hand side of (2.102) is zero in view of (2.95).

For large N we then have

$$\lim_{N \to \infty} \left[1 - \frac{k^2}{2N} \langle x^2 \rangle + o(N^{-1}) \right]^N = \exp\left[-\frac{k^2}{2} \langle x^2 \rangle \right]. \tag{2.103}$$

Inserting this result into (2.100) we eventually obtain the *central limit Gaussian theorem*

$$W_g(U) = \lim_{N\to\infty} W(U_N) = \frac{1}{\sqrt{2\pi\sigma^2}} \exp\left[-\frac{U}{2\sigma^2}\right] \quad (2.104)$$

with variance

$$\sigma^2 = \langle x^2 \rangle \quad (2.105)$$

which is equal to that of $p(x)$. Considering the variable X_N, we can say that for sufficiently large N it is a Gaussian variable with zero average value and variance given by $\langle X_N^2 \rangle \simeq N\sigma^2$. The central limit theorem thus gives the well known property of the random walk, that after N steps a distance $\simeq \sqrt{N}\sigma$ is covered.

We point out an important property of the Gaussian PDF often referred to as *stability* which is strictly related to the above limit theorem. A PDF $p(x)$ is said to be *stable* if, given two mutually independent variables X_1 and X_2 identically distributed according to $p(x)$, then the sum variable $Y = X_1 + X_2$, appropriately normalized (in the present case divided by $\sqrt{2}$), is again distributed according to the same PDF $p(x)$. The Gaussian PDF is not the only stable probability law, but it is probably the most important in physical applications.

There are two essential elements which have allowed us to obtain the Gaussian distribution for the sum of variables i.e., the *central limit theorem*:

1. The steps are mutually independent and identically distributed. This allows us to factorize the total probability distribution in (2.98) and to obtain a product of N identical terms in (2.100).
2. The second moment $\langle x^2 \rangle$ is finite permitting the expansion in (2.102) and (2.103).

Note that, in order to obtain the result (2.104), it is not necessary to have for each step the same PDF $p(x)$. It is sufficient that all the steps have only the same $\langle x \rangle = 0$ and the same finite $\langle x^2 \rangle > 0$.

It is possible to define random walks satisfying the first requirement above but not the second. A particularly interesting case is represented by the so-called Levy flights [178] (see Appendix B). For these walks $p(x)$ has a power-law tail at large x and the second moment $\langle x^2 \rangle$ diverges. The main effect of this modification is that the asymptotic end-to-end distribution is no longer Gaussian but has a so-called *fat* tail (i.e., slow power-law). However other limit theorems can be proved for this case leading to other stable asymptotic PDFs [178].

Weakening the first condition can also lead to deviations from the central limit Gaussian behavior. Let us analyze this point in more detail. Following a derivation analogous to that given above, it can be shown that if all the steps x_n are mutually independent, but not equally distributed i.e., all with zero mean but different finite variances $\langle x_n^2 \rangle$ such that $\sum_{n=1}^{N} \langle x_n^2 \rangle \sim N$ for large N and under suitable hypothesis on the higher moments of $\{x_n\}$, then for asymptotically large N the variable X_N is still Gaussian with zero mean but with variance

2.10 The Random Walk and the Central Limit Theorem

$$\sigma^2(X_N) = \sum_{n=1}^{N} \langle x_n^2 \rangle .$$

Since $\sigma^2(X_N) \sim N$ for large N, then U_N is Gaussian with the variance

$$\sigma^2(U_N) = \frac{\sum_{n=1}^{N} \langle x_n^2 \rangle}{N}$$

independent of N. In more general, if all $\{x_n\}$ are independent and $\langle x_n^2 \rangle < +\infty$ such that $\sigma^2(X_N) \sim N^\alpha$ with $\alpha > 0$, under suitable hypothesis on the tails of the PDFs $\{p_n(x_n)\}$ of the variables $\{x_n\}$, then U_N, redefined as $X_N/N^{\alpha/2}$, is Gaussian in the large N limit with asymptotic variance independent of N.

Another situation in which the first condition is relaxed is when the single steps x_n are identically distributed with zero mean and finite variance, but non-zero correlations are present i.e.,

$$\langle x_n x_m \rangle \neq 0 \text{ for some } n \neq m .$$

In this case the behavior of U_N (and also of X_N) for large N depends on how long the correlation of the steps persists i.e., it depends on the *correlation length* of the step-step correlation function $\langle x_n x_m \rangle$. In particular it can be shown [84, 186] that if the correlation length (expressed as a number of steps) is finite and N is much larger than it, then U_N is again a Gaussian variable with zero mean and finite variance. In this case, however, the variance is not simply given by $\langle x^2 \rangle$, but depends also on the correlation function. For example, if $\langle x_n x_m \rangle$ decays as $A \exp\left(-\frac{|n-m|}{\tau}\right)$ with finite correlation length $\tau > 0$, then for $N/\tau \gg 1$, the variable U_N can be considered with good approximation a Gaussian variable with a variance depending on $\langle x^2 \rangle$, A, and τ, diverging when one of these three parameters diverges.

To illustrate this last case we consider a further simple explicit example. Let us consider the set of random variables $\{x_1, x_2, \ldots, x_N\}$ all with zero mean and the same finite variance $\langle x^2 \rangle$. Moreover we suppose that we can partition this set into N/m (we take for simplicity N as an integer multiple of m) sub-sets $\{x_{(l-1)m+1}, x_{(l-1)m+2}, \ldots, x_{lm}\}$ with $l = 1, \ldots, N/m$, in such a way that two variables are correlated if they belong both to the the same sub-set, while they are not if they belong to two different sub-sets. It is clear that m represents the correlation length of this system. Moreover let us suppose for simplicity that $\langle x_{(l-1)m+i} x_{(l-1)m+j} \rangle = a \neq 0$, with $1 \leq i \neq j \leq m$ and $l = 1, \ldots, N/m$, where a is a constant independent of i, j and l. Clearly $|a| \leq \langle x^2 \rangle$. The variable X_N, as defined by (2.96), can be rewritten as

$$X_N = \sum_{l=1}^{N/m} y_l , \qquad (2.106)$$

where

$$y_l = \sum_{i=1}^{m} x_{(l-1)m+i} \text{ for } l = 1, \ldots, N/m ,\qquad(2.107)$$

form a set of independent variables with zero mean and common variance

$$\langle y_l^2 \rangle = m \langle x^2 \rangle + m(m-1)a .\qquad(2.108)$$

At this point we can apply the central limit theorem for independent variables to (2.106), obtaining that, when $N/m \to \infty$, the PDF of X_N is a Gaussian with zero mean and variance given by

$$\sigma^2(X_N) = N[\langle x^2 \rangle + (m-1)a] .$$

Consequently, the variable U_N defined by (2.97) is Gaussian with variance $\langle x^2 \rangle + (m-1)a$.

Coming back to the general case, it is possible to show that, if the correlation function $\langle x_n x_m \rangle$ has an infinite correlation length i.e., it decreases with $|n-m|$ as a sufficiently slow power-law, the PDF of X_N in the large N limit no longer converges to a Gaussian distribution, and its form strongly depends on the step-step correlation function.

2.10.1 Probability Distribution of Mass Fluctuations in Large Volumes

Let us now apply these results on the random walk to stationary stochastic mass density fields in d-dimensions with a well defined average density $\rho_0 > 0$. In complete analogy, we can say that, if the reduced density-density correlation function $C_2(r)$ has a finite correlation length r_c (i.e., if $C_2(r)$ decreases sufficiently rapidly as a function of r), then the mass $M(R)$ contained in a given spherical volume of radius R, with $R \gg r_c$ can be considered approximately a Gaussian variable with mean $\rho_0 V(R)$ (where $V(R)$ is the volume of the sphere of radius R), and variance proportional to $(R/r_c)^d$ with a coefficient depending on r_c and on the amplitude coefficient of $C_2(r)$. Note that, as a consequence, the "local" average density in the same volume, defined by $\overline{\rho}(R) = M(R)/V(R)$, is also Gaussian, but with mean ρ_0 and variance proportional to R^{-d}. Therefore in the limit $R \to \infty$ the PDF of $\overline{\rho}(R)$ converges to the Dirac delta-function $\delta\left(\overline{\rho}(R) - \rho_0\right)$. This shows that $\overline{\rho}(R)$ in the large R limit is a *self-averaging* variable with mean ρ_0.

On the contrary if $\rho_0 > 0$ is still well defined, but $C_2(r)$ has an infinite correlation length, decreasing slowly to zero, then the variable $M(R)$ cannot in general be considered Gaussian at any R, and the shape of its PDF will depend strongly on $C_2(r)$. As shown above, in this case, the variance of $M(R)$ is proportional to R^a with $d \leq a < 2d$. From this we can derive simply that the variance of $\overline{\rho}(R)$ is proportional to R^{a-2d}. This ensures, even in this "strongly correlated" case, the convergence of the PDF of $\overline{\rho}(R)$

to $\delta(\overline{\rho}(R) - \rho_0)$ for $R \to \infty$. This is another way of saying that, because of the normalization of $\overline{\rho}(R)$ with respect to $M(R)$, the convergence of its PDF to a delta-function is not governed by the central limit theorem, but by the *law of large numbers* [100], which has a range of validity much larger than the central limit theorem. In fact, it can be shown that for its validity it is sufficient only that $C_2(r) \to 0$ for $r \to \infty$, independently of the rate of decrease to zero (i.e., of the correlation length) of $C_2(r)$ itself. Due to this, the limit self-averaging shape of the PDF of $\overline{\rho}(R)$ is approximately reached when $R \gg \lambda_0$, where λ_0 is the homogeneity scale which has been defined above. However, in this strongly correlated case, there is no length beyond which we can say that $\overline{\rho}(R)$ is, with good approximation, Gaussian. For instance, at the critical point of a second order phase transition, where the long-range correlations yield diverging fluctuations, the situation is more complicated and the PDF is generally non trivial. A remarkable simplification occurs even in critical systems where the large scale fluctuations give rise to universality [148]: for macroscopic quantities this means that the shape of the PDF depends only on few general characteristics of the system (see [5, 197] and references therein). Many aspects of non-equilibrium PDF have been explored by [5, 197], resulting in a classification of non-Gaussian PDFs for macroscopic quantities in non-equilibrium steady states.

This discussion clarifies further the different roles of the correlation and the homogeneity scales in homogeneous stationary stochastic fields. A more extreme case will be presented in Chap. 4 when fractal mass distributions will be considered.

2.11 Gaussian Distribution as the Most Probable Probability Distribution

Here we show briefly that the Gaussian distribution for a random variable can be seen as the *most probable distribution* if, about the variable, we know only the mean value and the finite variance.

The derivation of this result is based on the use of the so-called Shannon entropy introduced in the context of *information theory* [174]. The concept of Shannon entropy is just fundamental in information theory. The entropy of a random variable is defined in terms of its probability distribution and can be shown to be a good measure of randomness or uncertainty. Given a set of events $\{\mathcal{E}_1, \mathcal{E}_2, \ldots, \mathcal{E}_n\}$ with respectively normalized probabilities $\{P_1, P_2, \ldots, P_n\}$ ($\sum_{i=1}^{n} P_i = 1$), the related Shannon entropy is defined as

$$S_d(P_1, P_2, \ldots, P_n) = -\sum_{i=1}^{n} P_i \log P_i \,. \tag{2.109}$$

It can be shown that the larger this entropy the larger is the *uncertainty* and the lower is the information about the set of events $\{\mathcal{E}_1, \mathcal{E}_2, \ldots, \mathcal{E}_n\}$ [239].

Clearly the definition given in (2.109) includes the case in which the set of events is the set of possible values of a numerical variable. Therefore it can be extended to the case in which the random variable X gets a continuous set of values $x \in \mathbb{R}$ with a probability density function $p(x)$. In this case (2.109) can be rewritten as[5]

$$S_c[p(x)] = -\int_{-\infty}^{+\infty} dx\, p(x) \log p(x) \,. \qquad (2.110)$$

Let us now consider a random variable X about which we know only the mean value

$$\langle X \rangle = \int_{-\infty}^{+\infty} dx\, x\, p(x) \qquad (2.111)$$

and the second moment

$$\langle X^2 \rangle = \int_{-\infty}^{+\infty} dx\, x^2\, p(x) \qquad (2.112)$$

which is taken to be finite. The question we want to address is: what is the most probable distribution for X given only this information? The usual prescription to answer this question is the so-called *maximum entropy principle* saying that we have to find the function $p(x)$ maximizing the Shannon entropy given by (2.110) and satisfying the conditions (2.111) and (2.112). Moreover we have to require that $p(x)$ satisfies the normalization condition

$$\int_{-\infty}^{+\infty} dx\, p(x) = 1 \,. \qquad (2.113)$$

This is done by looking for the function $p(x)$ which is a zero of the functional derivative of the *constrained* entropy:

$$\mathcal{F}[p(x)] = -\int_{-\infty}^{+\infty} dx\, p(x) \log p(x) - \lambda_1 \int_{-\infty}^{+\infty} dx\, p(x) - \lambda_2 \int_{-\infty}^{+\infty} dx\, x\, p(x) - \lambda_3 \int_{-\infty}^{+\infty} dx\, x^2 p(x) \,, \qquad (2.114)$$

---

[5] Note that the entropy S_d defined by (2.109) for a discrete random variable is always positive just as the thermodynamic entropy in physics. Instead the entropy S_c for a continuous variable, defined by (2.110), can be both positive and negative. This problem can be overcome by discretizing the phase space of the continuous variable in a discrete set of cells of size ϵ, and considering the "discrete" entropy S_d defined by (2.109) where $P_i \simeq p(x_i)\epsilon^d$ is now the probability that the continuous variable lies in the ith cell around x_i. It is simple to show [174] that for sufficiently small ϵ, the two entropies S_c and S_d differ by only a constant: $S_d \simeq S_c - d \log \epsilon$ where d is the dimensionality of the variable.

where λ_1, λ_2 and λ_3 are the Lagrange multipliers corresponding to the three conditions (2.111), (2.112), and (2.113). By taking the functional derivative and setting it equal to zero, we find the equation:

$$\frac{\delta \mathcal{F}[p(x)]}{\delta p(x)} \equiv -\log p(x) - 1 - \lambda_1 - \lambda_2 x - \lambda_3 x^2 = 0 , \tag{2.115}$$

which gives

$$p(x) = \exp\left[-\lambda_3 x^2 - \lambda_2 x - 1 - \lambda_1\right] , \tag{2.116}$$

which is a Gaussian distribution. The Lagrange multipliers can be found by requiring the equalities (2.111), (2.112) and (2.113). This gives finally

$$p(x) = \frac{1}{\sqrt{2\pi\sigma^2}} \exp\left[-\frac{(x - \langle X \rangle)^2}{2\sigma^2}\right] , \tag{2.117}$$

where $\sigma^2 = \langle X^2 \rangle - \langle X \rangle^2$. Finally, in order to verify that (2.117) is actually a maximum of the expression (2.114) and not only an extremum, it is sufficient to verify that the second functional derivative of (2.114) is negative definite. This can immediately seen to be the case as the second functional derivative is easily found to be $-1/p(x)$, which for any PDF $p(x)$ is negative for all x. This result can be simply extended to the case in which X is not a scalar but a vector of N components $X_1, \ldots, X_N$ about which we know $\langle X_i \rangle$ and $\langle X_i X_j \rangle$ for $i, j = 1, \ldots, N$. In this case we find the generalized N-dimensional Gaussian distribution given by (2.76).

2.12 Summary and Discussion

In this chapter we have defined and studied the main characteristics of *stationary stochastic processes* (SSP), focusing mainly on stochastic *mass density fields* $\hat{\rho}(\boldsymbol{r})$. In this context we have distinguished between (1) *continuous* fields and (2) discrete point-particle distributions which are usually called *stochastic point processes* (SPP). In particular we have analyzed the properties of *uniform or homogeneous* (i.e., non-fractal) mass density fields which in general can be characterized as a mass fluctuation field around a well defined average density $\rho_0 > 0$.

For the general class of SSP, we have defined the concepts of *ensemble* average, and discussed the hypothesis of ergodicity required to replace the ensemble average with the volume average. Moreover, we have introduced the concept of spatial correlation and of n-point correlation functions, focusing on the two-point correlation analysis, which characterize the influence that the knowledge of the values of the field at a point of the space has on the conditional probability of observing another value of the field at another spatial point. Restricting the analysis to fluctuation fields (i.e. defined with respect to the average $\rho_0 > 0$), we have introduced also to concept of *cumulants* and *connected* correlation functions. The two main lengths characterizing fluctuations and correlation properties of a homogeneous stochastic field are the *homogeneity* scale and the *correlation length*. The former marks the length scale beyond which fluctuations can be considered "small", while the latter characterize the rate of decrease in space of correlations between fluctuations. We have noted that these two distinct scales characterizing very different properties of stochastic fields are very often confused in the cosmological literature.

We have discussed also the general properties of *integrated fluctuations* (e.g., mass fluctuations in finite volumes) relating them to correlation functions.

For the SPP class we have presented a complete treatment of an explicit example: the *Poisson particle distribution*, which is the "most random" particle distribution possible. Then we have generalized the analysis to more general SPPs introducing the concept of *conditional densities* and conditional fluctuations, relating them again to correlation functions. In this context we have studied also the statistics of the distance between neighboring particles.

Also for the continuous SSP class, we have presented an explicit and paradigmatic example: the continuous *Gaussian field*. We have stressed the fundamental role it plays in the study of randomly fluctuating signals and phenomena, discussing the *central limit theorem*, and the relation between Gaussianity and *information theory*.

Finally we have presented the concepts of *scale-invariance* and *self-similarity* of the fluctuation field characterizing them in terms of the scaling properties of the two-point correlation function. These concepts will be developed in Chaps. 4–5.

3 The Power Spectrum and the Classification of Stationary Stochastic Fields

3.1 Introduction

In the previous chapter ordinary continuous SSP and discrete SPP have been characterized only through a real space description based on the n-point correlation functions. In particular a central role in the real space statistical analysis of such stochastic processes is played by the connected two-point correlation function $C_2(r)$ (or $\tilde{\xi}(r)$ for uniform stochastic mass density fields with well defined average density $\rho_0 > 0$). The analysis of a stochastic process limited to the two-point correlation properties is usually called *correlation theory*. In this chapter we introduce the Fourier conjugate characterization of the same kind of stochastic fields through the definition and the study of the so-called *power spectrum* (PS).

3.2 General Properties

3.2.1 Mathematical Definitions

First of all, let us introduce some mathematical definitions and results of Fourier analysis which we will use. Let us suppose that our stationary stochastic field is defined in a cubic volume V of size L ($V = L^d$, where d is the spatial dimension). Let us take an arbitrary function $f(r)$ defined in this volume. We define the Fourier transform (FT) of $f(r)$ in the volume V as:

$$\tilde{f}(\boldsymbol{k}) = \int_V d^d r\, f(\boldsymbol{r})\, e^{-i\boldsymbol{k}\cdot\boldsymbol{r}} \ . \tag{3.1}$$

This implies that

$$f(\boldsymbol{r}) = \frac{1}{V} \sum_{\boldsymbol{k}} \tilde{f}(\boldsymbol{k})\, e^{i\boldsymbol{k}\cdot\boldsymbol{r}} \ , \tag{3.2}$$

where the sum is over all the $\boldsymbol{k}$ whose components k_i satisfy $k_i = 2m\pi/L$ with $m = \ldots, -2, -1, 0, 1, 2, \ldots$. Usually, in mathematical textbooks, the FT $\tilde{f}(\boldsymbol{k})$ is equal to that of (3.1) multiplied by a factor V^{-1} in order to have (3.2) without this factor. We choose this different definition because with this choice it is not necessary to redefine the FT in the limit $V \to \infty$. In fact, in this limit, it is simple to show that above equations become directly the usual Fourier relations in the d-dimensional infinite Euclidean space:

$$\begin{cases} \tilde{f}(\boldsymbol{k}) = \int d^d r\, f(\boldsymbol{r})\, e^{-i\boldsymbol{k}\cdot\boldsymbol{r}} \\ f(\boldsymbol{r}) = \frac{1}{(2\pi)^d} \int d^d r\, \tilde{f}(\boldsymbol{k})\, e^{i\boldsymbol{k}\cdot\boldsymbol{r}} \end{cases} \quad (3.3)$$

In what follows, we introduce the following simplified notation for (3.3):

$$\begin{cases} \tilde{f}(\boldsymbol{k}) = FT\,[f(\boldsymbol{r})] \\ f(\boldsymbol{r}) = FT^{-1}\left[\tilde{f}(\boldsymbol{k})\right] \end{cases}. \quad (3.4)$$

Let us now consider the FT of a stochastic fluctuation field. From this chapter on, to simplify notation we will denote by $\rho(\boldsymbol{r})$ both the stochastic function $\hat{\rho}(\boldsymbol{r})$ (i.e., the infinite dimensional random variable) and its realization $\rho(\boldsymbol{r})$ (i.e., the value of the random variable). The stochastic fluctuation field is then denoted by $\delta\rho(\boldsymbol{r}) = \rho(\boldsymbol{r}) - \rho_0$ and its Fourier integral in the volume V by:

$$\delta\tilde{\rho}(\boldsymbol{k}; V) = \int_V d^d r\, \delta\rho(\boldsymbol{r})\, e^{-i\boldsymbol{k}\cdot\boldsymbol{r}}. \quad (3.5)$$

Since $\delta\rho(\boldsymbol{r})$ is real, we have that $\delta\tilde{\rho}(-\boldsymbol{k}; V) = \delta\tilde{\rho}^*(\boldsymbol{k}; V)$ (where a^* is the complex conjugate of the generic complex number a). Through the definition of $\delta\tilde{\rho}(\boldsymbol{k}; V)$ we can define the PS of the stochastic field $\rho(\boldsymbol{r})$ as:

$$S(\boldsymbol{k}) = \frac{\left\langle |\delta\tilde{\rho}(\boldsymbol{k}; V)|^2 \right\rangle}{V}, \quad (3.6)$$

where $\langle\ldots\rangle$ is the usual ensemble average. In several physical contexts, as in condensed matter theory, $S(\boldsymbol{k})$ is also called the *structure factor* [108]. In the mathematical context of SPP it is instead often called the *Bartlett spectrum* [16]. Since ergodicity is strictly valid only in the infinite volume limit, the ensemble average can be rigorously substituted by the volume average only in the limit $V \to \infty$. Therefore from now on we shall work directly in this limit:

$$S(\boldsymbol{k}) = \lim_{V \to \infty} \frac{\left\langle |\delta\tilde{\rho}(\boldsymbol{k}; V)|^2 \right\rangle}{V}. \quad (3.7)$$

The PS is the simplest statistical measure of the net contribution of the mode $\boldsymbol{k}$ to the realization of the stochastic process. Note that if the stochastic process $\rho(\boldsymbol{r})$ is statistically isotropic then $S(\boldsymbol{k}) \equiv S(k)$. In order to find the relationship between $S(\boldsymbol{k})$ and the connected two-point correlation function $C_2(r)$, let us consider the covariance between different modes directly in the limit $V \to \infty$:

$$\lim_{V \to \infty} \langle \delta\tilde{\rho}(\boldsymbol{k}; V) \delta\tilde{\rho}(\boldsymbol{k}'; V) \rangle = \int\int d^d r\, d^d r'\, \langle \delta\rho(\boldsymbol{r})\delta\rho(\boldsymbol{r}') \rangle\, e^{-i(\boldsymbol{k}\cdot\boldsymbol{r}+\boldsymbol{k}'\cdot\boldsymbol{r}')}$$

$$= (2\pi)^d \delta(\boldsymbol{k}+\boldsymbol{k}') \int d^d r\, C_2(r)\, e^{-i\boldsymbol{k}\cdot\boldsymbol{r}}, \quad (3.8)$$

where we have used the assumed statistical translational invariance of the SSP. Equations (3.7) and (3.8) imply that, for a stationary stochastic field, the PS is the FT of the reduced two-point correlation function:

$$S(\boldsymbol{k}) = \int d^d r\, C_2(\boldsymbol{r})\, e^{-i\boldsymbol{k}\cdot\boldsymbol{r}} \equiv FT\left[C_2(\boldsymbol{r})\right]. \quad (3.9)$$

Conversely, we can write:

$$C_2(\boldsymbol{r}) = \frac{1}{(2\pi)^d} \int d^d k\, S(\boldsymbol{k}) e^{i\boldsymbol{k}\cdot\boldsymbol{r}} \equiv FT^{-1}\left[S(\boldsymbol{k})\right]. \quad (3.10)$$

Rigorously speaking (3.9) is valid if $C_2(\boldsymbol{r})$ is strictly integrable over all space, otherwise it must be understood in the sense of distributions [100]. These results constitute one of the main points of the celebrated Wiener-Khinchin (or only Khinchin) theorem already introduced in Chap. 2 which states that [100, 230]:

- Given an arbitrary function $C_2(\boldsymbol{r})$, there exists at least one continuous SSP which has $C_2(\boldsymbol{r})$ as its connected two-point correlation function, if and only if its FT $S(\boldsymbol{k})$ is non negative at all $\boldsymbol{k}$. Moreover, if the SSP has finite one point variance $s^2 \equiv C_2(0)$, $S(\boldsymbol{k})$ must be integrable over all space; in other words $S(\boldsymbol{k})$ must be, apart from a normalization constant, a probability density function[1].
- For a SPP a similar condition holds, even though in this case it is only a necessary condition (the search for sufficient conditions is still an open problem): Given a $C_2(\boldsymbol{r})$, which at $r = 0$ is equal to $\rho_0 \delta(\boldsymbol{r})$, a necessary condition to have an SPP of average density ρ_0 having $C_2(\boldsymbol{r})$ as connected two-point correlation function is that its FT $S(\boldsymbol{k})$ is non-negative at any k and that $S(\boldsymbol{k}) \to \rho_0$ for $k \to \infty$.

For a statistically isotropic SSP in $d = 3$ (3.9) and (3.10) can be rewritten as

$$S(k) = 4\pi \int_0^\infty C_2(r) \frac{\sin(kr)}{kr} r^2 dr \quad (3.11)$$

$$C_2(r) = \frac{1}{2\pi^2} \int_0^\infty S(k) \frac{\sin(kr)}{kr} k^2 dk. \quad (3.12)$$

If the SSP $\rho(\boldsymbol{r})$ represents an homogeneous mass density field (i.e., with a finite homogeneity scale beyond which it can be considered spatially uniform

---

[1] In this context we include in the continuous SSP also the case in which, though $\lim_{r \to r'} \left\langle |\rho(\boldsymbol{r}) - \rho(\boldsymbol{r}')|^2 \right\rangle = 0$, the variance $C_2(0) = +\infty$, but this limit is approached continuously for $r \to 0$. The sub-case with $C_2(0) < +\infty$ can thus be called "strictly" continuous. While if $C_2(0) < +\infty$, integrability of $S(\boldsymbol{k})$ is required, for the weaker case $C_2(0) = +\infty$, approached continuously for $r \to 0$, we have only that $S(\boldsymbol{k})$ vanishes for $k \to \infty$.

to a good approximation), either continuous or discrete, with a well defined $\rho_0 > 0$, then the fluctuation field can be made dimensionless by writing

$$\delta_\rho(\boldsymbol{r}) \equiv \frac{\delta\rho(\boldsymbol{r})}{\rho_0} .$$

In cosmology this is called the *matter density contrast*. As shown in Chap. 2 the dimensionless connected two-point correlation function can be written as

$$\tilde{\xi}(\boldsymbol{r}) = \langle \delta_\rho(\boldsymbol{r}+\boldsymbol{r}_0)\delta_\rho(\boldsymbol{r}_0) \rangle = \frac{C_2(\boldsymbol{r})}{\rho_0^2} .$$

In this case the PS is usually redefined as

$$P(\boldsymbol{k}) = \frac{S(\boldsymbol{k})}{\rho_0^2} ,$$

in such a way that the relations (3.9) and (3.10) can be rewritten as

$$P(\boldsymbol{k}) = FT\left[\tilde{\xi}(\boldsymbol{r})\right] \tag{3.13}$$

$$\tilde{\xi}(\boldsymbol{r}) = FT^{-1}\left[P(\boldsymbol{k})\right] . \tag{3.14}$$

An analogous replacement of $S(k)$ with $P(k)$ and of $C_2(r)$ with $\tilde{\xi}(r)$ can be made in (3.11) and (3.12).

3.2.2 Limit Conditions

We limit here our discussion to isotropic SSPs. The limit conditions for the connected two-point correlation function $C_2(r)$ are:

- For $r \to 0$ we must have $C_2(r) \sim r^\alpha$ with $\alpha > -d$. This is a necessary condition in order that the integrated (i.e., mass) fluctuations have finite variance at any finite spatial scale (see (2.29)). This is a fundamental requirement in order for the density field to be a well defined stochastic process (because, if $\alpha < -d$ the FT of $C_2(r)$ diverges at any k). For a point-particle distribution (SPP) we always have also the *diagonal* contribution $\rho_0 \delta(r)$ to $C_2(r)$ which is relevant only at $r = 0$ (see Chap. 2).
- For $r \to \infty$ correlations must vanish i.e., $C_2(r) \sim r^\beta$ with $\beta < 0$ (in the case of a power-law behavior).

These conditions on $C_2(r)$ imply respectively the following properties for the PS $S(k)$:

- If the SSP is continuous (e.g., a Gaussian field), then $\lim_{k \to \infty} S(k) = 0$ i.e., $S(k) \sim k^\gamma$ with $\gamma < 0$ for a power-law behavior. If, moreover, it has finite one-point variance $C_2(0) < +\infty$ (i.e., "strictly" continuous with finite variance), then $\lim_{k \to \infty} k^d S(k) = 0$ i.e., $\gamma < -d$ for the power-law case.

In Chaps. 6–7 we will see that this condition is satisfied in cosmological models of the mass density field $\rho(\boldsymbol{r})$.

For a point-particle distribution (SPP), on the other hand, the presence of the delta-like diagonal term in $C_2(r)$ implies that $S(k \to \infty) = \rho_0$ (i.e., $P(k \to \infty) = 1/\rho_0$). However, even in this case, there is a similar constraint on the large k behavior: $\lim_{k\to\infty} |S(k) - \rho_0| = 0$ i.e., $|S(k) - \rho_0| \sim k^\gamma$ again with $\gamma < 0$.

- For $k \to 0$ one has $S(k) \sim k^\delta$ with $\delta > -d$.

By studying the FT relating $S(k)$ and $C_2(r)$, it is possible to write relations between the pairs of scaling exponents α, γ and β, δ. For instance, in $d = 3$ one has that, if $S(k) = Ak^\beta \exp(-k/k_c)$, then $\beta = -\delta - 3$ (the details of the derivation can be found in [184]). However, if $S(k) = k^\beta f(k)$ where $f(k)$ is analytic everywhere in $\boldsymbol{k}$ (note that $\exp(-k/k_c)$ is not analytic at $\boldsymbol{k} = 0$ because it is a function of the modulus of $\boldsymbol{k}$) and $f(0) = a > 0$, $C_2(r)$ decays at large r faster than any power (e.g., exponentially) for $\beta = 2m$ where $m = 0, 1, 2, 3, \ldots$. In fact in this case $S(k)$ is analytic everywhere and therefore its FT must decay faster than any power at large r.

3.3 The Power Spectrum for the Poisson Point Process and Other SPP

As seen in the previous chapter, in the case of a Poisson distribution of unitary mass point-particles, the two-point correlation function is

$$\tilde{\xi}(r) = \frac{\delta(\boldsymbol{r})}{\rho_0} . \tag{3.15}$$

Therefore, by applying directly (3.9), one finds the PS to be simply:

$$P(k) = \frac{1}{\rho_0} . \tag{3.16}$$

The interpretation of (3.16) is that in a Poisson point process every mode has the same weight, which is the signature of the complete absence of two-point correlations. This case is also called *white noise*.

Since for any stationary stochastic point process the function $\tilde{\xi}(r) = \delta(\boldsymbol{r})/\rho_0 + \xi(r)$ with $\xi(r)$ defined for $r > 0$, going to zero at $r \to \infty$, and integrable at $r = 0$, one always has

$$\lim_{k \to \infty} P(k) = 1/\rho_0$$

as in the Poisson case.

It is important to note that $P(0)$ has an important meaning when related to large scale mass fluctuations for any stationary mass density field. By definition

$$P(0) = \int d^d r \tilde{\xi}(r) \,, \qquad (3.17)$$

where the integral extends over all space. In the next section we will show that large scale mass fluctuations are related to the behavior of $P(k)$ around $k = 0$. In particular we will see that every stationary stochastic density field with $P(0) = A > 0$ but finite presents the same large scale behavior of the mass fluctuations as for a Poisson SPP with an effective density $\rho_0 = 1/A$. Therefore we name this class of SSP *substantially Poisson*. This class is characterized by having sufficiently short range and mainly positive two-point correlations, as for example in a homogeneous gas in equilibrium at high temperature.

3.4 The Power Spectrum and the Mass Variance: A Complete Classification

Let us analyze the relation between the PS and the mass-variance in both discrete or continuous stochastic mass density fields with a well defined positive average density ρ_0. We first rewrite the relations (2.27)–(2.28), generalizing them to the case in which we calculate the mass variance in a topologically more complex volume $\mathcal{V}$ of size V. To do this one introduces the window function $W_\mathcal{V}(r)$ defined as

$$W_\mathcal{V}(r) = \begin{cases} 1 & \text{if } r \in \mathcal{V} \\ 0 & \text{otherwise} \,. \end{cases} \qquad (3.18)$$

Therefore we can rewrite (2.27) as

$$\langle M(\mathcal{V}) \rangle = \int W_\mathcal{V}(r) \langle \rho(r) \rangle \, d^d r = \rho_0 V \,. \qquad (3.19)$$

and (2.28) as

$$\langle M^2(\mathcal{V}) \rangle = \int\!\!\int d^d r_1 d^d r_2 W_\mathcal{V}(r_1) W_\mathcal{V}(r_2) \langle \rho(r_1) \rho(r_2) \rangle \,, \qquad (3.20)$$

where all the integrals are over all space. The normalized variance

$$\sigma^2(\mathcal{V}) \equiv \frac{\langle M^2(\mathcal{V}) \rangle - \langle M(\mathcal{V}) \rangle^2}{\langle M(\mathcal{V}) \rangle^2}$$

is then given by

$$\sigma^2(\mathcal{V}) = \frac{1}{V^2} \int\!\!\int d^d r_1 d^d r_2 W_\mathcal{V}(r_1) W_\mathcal{V}(r_2) \tilde{\xi}(r_1 - r_2) \,. \qquad (3.21)$$

By writing $\tilde{\xi}(r)$ as the FT of the PS $P(k)$, one obtains

3.4 The Power Spectrum and the Mass Variance

$$\sigma^2(\mathcal{V}) = \frac{1}{(2\pi)^d} \int d^d k P(\mathbf{k}) |\tilde{W}_\mathcal{V}(\mathbf{k})|^2 \tag{3.22}$$

which is explicitly positive, and $\tilde{W}_\mathcal{V}(\mathbf{k})$ is the FT of $W_\mathcal{V}(\mathbf{r})$, divided by the volume defined by the window function itself, i.e.,

$$\tilde{W}_\mathcal{V}(\mathbf{k}) = \frac{1}{V} \int d^d r\, e^{-i\mathbf{k}\cdot\mathbf{r}} W_\mathcal{V}(\mathbf{r}) \tag{3.23}$$

with $V = \int W_\mathcal{V}(\mathbf{r}) d^d r$ by definition (therefore $\tilde{W}_\mathcal{V}(0) = 1$).

We now treat directly the case in which $\mathcal{V}$ is a sphere of radius R in the three-dimensional space which is relevant for the application to cosmology discussed in Chap. 6. The generalization to $d \neq 3$ dimensions is straightforward. Therefore let us consider the real sphere of radius R for which the FT of the window function is

$$\tilde{W}_R(\mathbf{k}) = \frac{3}{(kR)^3} (\sin kR - kR \cos kR) . \tag{3.24}$$

One then finds assuming statistical isotropy so that $P(\mathbf{k}) = P(k)$, the expression for the normalized mass variance in real spheres

$$\sigma^2(R) = \frac{9}{2\pi^2} \int_0^\infty dk\, \frac{(\sin kR - kR \cos kR)^2}{(kR)^6} k^2 P(k) . \tag{3.25}$$

Let us, without loss of generality, assume that $P(k) = A k^n f(k)$, where $A > 0$ and $f(k)$ a cut-off function chosen such that (i) $\lim_{k\to 0} f(k) = 1$, and (ii) $\lim_{k\to\infty} k^n f(k)$ is finite. We also require, as discussed in the previous section, $n > -3$ to have the integrability of $P(k)$ around $k = 0$. The condition (ii), which implies that at large wave-numbers $f(k) \sim k^{-\alpha}$ with $\alpha \geq n$, includes both the case of a continuous SSP (for which $\lim_{k\to\infty} P(k) = 0$) and the point process (for which $\lim_{k\to\infty} P(k) = 1/\rho_0)^2$. It is convenient to rescale variables putting $x = kR$ to rewrite (3.25) as

$$\sigma^2(R) = \frac{9A}{2\pi^2} \frac{1}{R^{3+n}} \int_0^\infty dx\, (\sin x - x \cos x)^2\, x^{n-4} f\left(\frac{x}{R}\right) . \tag{3.26}$$

By analyzing in detail this formula, we will obtain the following general relation between the large R behavior of $\sigma^2(R)$ and the small k behavior of $P(k)$ [91]:

$$\sigma^2(R) \sim \begin{cases} R^{-(3+n)} & \text{for } -3 < n < 1 \\ R^{-4} \log R & \text{for } n = 1 \\ R^{-4} & \text{for } n > 1 . \end{cases} \tag{3.27}$$

---

[2] Actually, as shown Sect. 3.5.1, there are also point processes with only a discrete translational invariance characterized by a PS with delta function contributions on periodic lattice sites. The present arguments can be extended easily to include this case.

Therefore for $n = 0$, as mentioned above, a Poisson-like behavior of mass fluctuations is obtained. This agrees with the fact that $n = 0$ implies

$$\int d^3r\, \tilde{\xi}(r) = A > 0$$

as in the Poisson case. On the other hand for $-3 < n < 0$ i.e., in the case where

$$\int d^3r\, \tilde{\xi}(r) = +\infty$$

we obtain a *super-Poisson* scaling of mass fluctuations, that is $\langle \Delta M^2(R) \rangle$ grows as a function of R faster than in a Poisson system. This is the characteristic behavior of the order parameter fluctuations of a thermodynamic system at the critical point of a second order phase transition (e.g., liquid-gas critical point) [148]. Finally, for $n > 0$, we obtain a *sub-Poisson* behavior of mass fluctuations. That is $\langle \Delta M^2(R) \rangle$ grows as a function of R slower than in a Poisson system. For this reason we call this kind of systems *super-homogeneous*. Other authors refer to this class of mass density fields as *hyper-uniform* [231]. Note that $n > 0$ implies that

$$\int d^3r\, \tilde{\xi}(r) = 0\, ,$$

which is a non-local balance condition on positive and negative density-density correlations. It is simple to see that this non-local condition implies a sort of stochastic long-range order. In fact, as explicitly shown in the following, a regular array of unitary mass particles (i.e., a lattice), has $n \to \infty$. Therefore the larger is n the more regularly density fluctuations are spatially distributed. It is important to note that for any $n \geq 1$ we have as a limiting behavior $\sigma^2(R) \sim R^{-4}$. This means that there is no stationary stochastic mass density field, either continuous or discrete, such that the quadratic mass fluctuation $\langle \Delta M^2(R) \rangle$ in a sphere of radius R grows slower than the surface of the sphere for large values of R. This last sentence can be shown to be true in any dimension [24].

In order to derive (3.27), we introduce the following assumption only to simplify the demonstration, without limiting the general validity of the final result:

$$f(k) = \begin{cases} 1 & \text{for } k \leq k_0 \\ \left(\dfrac{k}{k_0}\right)^{-\alpha} & \text{for } k > k_0\, , \end{cases} \quad (3.28)$$

where $k_0 > 0$ is the cross-over wave number. In this case this cross-over wave number plays also the role of the upper cut-off of the integral in (3.25), as it marks the transition to the tail of $P(k)$ guaranteeing the integrability for large k of the integrand of (3.25). This is equivalent to requiring the finiteness of mass-fluctuations for finite values of R.

3.4 The Power Spectrum and the Mass Variance

We can rewrite (3.26) as follows:

$$\sigma^2(R) = \frac{9A}{2\pi^2} \frac{1}{R^{3+n}} \left[\int_0^{k_0 R} dx \left(\sin x - x \cos x\right)^2 x^{n-4} \right.$$
$$\left. + (k_0 R)^\alpha \int_{k_0 R}^\infty dx \left(\sin x - x \cos x\right)^2 x^{n-4-\alpha} \right], \quad (3.29)$$

where we have put $kR = x$. Let us now analyze the scaling behavior of the oscillating factor of the window function:

$$(\sin x - x \cos x)^2 \sim \begin{cases} \frac{x^6}{9} & \text{for } x \ll 1 \\ x^2 & \text{for } x \gg 1 \,. \end{cases} \quad (3.30)$$

We now consider the case $k_0 R \gg 1$ to study the asymptotic behavior of the normalized mass variance for $R \to \infty$. First of all, let us analyze the behavior of the first integral in (3.29):

$$I_1 \equiv \int_0^{k_0 R} dx \left(\sin x - x \cos x\right)^2 x^{n-4} \,. \quad (3.31)$$

Since $n > -3$, I_1 is well behaved at its lower limit $x = 0$. It is simple to show that for $n < 1$, I_1 converges to a finite positive value when $R \to +\infty$, while it diverges as R^{n-1} if $n > 1$. This implies that for $k_0 R \gg 1$ we can write:

$$I_1 \sim \begin{cases} a > 0 & \text{for } -3 < n < 1 \\ \log R & \text{for } n = 1 \\ R^{n-1} & \text{for } n > 1 \,. \end{cases} \quad (3.32)$$

Let us now study the second integral of (3.29) in the same limit:

$$I_2 \equiv \int_{k_0 R}^\infty dx \left(\sin x - x \cos x\right)^2 x^{n-\alpha-4} \simeq \int_{k_0 R}^\infty dx\, x^{n-\alpha-2} \,.$$

Recalling that $\alpha \geq n$ (from the hypothesis of the finiteness of mass fluctuations), we can therefore write at large R

$$I_2 \sim R^{n-\alpha-1} \,, \quad (3.33)$$

which goes to zero for $R \to +\infty$[3]. Consequently, the following scaling behavior for $\sigma^2(R)$ is found:

$$\sigma^2(R) = \frac{9A}{2\pi^2} \frac{1}{R^{3+n}} \left[I_1 + (k_0 R)^\alpha I_2\right] \sim \begin{cases} R^{-3-n} & \text{for } -3 < n < 1 \\ R^{-4} \log R & \text{for } n = 1 \\ R^{-4} & \text{for } n > 1 \end{cases}$$
$$(3.34)$$

which is the result given in (3.27).

---

[3] Note that, given the behavior in (3.33), the demonstration can be formally extended to the wider range $\alpha > n - 1$, instead of our starting hypothesis $\alpha \geq n$ which includes all possible SSP and SPP.

We note the following important property of (3.34). Since, as seen above, $\sigma^2(R)$ is determined basically by the integral I_1, we can say that, in the limit of large R ($R \gg k_0^{-1}$) and $n < 1$, mass fluctuations in a sphere of radius R are dominated by modes of the stochastic process with $k \simeq 1/R$ i.e., with a wave length k^{-1} of the same order as the radius of the sphere. In fact, in this case, the behavior of I_1 is dominated by the region of the integrand around $x = 1$. On the other hand, for $n \geq 1$, by a similar argument, we can say that the mass fluctuation are determined by modes with $k \simeq k_0$ i.e., the integration cut-off of the PS, which are independent of R. In practice, if we take two spheres of radius $R \gg 1/k_0$ in a system with $n > 1$, the mass fluctuation between them can be considered as given by a Poisson-like fluctuation restricted to the last shell of thickness k_0^{-1} of the spheres.

Sometimes in the cosmological literature (e.g., [147]) the divergence of I_1 as R^{n-1} for $n > 1$ is considered as a simple mathematical pathology due to the fact that relative mass fluctuations are calculated in a spherical window function with a perfectly sharp boundary. For this reason this sharp window function in real space is often replaced by a smooth Gaussian window of the following type:

$$W_\mathcal{R}(\mathbf{r}) \sim e^{-r^2/R^2} \ .$$

This means that, from a randomly chosen center point, particles at distance r are counted with a Gaussian probability

$$p(r;R) \sim e^{-r^2/R^2}$$

instead of having $p(r;R) = 1$ if $r \leq R$ and $p(r;R) = 0$ if $r > R$. With such a modification of the window function I_1 converges for $R \to +\infty$ even for $n \geq 1$, being dominated by modes with $k \simeq 1/R$ ($x \simeq 1$) for any n. Consequently, with the use of this smoothed window function one obtains $\sigma^2(R) \sim 1/R^{3+n}$ also for $n > 1$. This difference is important in cosmology because, as we will discuss in Chap. 6, the form of the "primordial" PS is the so-called *Harrison-Zeldovich spectrum* with $P(k) \sim k$ at small k. The reason why the use of such a Gaussian window function leads to $\sigma^2(R) \sim 1/R^{3+n}$ also for $n > 1$, is because it models the edge of the sphere as smeared on the length scale of the whole radius. If, instead, the window function were smeared around the edge over a length independent of R we would get again $\sigma^2(R) \sim 1/R^4$ for every $n > 1$. In applications to observational data, the smearing of the edge of the window function can take into account some uncertainty in the measure of distance (e.g., the measure of the distance between our galaxy and the others). This uncertainty is, in principle, not strictly proportional to the distance itself. Therefore, while of course the use of smoothed window functions is valid mathematically, it misses an important point, which is that the limit behavior of the normalized mass variance, as $1/R^4$ for $n > 1$, has a very real physical meaning which has to do with the minimal intrinsic noise present in any stochastic mass distribution.

Our simple treatment and results are supported by rigorous mathematical studies about point processes. It has been shown [24] that there is no way in d-dimensions of distributing points in the space in such a way that $\sigma^2(R)$ decreases faster than $R^{-(d+1)}$. The problem of finding in $d > 1$ dimensions (in $d = 1$ the problem has been solved [231]) the point-particle distribution having the minimal $\sigma^2(R)$ at large R (i.e., not only the exponent which is $d+1$ but also the coefficient) in ordinary sharp spherical window functions is still an open issue [231]. Going back to the use of Gaussian windows, we can say that the window $W_R(\mathbf{r}) = \exp(-r^2/R^2)$ completely obscures this behavior for $n \geq 1$, giving an apparent behavior of a real space variance $\propto 1/R^{d+n}$. On the other hand the measure of $\sigma^2(R)$ in such a Gaussian window function can be very useful as a direct real space measure of the exponent n characterizing the small k behavior of the PS ($P(k) \sim k^n$) even for $n \geq 1$. However the application of this window function in finite samples can be problematic and affected by strong finite size effects obscuring the real intrinsic behavior of the system.

3.4.1 The Complete Classification of Mass Fluctuations versus Power Spectrum

The arguments in the previous section can be generalized to Euclidean spaces of any dimension d. Therefore supposing $P(k) = Ak^n f(k)$ as above, it is possible to proceed to the following classification for the scaling behavior of the normalized mass-variance:

$$\sigma^2(R) \sim \begin{cases} R^{-(d+n)} & \text{for } -d < n < 1 \\ R^{-(d+1)} \log R & \text{for } n = 1 \\ R^{-(d+1)} & \text{for } n > 1 \end{cases} \quad (3.35)$$

Therefore

- For $-d < n < 0$ (i.e., $P(0) \equiv \int d^d r \, \tilde{\xi}(r) = +\infty$), we have "super-Poisson" mass fluctuations typical of systems at the critical point of a second order phase transition (see Figs. 3.1–3.3);
- For $n = 0$ (i.e., $P(0) \equiv \int d^d r \, \tilde{\xi}(r) = A > 0$), we have Poisson-like fluctuations, and the system can be called *substantially Poisson*. This behavior is typical of many common physical systems e.g., a homogeneous gas at thermodynamic equilibrium at sufficiently high temperature;
- For $n > 0$ (i.e., $P(0) \equiv \int d^d r \, \tilde{\xi}(r) = 0$), we have "sub-Poisson" fluctuations, and for this reason, as aforementioned, we name this class of systems *super-homogeneous*. This behavior is typical, for example, of lattice-like point distributions where positively correlated regions are balanced by negatively correlated ones, in order to satisfy the condition $\int d^d r \, \hat{\xi}(r) = 0$. Therefore the condition of $P(0) = 0$ corresponds to a sort of underlying long-range order. Since this class of mass distributions play an important role in cosmology (see Chap. 6), we devote the next section and part of Chap. 7 to a deeper analysis using several examples from physics and geometry.

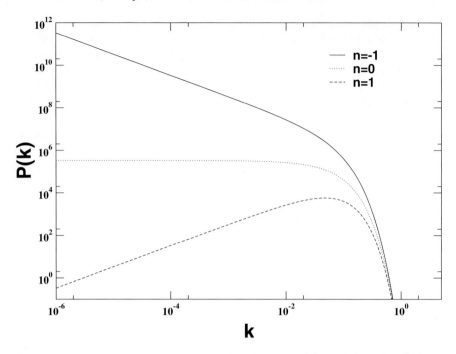

Fig. 3.1. Power spectra with a small k behavior as $P(\boldsymbol{k}) = Ak^n \exp(-k/k_c)$ for three values of $n = -1, 0, 1$ which correspond to critical, Poisson-like and super-homogeneous correlations in $d = 3$

3.5 Super-Homogeneous Mass Density Fields

As we will discuss in Chap. 6 current standard cosmological models for the distribution of mass fluctuations in the universe share approximately the same small k scaling behavior of $P(k)$, which is a direct consequence of the so-called *Harrison-Zeldovich* condition (see Chap. 6). This condition implies that at sufficiently small k the PS behaves as $P(k) \sim k$ at all cosmological epochs. Therefore it implies that the global mass distribution of the universe is *super-homogeneous* at large scales. As discussed in the previous section this means that these models are characterized by the most rapid decreasing behavior of normalized mass fluctuations with spatial scale: $\sigma^2(R) \sim R^{-4}$ in three dimensions (up to logarithmic corrections).

In this section we introduce some simple super-homogeneous point-particle distributions in order to illustrate the special properties of this class of mass distributions. The contrast with the Poisson behavior of mass fluctuations will be stressed.

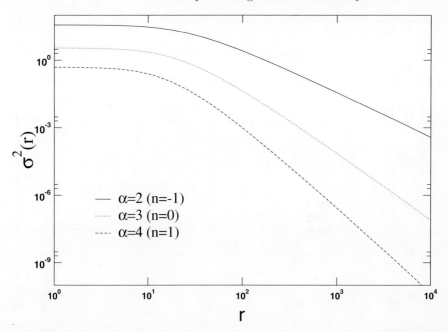

Fig. 3.2. The relative normalized mass variance for the three systems considered in Fig. 3.1: $\sigma^2(R)$ at large R behaves as $\sigma^2(R) \sim R^{-\alpha}$ with $\alpha = n + 3 = 2, 3, 4$ respectively for $n = -1, 0, 1$.

3.5.1 The Lattice Particle Distribution

The simplest case of a super-homogeneous mass distribution is a regular array of particles occupying all the sites of a Bravais lattice [253]. This can be considered the *most* uniform class of point processes in the following sense: For this kind of point-particle distribution it has been demonstrated rigorously that, for sufficiently large R,

$$\sigma^2(R) \sim R^{-d-1}, \tag{3.36}$$

which we have seen to be the lower limit on the scaling behavior for mass fluctuations at large scales. Moreover, in $d = 1$ the regular chain of point-particles has been shown to be the point process minimizing the amplitude pre-factor of (3.36). There is numerical evidence, but no rigorous demonstration, of the same property for particular lattices in higher dimensions [231].

We limit here the discussion to the case of a cubic lattice, but the results can be simply extended to more complex lattices. Let $\boldsymbol{R}$ be the vector position of the generic site of the lattice, and a the lattice spacing (or lattice constant). Therefore the microscopic density of particles of identical mass m placed on the lattice sites can be written:

$$\rho(\boldsymbol{r}) = m \sum_{\boldsymbol{R}} \delta(\boldsymbol{r} - \boldsymbol{R}), \tag{3.37}$$

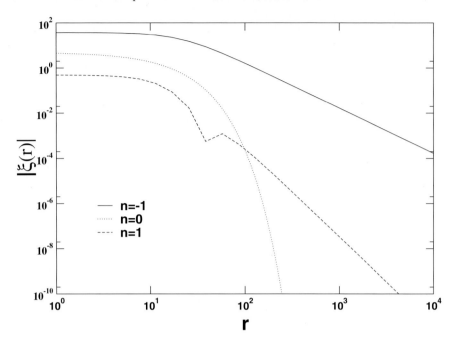

Fig. 3.3. Absolute value of the reduced two-point correlation function $\xi(r)$ for the three systems considered in Fig. 3.1. Note that for $n=-1$ $\xi(r) \sim r^{-2}$, while for $n=0$ it has an exponential decay corresponding to the absence of correlation. Instead for $n=1$ it has a *negative* power-law tail: $\xi(r) \sim -r^{-4}$

where the sum is extended to all the lattice sites. Clearly the average mass density is

$$\rho_0 \equiv \langle \rho(\boldsymbol{r}) \rangle = \frac{m}{a^d}. \tag{3.38}$$

Note that a lattice of particles cannot be seen as a genuine stochastic SPP. In fact, it is a deterministic particle distribution with continuous translational invariance broken to only a discrete translational invariance. Therefore its two-point correlation function depends on the coordinates of both the points and not only on their difference, and the Khinchin (elsewhere Wiener-Khinchin) theorem is not valid. Therefore, in order to evaluate the PS, we have to use the definition given by (3.7) which is valid also for nonstationary mass distributions. By performing the Fourier integral of (3.37) in a cubic volume V, taking the square modulus and sending V to infinity in the appropriate way (considering that in this limit the cubic lattice is invariant under lattice vector translations), it is simple to verify that (see Fig. 3.4):

$$P(\boldsymbol{k}) = (2\pi)^d \sum_{\boldsymbol{H} \neq \boldsymbol{0}} \delta(\boldsymbol{k} - \boldsymbol{H}), \tag{3.39}$$

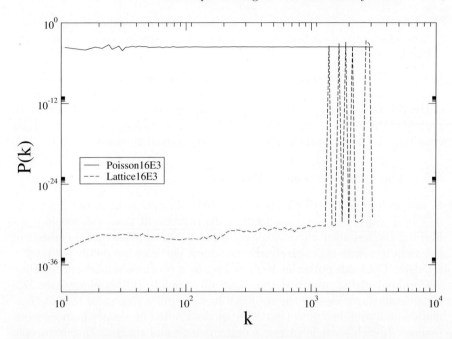

Fig. 3.4. Power spectrum of a cubic lattice together with the one of a Poisson distribution with the same number of points 16^3 in $d=3$

where the sum is over all the vectors of the *reciprocal* Bravais lattice with the exception of the vector **0**. About (3.39) we remark that

- the support of $P(\boldsymbol{k})$ (i.e., the region of the $\boldsymbol{k}$-space where $P(\boldsymbol{k}) > 0$) is discrete. This is due to the fact that the lattice is a deterministic point-particle distribution. All genuinely stochastic point processes give rise to a PS with a continuous support [198].
- $P(\mathbf{0}) = 0$, and around $\boldsymbol{k} = \mathbf{0}$ (more precisely in the first Brillouin zone [253]) the spectrum is completely flat, that is we can say that in the small k region it behaves as $P(\boldsymbol{k}) \sim k^{+\infty}$. As seen above this implies that $\sigma^2(R) \sim R^{-d-1}$ at sufficiently large scales.
- As mentioned above, the two-point correlation function of the lattice depends on the coordinates of both the points in a more complex way than only on their difference. This is because the continuous statistical translational invariance is broken. However the function $\tilde{\xi}(\boldsymbol{r})$ defined by

$$\tilde{\xi}(\boldsymbol{r}) \equiv FT^{-1}\left[P(\boldsymbol{k})\right] = a^d \sum_{\boldsymbol{R}} \delta(\boldsymbol{r}-\boldsymbol{R}) - 1 \qquad (3.40)$$

can be seen as a sort of connected two-point correlation function of the system in the following sense: If we define an *ensemble* of identical lattices differing one from each other only by a rigid random translation of the whole system, it is simple to show by direct calculation that the connected two-point correlation function defined by the *ensemble* average is given by (3.40).

In Chap. 7 it is shown that many other super-homogeneous particle distributions with a wide range of positive scaling exponents for the small k behavior of the PS can be obtained by perturbing appropriately this lattice system. In partcular this is the usual method employed in cosmology to generate initial conditions for numerical simulations of gravitational clustering.

3.5.2 The One Component Plasma

In this section we describe a classical statistical mechanical model characterized by a super-homogeneous particle distribution at thermodynamic equilibrium: the so-called *one component plasma* (OCP). This model consists of a system of identical positively charged point particles interacting through a repulsive Coulomb potential $V(r) = 1/r$, in a continuous uniform negative background giving overall charge neutrality usually called *jellium* (see [22] for a detailed overview of the subject). In what follows we show that at thermodynamic equilibrium the particles are distributed in a super-homogeneous manner. If each particle carries a unitary mass and charge, the microscopic mass density of the particles is given as usual by

$$\rho(\mathbf{r}) = \sum_i \delta(\mathbf{r} - \mathbf{r}_i) \qquad (3.41)$$

where $\mathbf{r}_i$ is the position of the ith particle. Consequently, the total microscopic charge density (including the uniform negative background) is

$$\rho_c(\mathbf{r}) = \sum_i \delta(\mathbf{r} - \mathbf{r}_i) - \rho_0 \qquad (3.42)$$

where

$$\rho_0 = \frac{1}{V} \lim_{V \to \infty} \int d^3 r \rho(\mathbf{r}) > 0, \qquad (3.43)$$

is the uniform charge density of the jellium which gives the global charge neutrality of the system.

Since the interaction potential is known, one can write down the Hamiltonian of the system and study the canonical ensemble averages (indicated as usual with $\langle \ldots \rangle$) of the main thermodynamic quantities [116]. In this way it has been shown [22] that this system can exists in two phases:

1) a low temperature ordered crystal solid phase, in which the particles are arranged in a lattice-like manner, and

2) a high temperature glassy disordered, but still super-homogeneous, phase.

Both phases are characterized by the fact that, in order to minimize the free energy, the positively charged particles arrange themselves in space in such a way as to screen the mutual interaction in the most efficient way at that temperature. If the temperature is sufficiently low this screening leads the particles to be arranged on a regular lattice array (the *ensemble* connected two-point correlation function is given by (3.40)). At high temperature the entropic contribution to the free energy is dominant and prevents the particle system from arranging itself in a perfect lattice order. However the particles are still spatially distributed in a super-homogeneous manner.

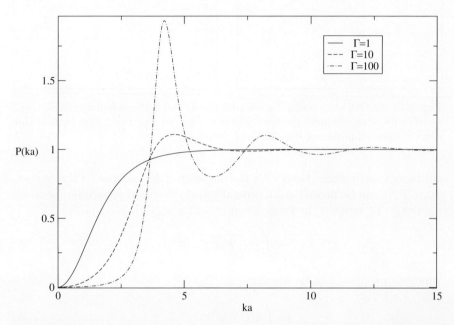

Fig. 3.5. Power spectrum (measured in units of the inter-particle separation a) of three different OCP models corresponding to three different value of the factor $\Gamma = (n^{1/3}kT)^{-1}$, which is defined as the ratio between the potential energy ($\simeq n^{-1/3}$) at the mean particle separation and their kinetic energy ($\simeq kT$). Larger values of Γ correspond to lower temperatures (or equivalently stronger coupling). The decay of the PS at small k reflects the fact that the distribution is more ordered, at large scales, than a typical Poisson configuration with the same average density. The oscillations for larger Γ correspond to the increase of the degree of order of the system as the temperature decreases

In this section we focus on this second "glassy" phase (see Figs. 3.5–3.6), in which the particle configurations can be described by a statistically isotropic and super-homogeneous SPP [155]. This implies also that the two-point correlation function is self-averaging and depends only on the modulus of the distance r between the two points i.e., it has the form:

$$\tilde{\xi}(r;T) = \frac{\delta(\boldsymbol{r})}{\rho_0} + \xi(r;T) , \qquad (3.44)$$

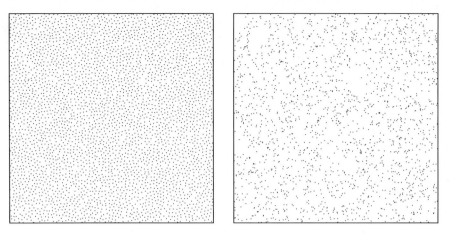

Fig. 3.6. An OCP-like configuration (*left panel*) and a Poisson distribution (*right panel*) with approximately the same number of points. Both are projections of thin slices of three dimensional distributions. (From [92])

where we have indicated explicitly the parametric dependence on the temperature T. It can be proved quite generally [155] that in this high temperature phase $\xi(r;T)$ satisfies the general condition for super-homogeneity

$$\int \tilde{\xi}(r;T) d^3r = 0$$

which, in our case, can be rewritten as:

$$\rho_0 \int \xi(r;T) d^3\mathbf{r} = -1 \ . \tag{3.45}$$

In the OCP context (3.45) is identified to as a sum rule which originates from the perfect screening of charge due to the long-range nature of the Coulomb potential. In order to see this in a simple way, let us suppose that an external infinitesimal charge density $\rho_{\text{ext}} = \epsilon e^{i\mathbf{k}\cdot\mathbf{r}}$ of very long wavelength is applied to the system. It creates an external electric potential

$$\Phi(\mathbf{r}) = \int \frac{\rho_{\text{ext}}(\mathbf{r}')}{|\mathbf{r}-\mathbf{r}'|} d^3\mathbf{r}' = \frac{4\pi}{k^2} \epsilon e^{i\mathbf{k}\cdot\mathbf{r}} \tag{3.46}$$

and a perturbation to the Hamiltonian

$$H_{\text{ext}} = \int \rho_c(\mathbf{r}) \Phi(\mathbf{r}) d^3\mathbf{r} = \epsilon \frac{4\pi}{k^2} \int \rho_c(\mathbf{r}) e^{i\mathbf{k}\cdot\mathbf{r}} d^3\mathbf{r} \ . \tag{3.47}$$

Using linear response theory [108], the induced charge is given by

$$\rho_{\text{ind}}(\mathbf{r}') = -\beta \langle \rho_c(\mathbf{r}') H_{\text{ext}} \rangle \tag{3.48}$$

where the average is over the unperturbed statistical ensemble and $\beta = (kT)^{-1}$ (where k is the Boltzmann constant). Thus, assuming that the applied charge is perfectly screened i.e., that the system responds with an induced charge density $\rho_{\text{ind}} = -\rho_{\text{ext}}$, we have, in the limit, $k \to 0$,

$$-\epsilon e^{i\mathbf{k}\cdot\mathbf{r}'} \simeq -\beta\epsilon \frac{4\pi}{k^2} \int \langle \rho_c(\mathbf{r}')\rho_c(\mathbf{r})\rangle e^{i\mathbf{k}\cdot\mathbf{r}} d^3\mathbf{r} \,. \tag{3.49}$$

Therefore the PS behaves, for $k \to 0$, as

$$P(k;T) \simeq \frac{k^2}{4\pi n^2 \beta} \,. \tag{3.50}$$

The behavior of the PS at small k is traceable through (3.46) as being simply, up to a factor β, the inverse of the FT (in the sense of distributions) of the repulsive $1/r$ potential. In a similar way it is simple to show that using a more general repulsive potential

$$\Phi(\mathbf{r}) = 1/r^a$$

with $-1 < a < 3$, the PS at small k behaves as $P(k) \sim k^{3-a}$ (i.e., super-homogeneous), while for any $a \geq 3$ one has a substantially Poisson small k behavior of the PS (with logarithmic corrections for $a = 3$). In particular for $a = 2$ the cosmological Harrison-Zeldovich spectrum for the PS is obtained [92] (see Chap. 6). We will discuss this point further in Chap. 7.

3.6 Further Analysis of Gaussian Fields

In this section we continue the analysis of continuous Gaussian fields, beyond the discussion given in Sect. 2.7.

We limit ourselves to *stationary* Gaussian fields, and present the discussion directly in the continuum. Statistical stationarity implies that the correlation matrix (see Sect. 2.7) between the two points $\mathbf{r}$ and $\mathbf{r}'$ is a function of only the difference $\mathbf{r} - \mathbf{r}'$. We will see that, in this context, the diagonalization procedure for the correlation matrix corresponds to the Fourier representation of the field, and to the introduction of the PS.

Let us call V the cubic volume in which the Gaussian SSP $\rho(\mathbf{r})$ is defined. Its probability density functional is given by (see Sect. 2.7)

$$\mathcal{P}[\rho(\mathbf{r})] \sim \exp\left[-\frac{1}{2}\int_V\int_V d^d r\, d^d r'\, \delta\rho(\mathbf{r}) K(\mathbf{r}-\mathbf{r}')\delta\rho(\mathbf{r}')\right] \,, \tag{3.51}$$

where $\delta\rho(\mathbf{r}) = \rho(\mathbf{r}) - \rho_0$ with $\rho_0 = \langle\rho(\mathbf{r})\rangle$ an arbitrary fixed constant (independent of r given the assumed stationarity of the process) giving the ensemble average of the field, and $K(\mathbf{r})$ is the so-called *correlation kernel*.

It is simple to extend to the *continuum* the arguments introduced in Sect. 2.7 to show that $K(\boldsymbol{r})$ must be positive definite i.e., such that

$$\int_V \int_V d^d r \, d^d r' \, f(\boldsymbol{r}) K(\boldsymbol{r} - \boldsymbol{r}') f(\boldsymbol{r}') > 0 \tag{3.52}$$

for any function $f(\boldsymbol{r})$ in the appropriate class. We will further clarify this point in passing to the Fourier space representation of the process. Any average $\langle \ldots \rangle$ of the stochastic field $\rho(\boldsymbol{r})$ is defined by the functional measure defined by (3.51). Let us denote by $F[\rho(\boldsymbol{r})]$ a generic function of the stochastic field at one or more spatial points. Formally, its average is defined to be

$$\langle F[\rho(\boldsymbol{r})]\rangle = \frac{\int D[\rho(\boldsymbol{r})] \, F[\rho(\boldsymbol{r})] \exp\left[-\frac{1}{2}\int_V\int_V d^d r\, d^d r'\,\delta\rho(\boldsymbol{r}) K(\boldsymbol{r}-\boldsymbol{r}')\delta\rho(\boldsymbol{r}')\right]}{\int D[\rho(\boldsymbol{r})] \exp\left[-\frac{1}{2}\int_V\int_V d^d r\, d^d r'\,\delta\rho(\boldsymbol{r}) K(\boldsymbol{r}-\boldsymbol{r}')\delta\rho(\boldsymbol{r}')\right]}, \tag{3.53}$$

where $\int D[\rho(\boldsymbol{r})] \ldots$ indicates the functional integral (also called *path integral*) and $D[\rho(\boldsymbol{r})]$ is the functional differential over the ensemble (for the general properties of functional integrals, see for instance [3]). By putting $F[\rho(\boldsymbol{r})] = \rho(\boldsymbol{r})\rho(\boldsymbol{r}')$ in (3.51), we obtain the complete two-point correlation function. Performing this average by using (3.53) may seem a difficult task. However, as we are going to show by passing to the Fourier representation, it is not necessary to perform it directly from (3.53) in order to find the correlation function $\langle \rho(\boldsymbol{r})\rho(\boldsymbol{r}')\rangle$ and its connected part.

As in Sect. 3.2, we can write

$$\delta\rho(\boldsymbol{r}) = \frac{1}{V}\sum_{\boldsymbol{k}} e^{i\boldsymbol{k}\cdot\boldsymbol{r}} \delta\tilde\rho(\boldsymbol{k}) , \tag{3.54}$$

and

$$K(\boldsymbol{r}) = \frac{1}{V}\sum_{\boldsymbol{k}} e^{i\boldsymbol{k}\cdot\boldsymbol{r}} \tilde K(\boldsymbol{k}) , \tag{3.55}$$

where $\delta\tilde\rho(\boldsymbol{k})$ and $\tilde K(\boldsymbol{k})$ are respectively the FT of $\delta\rho(\boldsymbol{r})$ and $K(\boldsymbol{r})$, and $\boldsymbol{k}$ runs over all the vectors whose components are any integer multiple of $(2\pi)/V^{1/d}$. Note that, as $K(\boldsymbol{r})$ is real and $K(\boldsymbol{r}) = K(-\boldsymbol{r})$, then also $\tilde K(\boldsymbol{k})$ is real and $\tilde K(\boldsymbol{k}) = \tilde K(-\boldsymbol{k})$. Recall that since $\delta\rho(\boldsymbol{r}) \in \mathbb{R}$, $\delta\tilde\rho(\boldsymbol{k})$ is a complex number satisfying

$$\delta\tilde\rho(-\boldsymbol{k}) = \delta\tilde\rho^*(\boldsymbol{k}) . \tag{3.56}$$

By substituting (3.54) and (3.55) in (3.51), we can write

$$\mathcal{P}[\rho(\boldsymbol{r})] \sim \exp\left[-\frac{1}{2V}\sum_{\boldsymbol{k}} \tilde K(\boldsymbol{k})|\delta\tilde\rho(\boldsymbol{k})|^2\right] =$$
$$\exp\left[-\frac{1}{V}\sum_{\boldsymbol{k}^+} \tilde K(\boldsymbol{k})|\delta\tilde\rho(\boldsymbol{k})|^2\right] = \prod_{\boldsymbol{k}^+} \exp\left[-\frac{1}{V}\tilde K(\boldsymbol{k})|\delta\tilde\rho(\boldsymbol{k})|^2\right] , \tag{3.57}$$

where $\sum_{\boldsymbol{k}^+}$ and $\prod_{\boldsymbol{k}^+}$ indicate respectively the sum and the product over half of the Fourier vectors $\boldsymbol{k}$ e.g., all the Fourier modes $\boldsymbol{k}$ with the first component limited to be positive or zero[4]. This corresponds to the application of the condition (3.56) imposing that $\delta\rho(\boldsymbol{r})$ be real. Let us call $Re\,[\delta\tilde\rho(\boldsymbol{k})]$ and $Im\,[\delta\tilde\rho(\boldsymbol{k})]$ respectively the real and the imaginary part of $\delta\tilde\rho(\boldsymbol{k})$, so that

$$|\delta\tilde\rho(\boldsymbol{k})|^2 = (Re\,[\delta\tilde\rho(\boldsymbol{k})])^2 + (Im\,[\delta\tilde\rho(\boldsymbol{k})])^2 \ .$$

Note that (3.57) shows that for *stationary* Gaussian fields the Fourier representation is equivalent to the linear transformation leading to the diagonalization presented in Sect. 2.7.

The last expression of (3.57) implies that all $\boldsymbol{k}$ belonging to the selected half of space are independent of each other, and that for each $\boldsymbol{k}$ the quantities $\{Re\,[\delta\tilde\rho(\boldsymbol{k})]\}$ and $\{Im\,[\delta\tilde\rho(\boldsymbol{k})]\}$ are two independent Gaussian random variables with zero mean and variance given by

$$\left\langle (Re\,[\delta\tilde\rho(\boldsymbol{k})])^2 \right\rangle = \left\langle (Im\,[\delta\tilde\rho(\boldsymbol{k})])^2 \right\rangle = \frac{V}{2\tilde{K}(\boldsymbol{k})} \ , \qquad (3.58)$$

which gives

$$\left\langle |\delta\tilde\rho(\boldsymbol{k})|^2 \right\rangle = \frac{V}{\tilde{K}(\boldsymbol{k})} \ . \qquad (3.59)$$

Since, by definition, the variance (3.58) must be positive, we conclude that $\tilde{K}(\boldsymbol{k}) > 0$ for each $\boldsymbol{k}$, and, by Fourier theory, this is equivalent to the condition (3.52). Given the definition (3.6) of the PS $S(\boldsymbol{k})$ of a SSP, we conclude that

$$S(\boldsymbol{k}) = \frac{1}{\tilde{K}(\boldsymbol{k})} \ . \qquad (3.60)$$

Equations (3.59) and (3.60) are the formulation of the Wiener-Khinchin (elsewhere only Khinchin) theorem for Gaussian fields. From (3.60), the connected two-point correlation function $C_2(\boldsymbol{r}-\boldsymbol{r}') = \langle \delta\rho(\boldsymbol{r})\delta\rho(\boldsymbol{r}') \rangle$ is calculated by simply evaluating (3.10) without explicitly performing the average (3.53).

Moreover, by using (3.57), it is simple to demonstrate that every cumulant of the process of order $n > 2$ vanishes. Consequently, any complete correlation function of order $n > 2$ is completely determined by ρ_0 and $C_2(\boldsymbol{r})$ and by using

---

[4] Actually, in this way, all the modes with the first component of $\boldsymbol{k}$ equal to zero are counted twice. However in the limit $V \to \infty$ this contribution vanishes as it corresponds to a set of k of measure zero.

the formulas of the cumulant expansion (see Sect. 2.3.1) with the cumulants of order larger than 2 set equal to zero[5].

Before passing to study the general problem of the statistical properties of the linear combinations of Gaussian variables, we note that the analysis just given provides a practical recipe to build a Gaussian field $\rho(\mathbf{r})$ with arbitrary connected two-point correlation function $C_2(\mathbf{r})$ with non-negative FT and average value ρ_0:

- First of all the FT of $C_2(\mathbf{r})$ has to be calculated. This gives the PS $S(\mathbf{k})$;
- The next step is given by choosing two independent Gaussian random numbers $a(\mathbf{k})$ and $b(\mathbf{k})$, for each $\mathbf{k}$ belonging to the half-space with positive first component and satisfying the Fourier condition in the cubic volume V (i.e., with components which are multiples of $2\pi/V^{1/d}$), with zero mean and variance $VS(\mathbf{k})/2$, for each $\mathbf{k}$ independently of the others;
- The third step consists in composing a and b in two conjugate complex numbers: $c(\mathbf{k}) = a(\mathbf{k}) + ib(\mathbf{k})$ and $c(-\mathbf{k}) = a(\mathbf{k}) - ib(\mathbf{k})$;
- The fourth step consists in evaluating the Fourier sum

$$\delta\rho(\mathbf{r}) = \frac{1}{V} \sum_{k} c(\mathbf{k}) e^{i\mathbf{k}\cdot\mathbf{r}} ,$$

where the sum is now over all the Fourier vectors.
- The last step consists simply in shifting the field by setting the mean equal to ρ_0:

$$\rho(\mathbf{r}) = \delta\rho(\mathbf{r}) + \rho_0 .$$

Finally, we note that the infinite volume limit $V \to \infty$ can be taken in a simple way, all the Fourier series being replaced by the appropriate Fourier integrals. This leads us to rewrite (3.57) as

$$\mathcal{P}\left[\rho(\mathbf{r})\right] \sim \exp\left[-\frac{1}{2(2\pi)^d} \int d^d k \, \tilde{K}(\mathbf{k}) |\delta\tilde{\rho}(\mathbf{k})|^2 \right] . \qquad (3.61)$$

Also in this limit we still have the relations

$$S(\mathbf{k}) = \frac{1}{\tilde{K}(\mathbf{k})} ,$$

and

$$\tilde{\xi}(\mathbf{r}) = FT^{-1}[S(\mathbf{k})] ,$$

which, together with the knowledge of ρ_0, give the complete description of the Gaussian field.

---

[5] In quantum field theory this recipe leads to the Wick theorem for Gaussian fields.

3.6.1 Real Space Composition of Gaussian Fields, Correlation Length and Size of Structures

Here we give another simple recipe to construct Gaussian fields with a given PS or two-point correlation function. We present this case in the *continuum* formulation, but the discretized version of it, useful for numerical applications, is evident. The recipe is the following (in d-dimensions):

- Take a *white* Gaussian field (also called simply *white noise*) $\phi(r)$ i.e., a Gaussian field whose correlation properties are the following:

$$\begin{cases} \langle \phi(r) \rangle = 0 \\ \langle \phi(r)\phi(r') \rangle = \sigma^2 \delta(r - r') \end{cases} \quad (3.62)$$

- Choose a *correlation factor* $F(r)$, and define the correlated Gaussian field $\psi(r)$ through

$$\psi(r) = \int d^d r' F(r - r') \phi(r') . \quad (3.63)$$

It is clear from this formula and from the properties of the *white* Gaussian field that $\psi(r)$ is statistically stationary. If, moreover $F(r)$ is a function only of $r = |r|$ then $\psi(r)$ is also statistically isotropic.
- From (3.62) and (3.63) it is simple to show by direct integration that

$$\langle \psi(r) \rangle = 0 , \quad (3.64)$$

and that the connected two-point correlation function is given by

$$C_2(r) = \sigma^2 \int d^d r' F(r') F(r' + r) . \quad (3.65)$$

As a consequence the PS is given by

$$S(k) = \sigma^2 |\tilde{F}(k)|^2 , \quad (3.66)$$

where $\tilde{F}(k) = FT[F(r)]$. From (3.66) and the Wiener-Khinchin theorem (see Sects. 3.6) we have that the function $F(r)$ must be such that $S(k)$ is integrable.

We now consider some simple explicit examples which show how the characteristic lengths of $F(r)$ can be related to the typical size of the structures in the field $\psi(r)$.

As a first example we set $\sigma = 1$ and take

$$F(r) = e^{-r/l} . \quad (3.67)$$

The field $\psi(r)$ defined by (3.63) will be correlated over a region of size l, and it can be seen from (3.65) that $C_2(r)$ will be exponentially damped for

$r > l$ in any dimension d i.e., l is the correlation length of the system. In particular in $d = 2$ for this choice of $F(\mathbf{r})$, from (3.66), we have

$$S(k) = \frac{4\pi^2 l^4}{[1+(kl)^2]^3} \, .$$

We can see that for $k < 1/l$ the PS $S(k)$ is approximately *white* (i.e., k-independent), meaning that the largest structures (i.e., correlated region) in the system are of size l. In Figs. 3.7 and 3.8 we show two pictures of this case in which the system volume has been discretized in 512×512 pixels. In the first case we have chosen $l = 14$ (i.e., much smaller than the sample size) and in the second one $l = 200$ (i.e., comparable with the sample size). In both cases the typical size of the over-densities (dark) and under-densities (light) is evidently given by the choice of l.

In the case that $F(\mathbf{r})$ (and, consequently, $C_2(\mathbf{r})$) has many different characteristic scales, or has no finite characteristic scale (i.e., "slow" power-law tailed $C_2(\mathbf{r})$), the relation of the typical size of over-densities (or under-densities) with the shape of $F(\mathbf{r})$ is much more complex [237]. For example if $F(\mathbf{r})$ presents m different characteristic scales as in

$$F(\mathbf{r}) = \sum_{i=1}^{m} A_i e^{-r/l_i} \, ,$$

the PS, in $d = 2$, will be given by

$$S(k) = \sum_{i=1}^{m} \frac{4\pi^2 A_i^2 l_i^4}{[1+(kl_i)^2]^3} \, .$$

It is clear from this formula that the typical scale of over-densities will depend in a complex way both on all the lengths l_i and on all the weights A_i.

3.7 Summary and Discussion

In this chapter we have presented the Fourier counterpart of the statistical analysis of the two-point correlation properties of an ordinary SSP, both in the continuous and in the discrete cases.

This kind of analysis is based on the introduction of the *power spectrum* (PS) $S(\mathbf{k})$ of the process, which measures the net contribution of each complex k-mode to the construction of the stochastic field in real space independently of its phase. It is simple to show that, in the case of spatial stationarity of the field, $S(\mathbf{k})$ is simply related to the connected two-point correlation function $C_2(\mathbf{r})$ by the FT operation. Therefore $C_2(\mathbf{r})$ and $S(\mathbf{k})$ contain the same information about the stochastic process. However, depending on the context, it may be more convenient or appropriate to concentrate on the analysis of one of the two.

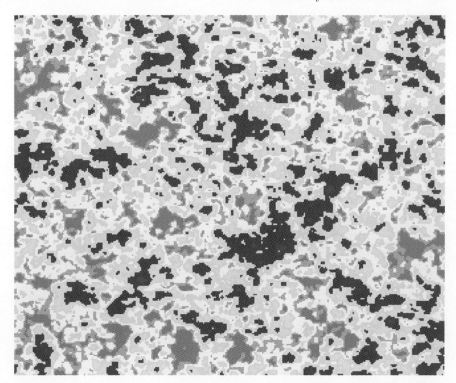

Fig. 3.7. Correlated Gaussian field with a single exponential cut-off as in (3.67). In this case $l = 14$ and the box size is $L = 512$ in units of the lattice spacing. The typical structures of over-densities (*dark*) and under-densities (*light*) have a characteristic size of order l

For instance, in the context of stochastic uniform (i.e., non-fractal) mass density fields (both continuous and discrete) in d-dimensions, a complete classification in terms of the scaling properties of the relative normalized mass variance $\sigma^2(R)$ in spheres at large scales R can be given most conveniently in terms of $S(\boldsymbol{k})$. In fact, while scaling conditions of $\sigma^2(R)$ are related to non-local integral properties of $C_2(r)$, it is simple to show that the same conditions can be translated into local scaling properties of $S(\boldsymbol{k})$ at small k. In particular, this kind of analysis of the PS at small k permits the classification of any uniform spatial mass distribution into only three classes:

- *Substantially Poisson*, if the large R behavior of $\sigma^2(R)$ is analogous to that of a Poisson SPP with appropriate average density i.e., $\sigma^2(R) \sim R^{-d}$. This case is found when $S(\boldsymbol{k}) \sim k^0$ at small k (small k white noise condition);
- *Critical* homogeneous mass densities i.e., mass distributions whose fluctuations are similar to those of thermodynamic critical phenomena (e.g. a simple fluid at the gas-liquid critical point). In this case $\sigma^2(R) \sim R^{-\alpha}$

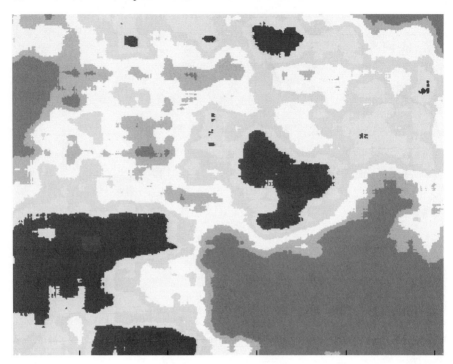

Fig. 3.8. Correlated Gaussian field with a single exponential cut-off. In this case $l = 200$ and $L = 512$. Structures have a characteristic size of order of the dimension of the box.

with $0 < \alpha < d$. This translates in a small k behavior of the PS given by $S(\mathbf{k}) \sim k^{\alpha-d}$ i.e., diverging in this limit.
- *Super-homogeneous* (or hyper-uniform) mass distributions. In this case $\sigma^2(R) \sim R^{-\alpha}$ with $\alpha > d$. This case can be seen as a condition of large scale *stochastic order*. It can be shown also that, in any case, $\alpha \leq d+1$ i.e., there is no stochastic mass distribution with $\alpha > d+1$. The related small k behavior of the PS is $S(\mathbf{k}) \sim k^\beta$ with $\beta = \alpha - d$ if $d < \alpha < d+1$, and $\beta \geq 1$ if $\alpha = d+1$.

This last class of mass distribution is very important in cosmology as we will discuss in more detail in Chap. 6. All standard models of the primordial mass density fluctuations fix, as the small k limit condition for the PS, $S(\mathbf{k}) \sim k$, which means a super-homogeneous distribution with the limit of the smallest possible real space fluctuations for a stochastic system. For this reason we have further illustrated this class of systems through the presentation of two explicit examples: (1) a regular lattice array of particles which shows well the link between the condition of *super-homogeneity* and large scale order; (2) an example from statistical physics known as "one component plasma". In the high temperature phase, it presents a PS behaving as $S(k) \sim k^2$ at small k. In this physical example the super-homogeneity condition for the fluctuations is linked to screening condition of long range interactions. In Chap. 7 we discuss further a generalization of this system which gives $S(k) \sim k$ at small k.

Finally we have completed the presentation of the continuous Gaussian stochastic fields started in Chap. 2, through the introduction of the Fourier representation, which makes simpler the explicit calculation of its statistical properties. Furthermore this Fourier representation leads to the introduction of a simple recipe for "building" a stationary Gaussian stochastic field with an arbitrary correlation function (or the equivalent PS).

4 Fractals

4.1 Introduction

In the last twenty years the introduction and the development of fractal geometry has allowed us to classify and study a large variety of structures in nature which are intrinsically irregular and self-similar [150, 151]. From a mathematical point of view this situation corresponds to the fact that these structures are *singular* at every point and at any scale. This property can be characterized in a quantitative way through the use of new mathematical and statistical methods whose application is usually called *fractal analysis*. However, given these subtle geometrical properties, it is clear that formulating a theory for the "physical" origin of these structures is going to be a very challenging task. This is actually one of the objectives of current activity in the field of modern statistical physics [79].

The main difference between the "popular" fractals like coast-lines, mountains, clouds, lightning etc. and the self-similarity shown by thermodynamical critical phenomena is that criticality at phase transitions occurs only with an extremely accurate fine tuning of the relevant thermodynamical parameters (e.g. temperature, pressure, etc.) at the *critical point* (as for the Ising model discussed in Chap. 2) [3, 148]. In the most common fractal structures observed in nature, fractal properties are instead developed spontaneously by the dynamical formation process, for a large range of the physical parameters of the system. However, before developing a complete physical description of the dynamical evolution and appearance of irregular and complex structures, one must have a clear way to characterize mathematically the relevant geometrical properties of self-similar distributions. This is the task of fractal geometry.

The fact that we are traditionally accustomed to describe physical systems in terms only of *regular* functions has crucial consequences on the type of questions we ask and on the methods we use to answer them. If one has never been exposed to the subtleness of strongly irregular structures, it is natural that mathematical regularity is not even questioned and it is an ordinary working hypothesis. Only after the aforementioned introduction and developments of *fractal geometry* has it become possible to test and determine if a physical process characterized by intrinsic fluctuations may be described by "smooth" mathematics or whether it requires a fractal description. From the point of view of stochastic processes and fields, many fractal mass distributions can be seen as special cases where *the ensemble average density ρ_0 is zero* and fluctuations are large at any scale. For this reason the estimation of statistical quantities, such as the average density in a finite sample, becomes then a subtle and complex task which should be considered with great care.

In this chapter we present and discuss some basic properties of fractal objects[1]. We do not intend here to present a mathematically rigorous introduction to the field of fractals, for which we address the reader to the book by Mandelbrot [151] and the one by Falconer [83]. In the Appendix B we explain some simple algorithms to generate fractal distributions, while in Chap. 5 we generalize the concept of fractals (point distributions) to strongly irregular measure densities (i.e. *multifractals*).

4.2 The Metric Dimension

The question which we would like to tackle is how we can describe and characterize irregular structures such as the ones shown in Fig. 1.2, Fig. 2.3, and in Figs. 4.1–4.4 (see Appendix B for explanation of the generation algorithms). The standard mathematical tools which are based on regular functions do not apply for these systems. Self-similarity, which implies that at smaller or larger scales the geometrical structure shows the same degree of *complexity* present at the original scale, makes it impossible to define a spatial derivative. As a result self-similarity is incompatible with mathematical regularity, with the exception of trivial cases such as a straight line. This lack of analyticity has deep consequences on the geometrical and statistical tools used to characterize these structures. It imposes radical changes of the mathematical perspective and of the description framework.

The *metric* dimension is the most important concept introduced to describe these intrinsically irregular systems. There is a large variety of definitions of the metric dimension but in the most of cases of interest they give the same result [83, 151]. Basically it measures the rate of increase of the "mass" of the set with the size of the volume in which it is measured. The most widely used definition for practical analysis is the *box dimension* or *capacity*. Let us focus on a set of points (i.e. geometrical structures) $S \subset \mathbb{R}^p$, where $\mathbb{R}^p$ is the p-dimensional space. The definition of the box dimension of a set $S \subset \mathbb{R}^p$ is given by

$$D_B = \lim_{\epsilon \to 0} \frac{\log N(\epsilon)}{\log(1/\epsilon)} \qquad (4.1)$$

where $N(\epsilon)$ is the minimum number of p-dimensional boxes of size ϵ needed to cover completely the set S. If we apply this definition on idealized smooth objects such as a line, a square or a cube we readily find respectively $N(\epsilon) \sim \epsilon^{-1}$,

---
[1] We have used, in the first part of this chapter, the Ph.D Thesis by A.P. Siebesma [213].

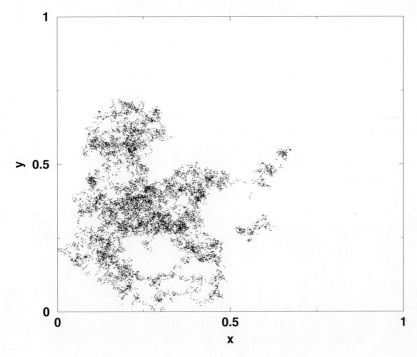

Fig. 4.1. Random walk in $d = 3$, projected on the x-y plane. The fractal dimension is $D = 2$

$N(\epsilon) \sim \epsilon^{-2}$, $N(\epsilon) \sim \epsilon^{-3}$ and thus for the dimension $D_B = 1, 2, 3$ respectively, as expected. Note that, if the set is covered with boxes of size ϵ (neighborhoods), one obtains a *coarse-grained* picture of the set up to a scale ϵ.

The box dimension can be interpreted as a simplified version of the *true* metric dimension: the *Hausdorff dimension* [83, 151]. To define the Hausdorff dimension of a subset $S \subset I\!R^p$, let us consider a covering of the set by p-dimensional neighborhoods the ith of which has a linear size ϵ_i. The Hausdorff dimension D_H is the critical dimension at which the Hausdorff measure $H_d(\epsilon)$ passes from zero to an infinite value:

$$H_d(\epsilon) = \inf \sum_i \epsilon_i^d \underset{\epsilon \to 0}{\longrightarrow} \begin{cases} 0 & \text{if } d > D_H \\ \infty & \text{if } d < D_H \end{cases} \qquad (4.2)$$

and where the infimum extends over all the possible coverings subject to the constraint that any $\epsilon_i \leq \epsilon$. It is quite easy to see that $D_B \leq D_H$ where the equality holds if the infimum of (4.2) implies a coverage of ϵ-neighborhoods all with the same linear size ϵ. As already mentioned, in almost every case of interest the equality holds.

The definition proposed by Mandelbrot for a fractal [151] is *"A fractal is a set for which the Hausdorff dimension strictly exceeds the topological*

104   4 Fractals

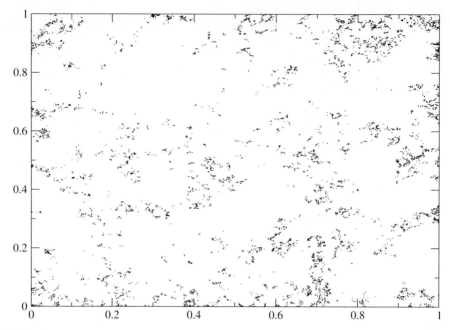

Fig. 4.2. Random trema dust in $d = 2$ with $D = 1.8$

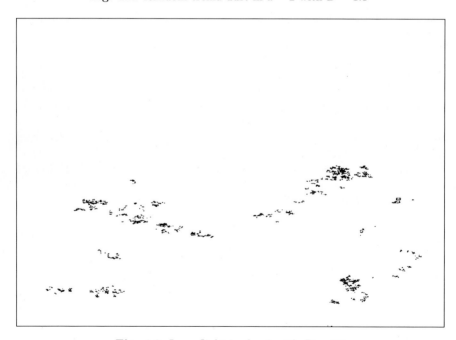

Fig. 4.3. Levy flight in $d = 3$ with $D = 1.0$

Fig. 4.4. Example of fractal growth phenomenon in the diffusion limited aggregation model, one of the most studied self-organized fractals in physics. A particle follows a Brownian motion in space until it has a collision with the central structure. At this point the particle sticks to the structure and another particle starts from a faraway point. The cluster represented in the figure contains 50.000 particles. The various colours are related to different physical properties (Elaboration and photo by B.B. Mandelbrot and C.J.G. Evertsz.)

dimension". The topological dimension can be simply defined as the number of independent directions in which one can move around a given point of the set. As a result the examples reported in Fig. 2.3, Fig. 1.2 in the Appendix B are indeed fractals. The Cantor set for instance has a topological dimension 0 since it consists of a set of isolated points while its Hausdorff dimension is strictly positive. Smooth idealized forms like a plane and a cube, where the topological dimension equals the Hausdorff dimension, are non-fractal and are called commonly *homogeneous* or *compact*. Whenever a set has a non-integer Hausdorff dimension it is a fractal. This is a sufficient but not a necessary condition. For example, the set of points visited by an ordinary random walk (see Chap. 2) has an integer Hausdorff dimension 2 while its topological dimension is 0 (if one considers the continuous trajectory the topological dimension is 1).

From here on we focus our attention on distributions of isolated points (or particles) representing a fractal or an homogeneous set which is the real subject of interest for cosmological applications. Strictly speaking a set of isolated particles, whose microscopic density can always be written as

$$\rho(\bm{r}) = \sum_i \delta(\bm{r} - \bm{r}_i) \, ,$$

where $\bm{r}_i$ is the position of the ith particle of the system, has a topological dimension equal to zero. However, given a connected fractal or homogeneous set of points of arbitrary metric and topological dimension, it can be represented with an arbitrary accuracy through a set of isolated points. In fact, one can always cover the original set with a grid of spacing Λ, and occupy with a single point at its center each cell of the grid whose intersection with the original set is not empty. In this way we obtain a set of isolated points (zero topological dimension) whose box dimension, for instance, is equal to that of the original set if measured using boxes with size $\epsilon > \Lambda$. It is clear that Mandelbrot's definition of a fractal for this kind of point distribution no longer applies because the metric dimension is larger than the topological one, even in the case in which the original set, from which the distribution of isolated points is obtained, is homogeneous (for instance, a surface in $d = 2$ or a volume in $d = 3$). Of course, the distribution of isolated points represents well the fractal properties of the original set only on scales larger than Λ, which for this reason is usually called the *lower cut-off*. This means that, given a distribution of isolated particles, as for example a galaxy survey, the determination of the fractal dimension is the main tool to study the degree of irregularity of the point distribution.

Another method used to determine the fractal dimension is the *mass versus radius method*. Loosely speaking, it consists in measuring the total properly defined mass M of an object, seen from an occupied point of the distribution, as a function of its linear size R. This corresponds to the most intuitive idea of fractality. If the mass M scales as

$$\lim_{R \to \infty} M(R) \sim R^D \tag{4.3}$$

then the exponent D is called the *mass-length dimension*. For self-similar systems D is usually equal to the box dimension, i.e. $D = D_B = D_H \leq d$ (where d is the spatial dimension) in most cases. In particular this definition is very useful in the case of the distribution of isolated particles (e.g. galaxy surveys) for which $M(R)$ can be measured, for example, by the number of points (galaxies) inside a sphere of radius R around a given point of the distribution, i.e. it is a *conditional quantity*. Clearly the amplitude of $M(R)$ in (4.3) depends on the lower cut-off Λ. Hereafter we mainly consider determinations of the fractal dimension via the average mass-length relation (MLR) and related tools (see also [102]).

4.3 Conditional Density

Many random fractal sets can be generated by stochastic algorithms which are spatially stationary, i.e. which generate point distributions whose statistical properties are translationally invariant [151, 153]. We will restrict our discussion to this kind of fractal. For such a fractal set any observer is a-priori statistically equivalent to any other one (i.e., the expected values are independent of his position)[2]. Moreover the algorithm generating the fractal set is also statistically isotropic. In this sense one might try to apply the statistical tools introduced for homogeneous mass distributions obtained for the SSP case. However, as shown by (4.3), a fractal with dimension $D < d$, has a mass (i.e. number of points) which grows more slowly than the volume; therefore the average density in any sample of volume V of the system vanishes when $V \to \infty$. This means that a fractal set of points is *asymptotically empty*. Consequently, if we take a randomly placed finite volume in an infinite fractal distribution of isolated particles it contains typically no points. This implies that only *conditional* correlation functions of the type $\langle \hat{\rho}(\boldsymbol{r}_1) \hat{\rho}(\boldsymbol{r}_2) \ldots \rangle_p$ of the stochastic density $\hat{\rho}(\boldsymbol{r})$ of points are well defined for a fractal distribution of points. The symbol $\langle \ldots \rangle_p$ adopted for the average has the same meaning as in Chap. 2: in principle it is the conditional ensemble average over all the possible realizations $\rho(\boldsymbol{r})$ of the density of points with the functional probability density $\mathcal{P}[\rho(\boldsymbol{r})]$. Supposing the *ergodicity* of the stochastic process or algorithm generating the fractal point distribution, if only one realization of the fractal is available, the average can be performed

---

[2] In this sense a fractal satisfies a weaker version of the cosmological principle (which states that all points in space should be equivalent), the so-called *conditional cosmological principle* [151], where the isotropy of the structure plus the non-analytical character of the distribution ensure the equivalence of all points without imposing strict homogeneity (or uniformity) of the distribution (see [18, 47, 151, 223] for further discussion).

as a conditional volume average given by (2.49) in Chap. 2. If this average is restricted to the points inside a finite portion of the whole infinite fractal, we have only a statistical estimator of the correlation function.

We can define the property of statistical isotropy in a more precise way. Let us focus on the three-dimensional case. Let $N(\hat{\phi}, \Omega; r)$ be the number of points contained in a portion of a sphere of radius r, which covers a solid angle Ω in the direction $\hat{\phi}$. We define as a statistically isotropic fractal, a structure for which [47, 88, 90, 151, 223]

$$\langle N(\hat{\phi}, \Omega; r)\rangle_p = B \frac{\Omega}{4\pi} r^D , \qquad (4.4)$$

with B independent of $\hat{\phi}$ and Ω. Clearly, because of the intrinsically irregular nature of a fractal, characterized by both large voids and structures (i.e. clusters of points), the behavior of $N(\hat{\phi}, \Omega; r)$ from a single observer presents sensible deviations from (4.4) at all scales. In fact, the presence of voids and structures at any scale creates local anisotropies which are erased only in the conditional average over all possible observers. The statistical properties of these deviations depend, in general, on the specific morphological properties of the distribution considered, and not only on the fractal dimension, as we are going to discuss below [88, 151, 153, 223].

As aforementioned, when one considers a self-similar point distribution, fractal properties are defined above a lower cut-off Λ. In a distribution of isolated particles the amplitude B in (4.4) is a measure of Λ. More precisely, since B is the average number of points (particles) in the unitary sphere around an occupied point (see Sect. 4.3.3), it is directly a measure of Λ^D [47, 88, 90, 223]. A point distribution can also have an upper cut-off R_s which is the spatial scale up to which self-similarity is detectable. For instance, in the case of a fractal set in a finite volume, R_s is the typical size of this volume. In particular it may be the scale marking a cross-over from fractal to non-fractal uniform behavior: in this case R_s may be identified with λ_0, the homogeneity scale defined in Chap. 2. Therefore, usually, when fractality is detected in real geometrical structures and/or point distributions (4.4) is valid in the finite range of scale $\Lambda < r < R_s$. Given (4.4) it is simple to see that the conditional average is expressed as [47, 223]

$$\langle \hat{\rho}(\mathbf{r})\rangle_p \equiv \Gamma(r) = \frac{DB}{4\pi} r^{D-3} . \qquad (4.5)$$

We use both the symbols $\langle \hat{\rho}(\mathbf{r})\rangle_p$ and $\Gamma(r)$ for the conditional average density to match with the notations used by [47, 190, 223]. Equation (4.5) can be generalized to spaces with d-dimensions:

$$\Gamma(r) = A r^{-\gamma} \quad \text{with} \quad \gamma = d - D , \qquad (4.6)$$

with $D < d$ and A is a constant. The exponent γ is called the *co-dimension* of the fractal. Operatively, the conditional average density in (4.5) represents the average density of particles measured in a shell of finite thickness Δr at distance r from an occupied point (with $\Delta r \ll r$), and then averaged over all the points contained in the sample (see Chap. 9 and Appendix F).

We may use also the integrated conditional density $\Gamma^*(r)$ which is defined as

$$\Gamma^*(r) = \frac{3}{4\pi r^3} \int_{C(r)} \Gamma(r')d^3r' , \qquad (4.7)$$

where $C(r)$ is the sphere of radius r. In practice the latter quantity measures the conditional density in a sphere of radius r rather than in a spherical shell as for $\Gamma(r)$. For this reason it smooths out the rapidly varying fluctuations of $\Gamma(r)$ estimated in a finite sample and it is a more stable quantity.

4.3.1 Conditional Density and Smooth Radial Particle Distributions

In this section we show that a power-law average conditional density can be found not only in fractal particle distributions, but also in finite but large samples of smooth spherically symmetric (radial) particle systems [36]. This implies that, even though the functions $\Gamma(r)$ and $\Gamma^*(r)$ are the primary tools to investigate the fractal and self-similar properties of a distribution of particles with the same mass, one should analyze also other properties as, for instance, the box-counting dimension, three-point correlations or the conditional variance in spheres (see below) to test whether the power-law correlations correspond to fractality.

Let us consider a unitary mass particle distribution in d-dimensions

$$\rho(\boldsymbol{r}) = \sum_i \delta(\boldsymbol{r} - \boldsymbol{r}_i) ,$$

defined by the following radial anisotropic Poisson algorithm in a spherical volume of radius R_s. Let us partition the total volume into small volume elements dV, and let us occupy the volume element around the point $\boldsymbol{r}$ with a particle with a radial probability $A/r^\alpha dV$ with $0 < \alpha < d$ (we choose $dV(r)$ so that $A/r^\alpha dV(r) \ll 1$ everywhere in the volume) or leave it empty with the complementary probability $1 - A/r^\alpha dV$. Clearly, at small r, say below a certain $\lambda \ll R_s$ (lower cut-off), we have to cut-off to a constant value the radial probability in order to have it well defined in any volume element. Moreover, we choose $0 < \alpha < d$ in order to have an infinite number of point-particles in the infinite volume limit $R_s \to \infty$. Note that there is no correlation between the probabilities of occupying two different volume elements. This corresponds to an anisotropic radial (or radial density profile) generalization of the Poisson distribution introduced in Chap. 2. Clearly the unconditional density behaves as

$$\langle \rho(\boldsymbol{r}) \rangle = \frac{A}{r^\alpha} . \qquad (4.8)$$

We now use a simple argument to obtain an estimate of $\Gamma^*(r)$ for $\lambda \ll r \ll R_s$. The results found for $\Gamma^*(r)$, apart from differences in the numerical coefficients, hold also for $\Gamma(r)$. Let us suppose, for simplicity, that R_s is an integer multiple of r, and partition the space into a sequence of spherical shells of thickness $2r$, i.e. the nth shell covers the distance between $(2n-2)r$ and $2nr$ from the origin. Moreover, we will consider the density in each shell to be constant, in such a way that in the nth shell its value is given by $\langle \rho(2nr) \rangle$. The result obtained with the following argument are valid also for R_s/r non-integer and in agreement with that obtained by a more rigorous treatment. If $n \gg 1$ a generic particle belonging to the nth shell will observe, within a distance r from it, a number of particles

$$N_n(r) \simeq C(r)\frac{A}{(2nr)^\alpha} = \frac{\Omega_d A}{2^d} n^{-\alpha}(2r)^{d-\alpha} \qquad (4.9)$$

where $C(r) = \Omega_d r^d$ is the volume of the d-dimensional sphere of radius r (and Ω_d is the whole solid angle). Moreover there are

$$M_n(r) \simeq \Omega_d d(2nr)^{d-1} 2r \frac{A}{(2nr)^\alpha} = \Omega_d A d n^{d-\alpha-1}(2r)^{d-\alpha} \qquad (4.10)$$

of such "observers" in the same nth shell. More rigorously, since the average density of particles in a shell is not constant (4.8), one should write

$$M_n(r) = \frac{\Omega_d A d}{d-\alpha} \int_{(2n-2)r}^{2nr} dr\, r^{d-\alpha-1} \qquad (4.11)$$

which is well approximated by (4.10) for $n \gg 1$. Therefore the average conditional number of particles seen within a distance r by another particle of the system is

$$\langle N(r) \rangle_p \simeq \frac{\sum_{n=1}^{R_s/r} M_n(r) N_n(r)}{\sum_{n=1}^{R_s/r} M_n(r)}. \qquad (4.12)$$

Note that $\sum_{n=1}^{R_s/r} M_n(r)$ is the total number of particles in the system, which, instead of performing the sum using the (4.11) or (4.10), may be directly evaluated by integrating in the sphere of radius R_s (4.8). In this way we obtain

$$\sum_{n=1}^{R_s/r} M_n(r) = \frac{d}{d-\alpha}\Omega_d A R_s^{d-\alpha}. \qquad (4.13)$$

We can now evaluate the numerator of (4.12) by using (4.9) and (4.10). This leads to

$$\langle N(r) \rangle_p \simeq \frac{\Omega_d A(d-\alpha)}{2^d} R_s^{\alpha-d}(2r)^{2(d-\alpha)} \sum_{n=1}^{R_s/r} n^{d-2\alpha-1}. \qquad (4.14)$$

The next step consists in approximating the sum in (4.14) with the following integral:

$$\sum_{n=1}^{R_s/r} n^{d-2\alpha-1} \simeq \int_1^{R_s/r} dx\, x^{d-2\alpha-1} \,. \qquad (4.15)$$

At this point we have to distinguish two cases (see Figs. 4.5–4.6):

1. $0 < \alpha \le \frac{d}{2}$ (see Fig. 4.5): in this case the integral of (4.15) is divergent for $R_s \to \infty$ and r fixed. Therefore in the limit of large R_s/r, we can approximate it as

$$\int_{1*}^{R_s/r} dx\, x^{d-2\alpha-1} \simeq \frac{1}{d-2\alpha} \left(\frac{R_s}{r}\right)^{d-2\alpha} \,. \qquad (4.16)$$

This leads to

$$\langle N(r) \rangle_p \simeq \frac{\Omega_d A(d-\alpha)}{2^{2\alpha}(d-2\alpha)} R_s^{-\alpha}(2r)^d \,. \qquad (4.17)$$

Since by definition $\Gamma^*(r) = \langle N(r) \rangle_p / C(r)$, we can write

$$\Gamma^*(r) \simeq \frac{2^{d-2\alpha} A(d-\alpha)}{d-2\alpha} R_s^{-\alpha} \,. \qquad (4.18)$$

Thus we arrive at the result that, if $0 < \alpha < \frac{d}{2}$, the conditional density is independent of r and depends on the upper cut-off R_s (i.e. the size of the system) following the scaling relation $R_s^{-\alpha}$, i.e. it goes to zero in the infinite volume limit with the same exponent as the average radial density. In the case $\alpha = d/2$ the integral of (4.15) diverges only logarithmically, leading to logarithmic corrections to this result.

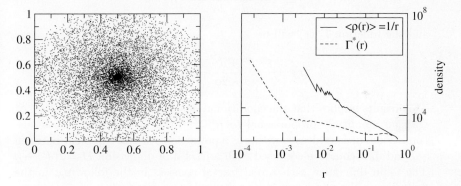

Fig. 4.5. *Right panel*: Density profile in $d = 3$, with a power-law density decay (as $1/r$) from the center. In the *left panel* is shown the behavior of the density computed from the center (*solid line*) and of the conditional average density: $\langle n(r) \rangle_p$ shows an almost flat behavior as a function of scale (which corresponds to homogeneity)

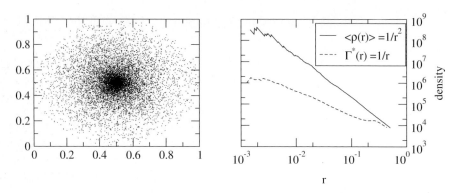

Fig. 4.6. *Right panel*: Density profile in $d = 3$, with a power-law density decay (as $1/r^2$) from the center. In the *left panel* is shown the behavior of the density computed from the center (*solid line*) and of the conditional average density: $\langle n(r) \rangle_p$ shows a power-law behavior with the exponent predicted by (4.21)

2. $\frac{d}{2} < \alpha < d$ (see Fig. 4.6): in this case the distribution becomes empty more rapidly than in the first case, and the integral of (4.15) converges to a constant in the limit of $R_s \to \infty$ and r fixed. Therefore in the limit of large R_s/r we can make the approximation:

$$\int_1^{R_s/r} dx\, x^{d-2\alpha-1} \simeq \frac{1}{d-2\alpha}\, . \tag{4.19}$$

As a consequence, the conditional average number of particles is

$$\langle N(r) \rangle_p \simeq \frac{2^{d-2\alpha} \Omega_d A(d-\alpha)}{d-2\alpha} R_s^{\alpha-d} r^{2(d-\alpha)}\, , \tag{4.20}$$

and the function $\Gamma^*(r)$ will be

$$\Gamma^*(r) \simeq \frac{2^{d-2\alpha} A(d-\alpha)}{d-2\alpha} R_s^{\alpha-d} r^{d-2\alpha}\, . \tag{4.21}$$

This results implies that the conditional density decays with r as $\sim r^{d-2\alpha}$ with an amplitude proportional to $R_s^{\alpha-d}$. Since in this case the average conditional density decays with r as a power-law with an exponent between $-d$ and 0, it could be confused with a fractal of dimension $D = 2(d-\alpha)$. The main difference with a real fractal consists in the following: in the fractal case the amplitude does not depend on the system size, while in the smooth radial particle distribution it goes to zero when the size of the system goes to infinity. Finally, we notice that in the case where $\frac{d}{2} < \alpha < d$, since the integral (4.19) converges in the infinite volume limit, the behavior of the conditional density is dominated by the first shells. Therefore in a more rigorous approach, we should separate these first contributions from the rest and analyze them more carefully. This would lead to a difference in the numerical coefficients in (4.21), but leaves the scaling in r and R_s unchanged.

This specific problem of how to distinguish between a radial density profile and a fractal set can be very relevant in cosmology, in particular for the interpretation of clustering in N-body simulations of gravitational clustering. Indeed in this context a power-law behavior of the two-point correlation function is observed in a finite range of scales whether this corresponds to a fractal nature of structures, rather than to a number of several spherical symmetrical density profiles, is a point which must be carefully investigated. We consider an example of such a situation in the following section.

4.3.2 Statistically Homogeneous and Isotropic Distribution of Radial Density Profiles

We consider distributing N points in a correlated manner using a prescription similar to the one used in [166, 167]. Specifically we distribute points in N_c "clusters", each cluster containing N_g points. Particles are attached to each cluster with a probability which depends on the distance from the center so that each cluster has a radial density profile $\rho(r) \sim r^{-1}$. The free parameter is the average mass (number of particles) per cluster $\langle N_g \rangle$. This implies that the number of "clusters", which are randomly distributed, is $N_c = N/\langle N_g \rangle$ and their average separation is given by $\langle \Lambda \rangle = 0.55(1/N_c)^{1/3}$. The clusters can have a distribution of masses (number of points inside) which is described, for example, by an exponential behavior

$$p(N_g) \sim \exp(-N_g/\langle N_g \rangle)$$

and have a spatial extension of about $\langle \Lambda \rangle$. This means that the N_g points belonging to the ith cluster are distributed in a spherically symmetric way with a uniform probability to find a point in the interval $[0, \langle \Lambda \rangle]$ from the cluster's center. This represents the situation where each cluster has a spherically symmetric density profile with $\rho(r) \sim r^{-2}$ (see Fig. 4.7–4.8) In Fig. 4.9 we show the behavior of the conditional density for various realizations of such a distribution, obtained by changing the average mass of each cluster $\langle N_g \rangle$. We note that such distributions are useful to capture the essential elements of the so-called "halo" models in cosmology for the large scale distribution of matter (see e.g. [50] for a review).

4.3.3 Nearest Neighbor Probability Density for Radial and Fractal Point-Particle Distributions

As shown in Chap. 2, in the case of particle distributions with no correlations between the positions of different particles (i.e. generalized Poisson distributions), the PDF of the first nearest neighbor (nn) distance is given by (2.71).

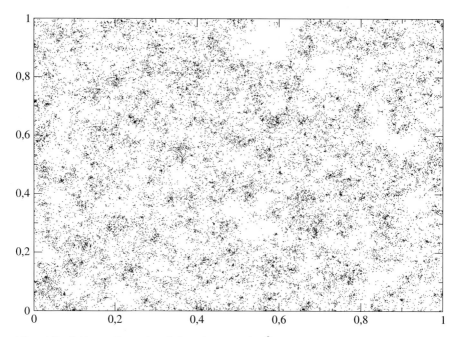

Fig. 4.7. A three-dimensional distribution of 32^3 points with the prescription roles discussed in the text. Clusters have an average mass of $\langle N_g \rangle = 20$. This is an orthogonal projection on the x-y plane

For a radial Poisson particle distribution one can study rigorously the probability distribution of the distance between a particle and its first nearest neighbor (nn). This quantity is relevant for applications in astrophysics, for instance in the study of the statistical properties of the gravitational force in radial distributions of massive objects (e.g. cluster of stars). In fact, as it will be shown in Chap. 14, the contribution to the force felt by a particle of the system due to its first nn is, in many cases of interest, the dominant one, and it is directly determined by the statistics of the nn distance. Furthermore, as explained below, the result found for smooth radial distributions is useful also for the statistics of the same quantity for fractal particle distributions.

For a three-dimensional radial Poisson distribution with average spatial density $\langle \rho(\mathbf{r}) \rangle = Ar^{-\alpha}$ in a finite volume of size R_s, we have seen in the previous subsection that $\Gamma(r) = Cr^{-\gamma}$ with $\gamma = 0$ and $C \sim R_s^{-\alpha}$ if $0 < \alpha \leq 3/2$, while $0 < \gamma = 2\alpha - 3 < 3$ and $C \sim R_s^{\alpha-3}$ if $3/2 < \alpha < 3$. At this point, since there is no correlation between the positions of different particles in order to find the PDF of the nn distance we can repeat the reasoning presented in Chap. 2 leading to that of an isotropic Poisson distribution. This results again in (2.71) with the appropriate $\Gamma(r)$, i.e.

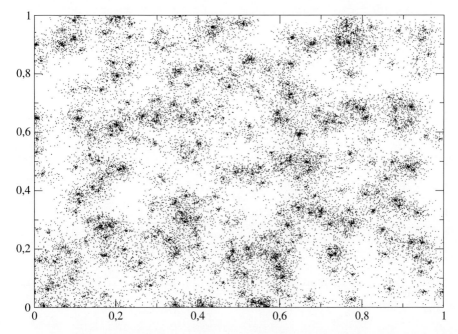

Fig. 4.8. As in the previous figure, but now with $\langle N_g \rangle = 50$

$$\omega(r) = 4\pi C r^{2-\gamma} \exp\left(-\frac{4\pi C}{3-\gamma} r^{3-\gamma}\right) . \tag{4.22}$$

As expected, this function satisfies the normalization condition

$$\int_0^\infty dr \, \omega(r) = 1 ,$$

and can be extended easily to any dimension d. Given the PDF (4.22), it is simple to calculate the average nn distance through the formula

$$\langle \Lambda \rangle = \int_0^\infty dr \, r \, \omega(r) . \tag{4.23}$$

In our case this gives

$$\langle \Lambda \rangle = \left(\frac{3-\gamma}{4\pi C}\right)^{\frac{1}{3-\gamma}} \Gamma_e\left(\frac{4-\gamma}{3-\gamma}\right) , \tag{4.24}$$

where $\Gamma_e(x)$ is the Euler gamma function.

As we will discuss in Chap. 14 (4.22) works well also in the case of a stochastic fractal particle distribution, at least in the range of small distances $r < \langle \Lambda \rangle$. As shown in previous sections, for a fractal distribution we have again $\Gamma(r) = C r^{-\gamma}$ with $\gamma = 3 - D$ and $C = \frac{DB}{4\pi}$ which is independent of

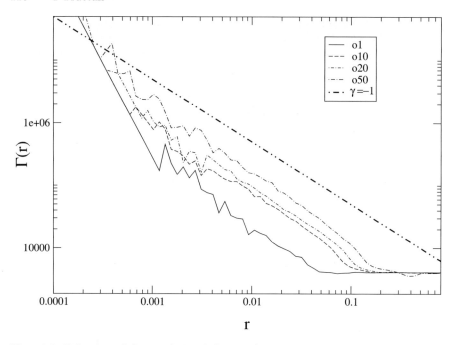

Fig. 4.9. Behavior of the conditional density for various realizations of a statistically homogeneous and isotropic distribution of radial density profiles (see text), obtained by changing the average mass of each cluster (from left to right the average mass for cluster increases from 1 to 50). Note that because $\langle \rho(r) \rangle \sim r^{-2}$ we get $\Gamma(r) \sim r^{-1}$

R_s as shown in previous sections. Therefore, from (4.24) we can write the following approximation for the average nn distance:

$$\langle \Lambda \rangle \simeq \left(\frac{1}{B}\right)^{\frac{1}{D}} \Gamma_e \left(1 + \frac{1}{D}\right) . \tag{4.25}$$

4.4 The Two-Point Conditional Density

As already mentioned, the fractal dimension is the primary global statistical quantity to measure in an intrinsically irregular structure displaying fluctuations at all scales. More information, about the morphology, can be obtained by studying other many-point statistical properties. The average MLR gives us the density of points in spheres or spherical shells at distance r and thickness Δr from an occupied point. However this statistical tool is not able to account for other more complex morphological properties such as, for example, the void size distribution around a point of the system, and other measures of the *degree of isotropy* around a single point. In order to analyze

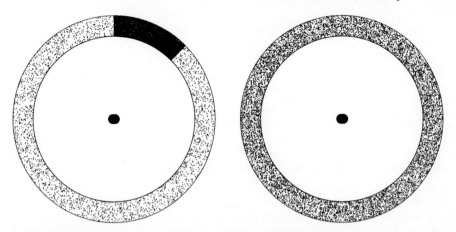

Fig. 4.10. A simplified scheme for an anisotropic (*left panel*) versus isotropic (*right panel*) point distribution (as seen from an "average" point which is the center of the spherical shell). The conditional density fixes the number of points which there are, on average, in a given shell in a given shell at distance r from an occupied point (which, in this example, is fixed to be the same in both spherical shells). The two-point conditional density can discriminate between the two situations shown in the figure

these properties, it is necessary to use at least the conditional two-point correlation function. In Fig. 4.10 a simple schema is given which illustrates this point: three-point analysis is needed to characterize morphological properties and quantify the anisotropy of structures. In the context of stochastic fractal particle distributions the principal three-point correlation function is the two-point conditional density. Through it we can describe the "level" of isotropy of the distribution around a generic point of the system, and the effect of self-similar fluctuations on various physical properties such as, for instance, the gravitational force exerted by the system on a fixed point (see Chap. 14). Such a quantity plays a basic role in the determination of the average quadratic gravitational force acting on a point (see Chap. 14) and on many quantities measurable in galaxy samples (see Chap. 8).

It is possible to show that statistical translation and rotational invariance, together with fractal scale-invariance, lead to the following ansatz, which is verified rigorously for the fractal point distribution generated by a random walk and numerically for many other fractal particle distributions [34]:

$$\Gamma^{(2)}(\boldsymbol{r_1}, \boldsymbol{r_2}) \simeq \langle \hat{\rho}(r_1) \rangle_p \langle \hat{\rho}(r_2) \rangle_p \; \mathcal{L}\left(\frac{r_1}{r_2}, \theta\right) , \qquad (4.26)$$

where $\Gamma^{(2)}(\boldsymbol{r_1}, \boldsymbol{r_2})$ is the non-diagonal part of the two-point conditional density $\langle \hat{\rho}(r_1)\hat{\rho}(r_2) \rangle_p$ as defined in Sect. 2.5.1 (2.53–2.55), and

$$\mathcal{L}\left(\frac{r_1}{r_2},\theta\right) = 1 + g\left(\frac{r_1}{r_2},\theta\right) \tag{4.27}$$

with

$$\lim_{r_1/r_2 \to 0,\, r_2/r_1 \to 0} g\left(\frac{r_1}{r_2},\theta\right) = 0. \tag{4.28}$$

This approximation will be useful in Chap. 8 and Chap. 14.

4.5 The Conditional Variance in Spheres

Let us now consider the conditional variance of the number of points $N(R)$ in a generic sphere $C(R)$ of radius R and its relation with the three-point correlation function, i.e. with the two-point conditional density. The normalized conditional variance is defined as (2.58)

$$\sigma_p^2(R) = \frac{\langle N^2(R)\rangle_p - \langle N(R)\rangle_p^2}{\langle N(R)\rangle_p^2} \equiv \frac{\langle \Delta N^2(R)\rangle_p}{\langle N(R)\rangle_p^2}. \tag{4.29}$$

Using (4.26) we obtain

$$\langle N(R)^2\rangle_p = \int_{C(R)}\int_{C(R)} \langle \hat\rho(r_1)\rangle_p \langle \hat\rho(r_2)\rangle_p \left[1 + g\left(\frac{r_1}{r_2},\theta\right)\right] d^3r_1 d^3r_2$$

$$= \langle N(R)\rangle_p^2 + \int_{C(R)}\int_{C(R)} \Gamma(r_1)\Gamma(r_2) g\left(\frac{r_1}{r_2},\theta\right) d^3r_1 d^3r_2 \tag{4.30}$$

which shows that for a fractal particle distribution for which $\Gamma(r) \sim r^{D-3}$, if $g\left(\frac{r_1}{r_2},\theta\right)$ is not exactly zero, we have

$$\langle \Delta N^2(R)\rangle_p \propto \langle N(R)\rangle_p^2 \tag{4.31}$$

at all scales. This is due to the fact that, because of the scale-invariance of the fractal structure, g is a function only of the ratio r_1/r_2 [88, 90, 223]. Note again that the value of constant of proportionality in (4.31) is related to three-point properties and can be smaller than one: here what matters is the behavior as a function of scale.

We stress the basic difference in this respect between the behavior of a fractal distribution and that of an homogeneous (or uniform) one (i.e. with, as discussed in Chap. 2, a well defined mean density $\rho_0 > 0$), no matter what the nature of correlations. In fact, as we have seen in Chap. 2, in this latter case we have always that $\sigma_p^2(R) \to 0$ at large scales. Therefore the behavior given by (4.31) is typical only of fractals and it is a direct effect of their intrinsic irregularities (while being still statistically stationary and isotropic).

4.6 Corrections to Scaling

Generally, for a stochastic fractal the average conditional mass-length relation is *exactly* a power-law function (from (4.4)). As usual by "average" is meant the *ensemble* average, or in the case in which ergodicity applies (usually the case for statistically stationary and isotropic stochastic fractals), a *volume* average in an infinite sample. In the case in which this average in a single sample is limited to a finite number of observers, and in particular for a single observer, there can be important fluctuations at any scale around the power-law behavior in the *estimators* of $\Gamma(r)$ and $\Gamma^*(r)$. In fact, as a fractal is an intrinsically critical system with strong correlations on all scales, we expect that the conditional density seen by a single point-particle shows large fluctuations changing, for example, with the direction of observation $\hat{\phi}$ in (4.4). These fluctuations, because of self-similarity, are present at any scale as shown by (4.31) and are correlated in angle. This implies that the quantity

$$\langle \Delta^2 N(|\hat{\phi}_1 - \hat{\phi}_2|; \Omega, r)\rangle_p = \left\langle \left(N(\hat{\phi}_1; \Omega, r) - N(\hat{\phi}_2; \Omega, r)\right)^2 \right\rangle_p \quad (4.32)$$

depends on the angular distance between $\hat{\phi}_1$ and $\hat{\phi}_2$, and that it is of the same order as $\langle N^2(\hat{\phi}, \Omega; r)\rangle_p$. In other words, one has to consider that the MLR from a single observer is determined by a sequence of fluctuations which are present at all scales. For example one encounters, at any scale, structures followed by large voids. These same fluctuations affect the behavior in r of $N_i(\hat{\phi}, \Omega, r)$ as measured from the ith point only. About this last fact we limit here the discussion to the case in which $\Omega = 4\pi$ is the whole solid angle, replacing $N_i(\hat{\phi}, \Omega, r)$ simply by $N_i(r)$. However the results can be extended easily to include the case of a smaller Ω. We can quantify these effects as a *modulating term* around the expected average power-law behavior given by (4.4). Therefore, in the observations from a single point in general one can write

$$N_i(r) = Br^D f_i(r) \,. \quad (4.33)$$

where $f_i(r)$, the multiplicative correction to scaling, is a positive limited function oscillating around 1, in such a way that the average over all the observers gives (4.4): If we perform the ensemble conditional average (or the volume one in the case of ergodicity and an infinite sample) of this fluctuating term, we can smooth out its effect and, as written above, we have that

$$\langle f_i(r)\rangle_p = 1 \,, \quad (4.34)$$

at all r. Thus the conditional density, averaged over all points of the infinite sample, has a single power-law behavior. Hereafter we omit, for simplicity, the subscript i.

Given (4.33) we may write (4.29) as

$$\sigma_p^2(r) = \langle (f(r) - 1)^2\rangle_p = \langle f(r)^2\rangle_p - 1 \,, \quad (4.35)$$

where the last equality follows considering (4.34). An example is shown in Fig. 4.11.

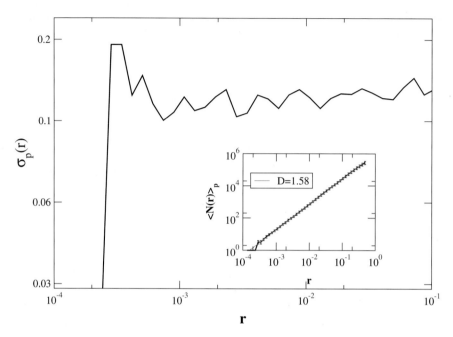

Fig. 4.11. Conditional variance for a two-dimensional Sierpinski carpet (see Fig. 4.12). The average has been performed over 100 random observers located on points of the distribution. In the *inset panel* is shown the behavior of the average MLR together with its errors

4.6.1 Correction to Scaling: Deterministic Fractals

As an interesting example of the multiplicative correction $f(r)$ to the average power-law behavior of the conditional density, we consider the case of deterministic fractals. In this case *self-similarity* is a concept stronger than in stochastic fractals. In fact, it does not simply imply the scale-invariance of the correlation functions, but also the fact that any finite portion of the system, if *appropriately* magnified, reproduces exactly the whole structure[3]. All the main methods to compute the fractal dimension D of a point set define it through the asymptotic slope of a suitable curve (for example, the average conditional density) in a log-log plot. In general, thermodynamic

---

[3] Apart from the same magnification of the lower cut-off if the original structure is not exactly a fractal but a so-called *pre-fractal* obtained by a finite number of iterations of the deterministic fractal algorithm (see Appendix B).

4.6 Corrections to Scaling

self-similar (i.e. critical) systems are characterized by an invariance with respect to arbitrary magnifying factors, i.e. continuous self-similarity. However a system could be invariant under a discrete set of dilation only: in such a situation it is usual to say that it exhibits *discrete scaling invariance* (DSI) [194, 215]. There are several examples in statistical physics of systems showing DSI [11, 30, 215]. Formally DSI leads to a complex exponent of the scaling quantities and log-periodic oscillations. Let us consider this point in more detail.

From a mathematical point of view, self-similarity (see Sect. 2.8) implies that a rescaling of a length r by a factor b

$$r' = br \tag{4.36}$$

leaves the correlation functions of the system, which describe its statistical behavior, unchanged up to a rescaling that depends on b and not or r. In particular for the conditional average number of points inside a sphere of radius r, one has

$$\langle N(r') \rangle_p = \langle N(br) \rangle_p = \langle N(r) \rangle_p A(b) \tag{4.37}$$

where A is a pre-factor that does not depend on r. This means that a rescaling of r gives rise exactly to the same function if $\langle N \rangle_p$ is also rescaled by the appropriate constant A. As we have already seen, if we require (4.37) to be satisfied for any value of b, the allowed functions are simple power-laws so that

$$\langle N(r) \rangle_p = B r^D \tag{4.38}$$

where $A(b) = b^D$. This is the case for most thermodynamic critical systems and for many stochastic fractals.

However there are cases such as *deterministic* fractals, e.g. the Sierpinski gasket (see Fig. 4.12 and Appendix B), where the condition of self-similarity is satisfied only for a discrete set of values of b, even though $N(r)$ is averaged over all the points of the set. If $N(r)$ is measured from a single observer this effect is quite evident. In the case of the Sierpinski gasket we have $b = 2^n$ where n can take any integer value. Quite generally, deterministic fractals satisfy self-similarity for a set of scale factors b defined by

$$b(n) = b_o^n \quad (n = 1, 2, 3, \ldots), \tag{4.39}$$

where b_0 is a characteristic integer number. In this case (4.37) only holds at discrete values of b

$$N(r') = N(b_o^n r) = A(b_o^n) N(r). \tag{4.40}$$

Attempting to solve (4.40) with functions of the type

$$N(r) = f(r) r^D \tag{4.41}$$

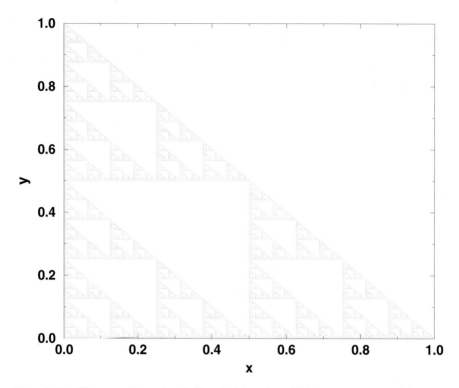

Fig. 4.12. The two dimensional Sierpinski gasket. This set is generated by applying an iterative fragmentation algorithm. At each iteration the set is divided into 2^2 boxes, of which only one box is removed. The fractal dimension is then $D = \log(3)/\log(2) = 1.585$. In this example we have iterated the algorithm 10 times

by finding the conditions to be satisfied by $f(r)$, we have

$$N(r') = N(b_o^n r) = f(b_o^n r)(b_o^n)^D r^D \ . \tag{4.42}$$

If we want to have (4.40) satisfied with $A(b_o^n) \sim (b_o^n)^D$, we have to set

$$f(b_o^n r) = f(r) \ . \tag{4.43}$$

Let us now consider the variable $y = \log(r)$. We have that

$$y' = \log(r') = n \log(b_o) + y \ . \tag{4.44}$$

If we call $g(y) = f(r = e^y)$, we can write

$$g(n \log(b_o) + y) = g(y) \ . \tag{4.45}$$

This relation is satisfied by all functions which are periodic with respect to the variable $y = \log(r)$ with period $\log(b_o)$. This means that in the case of

4.6 Corrections to Scaling

discrete scaling we expect a power-law modulated by a periodic function of $\log(r)$. A nice example can be found in [30]. It is possible to show that the periodic modulation of the power-law can be described by introducing an imaginary part of the fractal dimension. In fact, we can set, for example, in (4.45)

$$f(r) = 1 + a\sin(D_I \log(r)) \tag{4.46}$$

and

$$N(r) = Br^{D_R} f(r) . \tag{4.47}$$

It is then possible to write

$$N(r) = Br^{D_R} Re(1 + ar^{iD_I}) , \tag{4.48}$$

where $Re(x)$ is the real part of x. For this reason D_I is often called the *imaginary part* of the fractal dimension of the set.

It is evident from the previous discussion that if the period of oscillations around the average power-law behavior is large enough (i.e. comparable with the sample size), and the amplitude is non-negligible, a clear determination of the fractal dimension becomes an extremely difficult task via the unaveraged MLR.

As an example, we consider now the determination of the MLR from one point for the case of the Sierpinski gasket of Fig. 4.12. We compute the MLR from the vertex point, and then we plot $f(r)$ defined as

$$f(r) - 1 = \frac{N(r) - \langle N(r) \rangle_p}{\langle N(r) \rangle_p} . \tag{4.49}$$

The oscillatory components can be clearly seen in Fig. 4.13. The period in logarithmic space is $p = 2\pi/\log_{10}(2) \sim 21$, which means that in passing from $\log(r)$ to $\log(r) + p$, $f(r)$ takes the same value. The amplitude of oscillations is rather small with $a = 0.1$. We consider now a generic randomly chosen point of the set and we compute numerically $f(r)$. The result is shown in Fig. 4.14. In this case the amplitude is larger ($a \sim 1$) and it is more difficult to identify a single component in the oscillations. Such a situation requires a generalization of (4.46) to include further harmonics.

Let us now consider the conditional variance (Chap. 2) of a set which presents log-periodic corrections to scaling of the type:

$$f(r) = 1 + a\sin(b\log(r) + \phi) . \tag{4.50}$$

Then from (2.58) (4.29) we obtain that the conditional variance is given by

$$\sigma_p^2(r) = \langle (a\sin(b\log(r) + \phi))^2 \rangle = \frac{a^2}{2} . \tag{4.51}$$

Hence $\sigma_p^2(r)$ is related to the *amplitude* of the log-periodic oscillation and ultimately to three-point properties (see above). This means that $\langle \Delta N(r)^2 \rangle_p \approx$

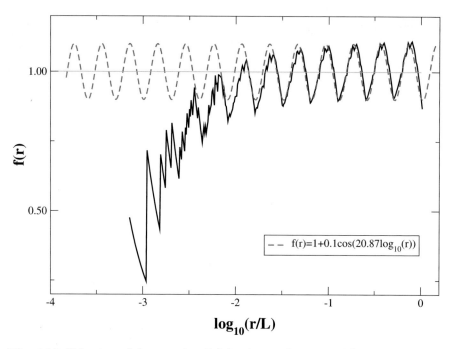

Fig. 4.13. Behaviors of the quantity $f(r)$ for the two dimensional Sierpinski gasket. In this case the origin is the vertex of the set (i.e. the point (0,0) in the previous figure). The oscillatory component can be clearly seen with period $= 2\pi/\log_{10}(2) \approx 20.87$. The amplitude of the fluctuations is about 0.1. At small scales the signal is dominated by shot noise. Such oscillations correspond to the sequence of voids of increasing size which can been seen in the previous figure

$\langle N(r) \rangle_p^2$ at any scale, and hence the set is characterized by persistent log-log fluctuations around the average power-law. On the other hand the typical logarithmic size of voids is defined by b^{-1}. The fact that the conditional variance is constant as a function of scale is another very particular property of fractals. Note again that the value of a is not related in a simple manner to the value of the fractal dimension.

4.6.2 Correction to Scaling: Random Fractals

As shown above, the presence of persistent oscillations, in the case of a deterministic fractal, comes directly from the *strong* self-similarity of the set. Also in the case of a stochastic fractal, the MLR from a single point belonging to the set shows persistent oscillation with respect to the average power-law. Unlike the case of deterministic fractals, these oscillations are not periodic. In the general case of a stochastic fractal, we can write

$$f(r) \equiv f(\log(r)) = \int_{-\infty}^{\infty} d\omega A(\omega) e^{i\omega \log(r)} \qquad (4.52)$$

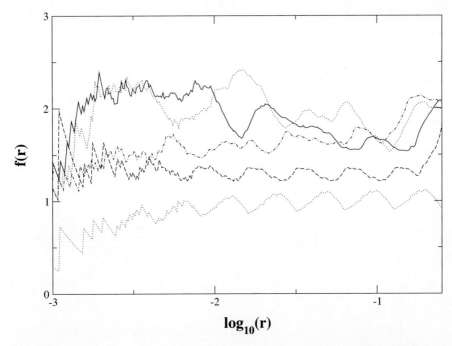

Fig. 4.14. Behaviors of $f(r)$ (see text) for the two dimensional Sierpinski gasket as determined from various randomly chosen points of the structure. The oscillating term can be described as a sum of different components. The amplitude of the fluctuations is larger than in the previous case (*dotted line*) and it is systematically higher

where $A(\omega) \in L^2(-\infty, +\infty)$ (and such that $f(r) > 0$ at all r) is the spectrum of the intrinsic fluctuations of the stochastic fractal (clearly $A^*(\omega) = A(-\omega)$ as $f(r)$ is real). By the definition of $f(r)$

$$\sigma_p^2(r) \sim \int d\omega \, |A(\omega)|^2 \ . \tag{4.53}$$

Hence in this case there is no preferred period of oscillations: Fluctuations are the superposition of log-waves of different amplitude and frequency. This is clearly a simplified treatment of the problem, and other concepts, such as "porosity", have been introduced to describe fully this case. However our motivation here is to look for a simple scheme which allows us to compute some observational quantities in the case of three-dimensional distributions of cosmological objects as for instance galaxies: for this purpose the approximation given by (4.52) is quite satisfactory.

The log-oscillations may produce fluctuations in the measurement of the dimension through the MLR from a single or a finite number of observers. Such an effect is particularly important especially when oscillations have a large amplitude or a long-range period in logarithmic scale, compared to the range of scales covered by the available samples. To give a simple example of log-periodic oscillations in a random fractal, we have generated an artificial sample with the random Cantor set (see Appendix B). In this case we have used a fragmentation order equal to 5 and we have iterated the algorithm 12 times. In Fig. 4.15 we show the behavior of $f_i(r)$ (4.49) related to the

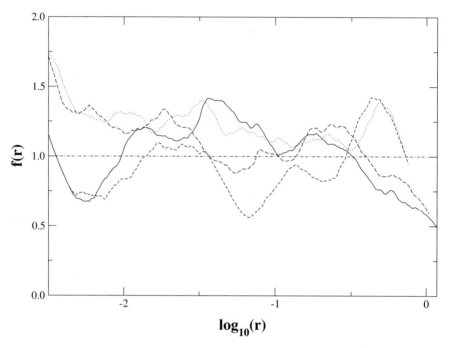

Fig. 4.15. Behavior of $f(r)$ for a stochastic fractal (random Cantor set with $D = 1.58$). The amplitude is larger with respect to the case of the Sierpinski gasket and there is no particular dominating frequency of oscillation

MLR from several randomly chosen points of the set. Oscillations are present but there are waves of different logarithmic frequencies and it is not possible to clearly identify a dominating frequency. The relative amplitude of the oscillations is of order 1. Since this fractal sample is finite we find that oscillations are still present, also for the average conditional density, but their amplitude is reduced by a factor of ten, so they are almost negligible in the average case.

4.7 Fractal with a Crossover to Homogeneity

We now discuss a simplified approach to describe the change of the statistical properties of a point-particle distribution which has a fractal behavior up to a scale λ_0 at which it has a crossover to homogeneity. Let us consider the case of a sharp crossover of the type

$$\Gamma(r) = \begin{cases} \frac{DB}{4\pi} r^{D-3} & \text{for } r < \lambda_0 \\ \frac{DB}{4\pi} \lambda_0^{D-3} & \text{for } r \geq \lambda_0 \,. \end{cases} \quad (4.54)$$

By neglecting the effect of the lower cut-off $\langle \Lambda \rangle$, we obtain that a possible estimator (see Chap. 9 and Appendixes F-G for more detailed discussion of estimators in finite samples) of the average density in a spherical sample of radius R_s is

$$\Gamma_E^*(R_s) = \frac{3}{4\pi R_s^3} \int_0^{R_s} \Gamma(r) d^3 r = \frac{3}{R_s^3} \frac{DB}{4\pi} \left[\frac{\lambda_0^D}{D} + \frac{\lambda_0^{D-3}}{3} \left(R_s^3 - \lambda_0^3 \right) \right] \,, \quad (4.55)$$

where the suffix E refers to the estimated quantity. Clearly we have that in the infinite volume limit

$$\lim_{R_s \to \infty} \Gamma_E^*(R_s) = \Gamma(R_s) \,. \quad (4.56)$$

An estimator of $\xi(r)$ can be then written as

$$\xi_E(r) = \frac{\Gamma_E(r)}{\Gamma_E^*(R_s)} - 1 \,. \quad (4.57)$$

We thus have $\xi_E(r) < 0$ for $r > \lambda_0$ and with a power-law behavior only for $r \ll \lambda_0$. Note that the fact that $\xi_E(r)$ is negative comes only as a consequence of the finite sample estimation and does not correspond to a real "anti-correlation". Indeed, in the infinite volume limit, one may see that the system is clustered and correlated for $r \leq \lambda_0$ and unclustered and uncorrelated at larger distances: for this reason the asymptotic $\xi(r)$ must be identically zero at scales larger than λ_0. This kind of finite-size effect must be carefully considered in the analysis of real measurements. While the approximation of (4.54) is rather rough, it turns out to be very useful in many practical cases because the residual two-point correlation in the regime $r \gg \lambda_0$ can be negligible with respect to Poisson fluctuations. For this reason the distribution is practically equivalent to an uncorrelated one for $r > \lambda_0$ as described by (4.54).

4.8 Correlation, Fractals and Clustering

In this section we relate the correlation properties of a particle distribution to its clustering features. Let us consider a particle distribution in a d-dimensional sample of volume V. We can distinguish different cases:

- The particles are distributed in a homogeneous (i.e. uniform) way in the volume (see Chap. 2), i.e., the fractal dimension of the distribution is $D = d$, and the homogeneity scale is smaller and independent of the system size $R_s = V^{1/d}$. In this case the quantity V/N is sample independent and gives the *specific volume*, i.e. the volume per particle. Therefore

$$l = \left(\frac{V}{N}\right)^{1/d}$$

gives the average distance between different particles. It is a well defined quantity and is independent of the sample size R_s. Another important distance which is well defined for this distribution is the average nn distance $\langle\Lambda\rangle$ introduced both in Chap. 2 and in Sect. 4.3.3. As illustrated by the approximation given by (2.71) the PDF of the nn distance, and as a consequence its average value, are, at least as a first approximation, functions of the average conditional density $\Gamma(r)$. In particular the more positive is the function $\Gamma(r)$ is at small distances, the smaller is the average nn distance. Therefore the ratio

$$\lambda_{cl} = \frac{l}{\langle\Lambda\rangle} \geq 1 , \qquad (4.58)$$

is a good parameter to characterize the degree of clustering in a distribution of particles. The particle configuration for which the clustering parameter λ_{cl} is the smallest possible is a cubic lattice for which it takes the value 1. A completely uncorrelated statistically stationary Poisson distribution has the smallest degree possible of correlation since particles are distributed completely at random in the volume. In this case, in three dimensions (see Sect. 4.3.3), we have

$$\lambda_{cl} = \left(\frac{4\pi}{3}\right)^{1/3} \Gamma_e^{-1}\left(\frac{4}{3}\right) > 1 .$$

The fact that in the Poisson case λ_{cl} takes a larger value than in a lattice can be simply explained by noticing that at short non-zero distances the correlation function for a lattice is negative (see Chap. 3). In general the larger is the correlation at short distances, the larger is the clustering coefficient λ_{cl} and the stronger the clustering.
- In a fractal particle distribution with $D < d$ the situation is completely different and, clearly, clustering is much stronger than in any homogeneous system. As we have already seen above, a fractal is asymptotically empty in the sense that the conditional average number of particles $\langle N(R_s)\rangle_p$ in a finite spherical volume of size R_s grows more slowly than the volume, i.e.

$$\langle N(R_s)\rangle_p = BR_s^D \text{ with } D < d ,$$

where B is an intrinsic coefficient related to the lower cut-off of the particle system. This implies that the average distance between particles l, as defined above, is a function of the sample size increasing as $\sim R_s^{1-D/d}$ and diverging in the infinite volume limit. On the other hand the average nn distance is well defined and depends only on B and D, and other intrinsic properties of the fractal system such as the lacunarity (see Sect. 4.3.3). Therefore in a finite sample the parameter λ_{cl} defined by (4.58) depends on the sample size and diverges in the infinite volume limit. This means that the degree of clustering of a fractal particle distribution is *infinitely larger* than in any homogeneous distribution.

On the other hand, the approximation (4.25) allows us to compare the degree of clustering of two fractal samples with the same number of particles N in the same spherical volume V. This equation shows that, neglecting the variation of the value of the Euler gamma function at different D, in this case the average nn distance is smaller in a distribution with a smaller fractal dimension . This is simply verified by considering that $B = N/V^{D/d}$ with N, V, d fixed equal for both fractals (see Fig. 4.16). Consequently, in general the smaller the fractal dimension the higher is considered the degree of clustering. This is illustrated by the four different cases in $d = 2$ in Fig. 4.17.

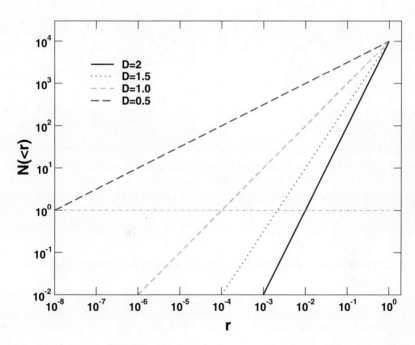

Fig. 4.16. The "clustering" can be described by the fractal dimension. In this figure is shown the average distance between nearest neighbors $\langle \Lambda \rangle$ as a function of the fractal dimension ($D = 0.5, 1.0, 1.5, 2.0$). The number of points and the sample volume are the same in all cases

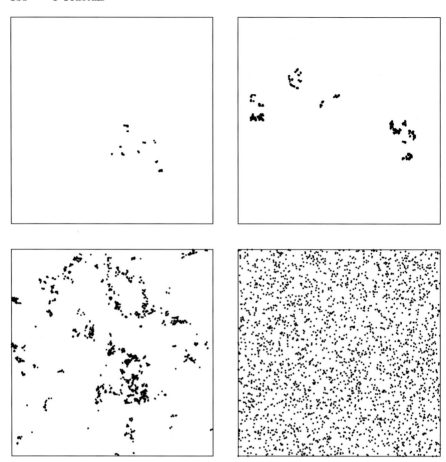

Fig. 4.17. For the same number of point-particles in the same volume, the fractal dimension is a measure of the "intensity" of clustering for fractal objects. In these simulations we have fixed the number of points as well as the volume of the sample. The fractal dimension takes the values $D \simeq 0.5, 1.0, 1.5, 2$, in $d = 2$, from *top left* to *bottom right*. The clustering increases when the fractal dimension decreases

4.9 Probability Distribution of Mass Fluctuations in a Fractal

In Chap. 2 we have seen that, for homogeneous (i.e. spatially uniform) mass density fields in d dimensions with a well defined $\rho_0 > 0$, the mass $M(R)$ contained in a sphere of radius R is a stochastic variable, whose behavior can be summarized as follows:

- if the correlation length r_c of the system, as defined in Sect. 2.3.2, is finite (i.e. if the connected correlation function $\tilde{\xi}(r)$ decreases sufficiently rapidly to zero), then the central limit theorem can be applied, and for $R \gg r_c$

4.9 Probability Distribution of Mass Fluctuations in a Fractal

the stochastic mass $M(R)$ can be considered as a Gaussian variable with mean $\rho_0 V(R)$ (where $V(R) \sim R^d$ is the volume of the sphere of radius R) and variance proportional to $(R/r_c)^d$;

- if $r_c \to \infty$ (i.e. $\tilde{\xi}(r)$ decreases slowly to zero as in a thermodynamic critical phenomenon) the central limit theorem is no longer valid, and consequently there is no finite scale beyond which $M(R)$ can be considered to have a Gaussian behavior. Its mean is again $\rho_0 V(R)$, but its variance is proportional to R^a with $d < a < 2d$.

In both cases, the "local" average density defined by

$$\bar{\rho}(R) = \frac{M(R)}{V(R)}$$

can be considered *self-averaging* (i.e. with a PDF $\delta\left(\bar{\rho}(R) - \rho_0\right)$) for $R \gg \lambda_0$, where λ_0 is the homogeneity scale (e.g. defined by $\xi(r) \leq 1$ for $r \geq \lambda_0$). In fact, because of the normalization of $\bar{\rho}(R)$ with respect to $M(R)$, the asymptotic behavior of its PDF is governed by the law of large numbers, for the application of which it is sufficient that $\tilde{\xi}(r) \to 0$ for $r \to \infty$. However, despite this fact, in the case in which $r_c \to \infty$, there is no spatial scale beyond which $\bar{\rho}(R)$ displays a Gaussian behavior.

For a fractal mass density the situation is even more radical. As shown in this chapter, a fractal mass distribution is characterized by being asymptotically empty, but shows large *conditional* fluctuations decreasing slowly on average in space because of strong correlations. Therefore both the correlation length and the homogeneity scale diverge for a fractal, and $M(R)$, which on average is non-zero only as a conditional quantity (i.e. as seen from a spatial point occupied by a particle of the system), is strongly fluctuating at any scale. We have seen in fact that $\langle M(R) \rangle_p = BR^D$ (where $D < d$ is the fractal dimension and B a coefficient related to the lower cut-off of the density field) and

$$\langle M^2(R) \rangle_p = CR^{2D}$$

with $C > B^2$. This means that relative fluctuations in $M(R)$ are persistent and statistically self-similar at any scale. Therefore since $r_c \to \infty$, as in a homogeneous (uniform) critical system, the central limit theorem does not hold, and consequently $M(R)$ does not in general have a Gaussian behavior at any R. Moreover, since in a fractal also $\lambda_0 \to \infty$, there is no finite length scale beyond which the PDF of the *conditional* "local" density

$$\bar{\rho}(R) = \frac{M(R)}{V(R)}$$

can be well approximated by the self-averaging PDF $\delta\left(\bar{\rho}(R) - \rho_0\right)$ where $\rho_0 = 0$ is the asymptotic average mean value. This is true even though the law of large numbers can be shown to imply asymptotically that

$$\lim_{R \to \infty} p(\bar{\rho}(R)) = \delta(\bar{\rho}(R)), \tag{4.59}$$

where $p(\bar{\rho}(R))$ is the PDF of the variable $\bar{\rho}$ at distance R from the occupied origin of coordinates. The problem is that at any R larger than the lower cut-off of the system we have

$$\langle \bar{\rho}(R) \rangle_p \sim R^{D-d}$$

and

$$\langle \bar{\rho}^2(R) \rangle_p \sim R^{2(D-d)},$$

which imply the asymptotic limit (4.59), but at any finite R the estimate of the mean gives an infinite relative error with respect to the limit and with relative fluctuations of order one.

In the case of a fractal, because of the intrinsic self-similarity, an important role in its statistical characterization is played by the study of the statistics of the multiplicative mass fluctuations from an occupied point $f(R) = M(R)/\langle M(R) \rangle_p$. In fact, as we have seen, in a stochastic fractal the statistical properties of $f(R)$ do not depend on R, while in a uniform mass distribution its PDF goes toward the self-averaging form $\delta(f(R) - 1)$ when R increases beyond the homogeneity scale λ_0.

The analysis of all these quantities can be very useful in studying the statistical properties of observational samples of point-particle distributions, as, for example, those given by galaxy redshift surveys. Studying the statistics, such as the Gaussianity or the self-averaging properties, of integrated quantities such as $M(R)$ or $\bar{\rho}(R)$ at finite scales, or the change in the PDF of $f(R)$, can give much insight into the correlation properties of the mass density field studied and help to classify it and characterize its properties correctly.

4.10 Intersection of Fractals

Let us suppose we have two independent stochastic fractal sets in the same space. An interesting question, which can be relevant in certain problems, is whether they intersect one each other, and in this case what the main features are of this intersection [213]. The answer was given by Mandelbrot [151] as a simple rule which states that the *co-dimension* of the intersection of two independent fractal sets S_1 and S_2, with dimension D_1 and D_2 respectively, embedded in the d-dimensional space is equal to the sum of the *co-dimensions* of the two sets:

$$d - D_I = (d - D_1) + (d - D_2), \tag{4.60}$$

where D_I is the fractal dimension of the intersection set: $D_I \equiv D(S_1 \cap S_2)$. Hence we can write for the dimension D_I:

$$D_I = D_1 + D_2 - d. \tag{4.61}$$

4.10 Intersection of Fractals

Note that this rule holds also in the case of the intersection of two stochastic distributions of isolated points, if we use the following rule to determine the intersection. Let us suppose we know the lower cut-offs Λ_1 and $\Lambda_2 < \Lambda_1$ of the two distributions. At this point we can cover the space with a grid of cells of size Λ_1, and define the intersection as the set of cells in which both points of the first and of the second point distributions appear.

A simple explanation of (4.61) can be given as follows. Consider two independent stochastic fractal sets S_1 and S_2 embedded in the d-dimensional space. Let us divide the space into boxes of size ϵ (in the case of point distributions ϵ must taken to be larger than the lower cut-off). If we consider a particular box, the probability p_i that it contains points of the set S_i is equal to $N_i(\epsilon)/N_{tot}$, i.e. the number of boxes containing points of the set divided by N_{tot}, the total number of boxes. Since for a fractal of dimension D_i we have that $N_i(\epsilon) \sim \epsilon^{-D_i}$ we can write:

$$p_i(\epsilon) = \frac{N_i(\epsilon)}{N_{tot}} \propto \epsilon^{d-D_i} \,. \tag{4.62}$$

Because of the randomness of the fractal this argument holds for each box. In the hypothesis of mutual independence of the two intersecting fractals, we readily find the probability $p_I(\epsilon)$ that a box contains points of both fractal sets:

$$p_I(\epsilon) = p_1(\epsilon)p_2(\epsilon) \propto \epsilon^{2d-D_1-D_2} \,. \tag{4.63}$$

In the limit of small ϵ we find for the total number of boxes N_I containing points of both sets:

$$N_I(\epsilon) = p_I(\epsilon)N_{tot} \propto \epsilon^{d-D_1-D_2} \,. \tag{4.64}$$

By taking into account the definition of the "box dimension", we find:

$$D_I = D_1 + D_2 - d \,. \tag{4.65}$$

This rule is the generalization of the rule valid for regular geometrical manifolds, e.g. the intersection of two non-parallel planes ($D_1 = D_2 = 2$) in three-dimensional space ($d = 3$) is a line ($D_I = 1$). The case in which $D_I < 0$ (for example by intersecting a fractal with $D_1 < 2$ in $d = 3$ with a line $D_2 = 1$) must be interpreted in the following way: the real fractal dimension of the intersection set is zero, but we can distinguish such sets of zero dimension by different scalings of $N_I(\epsilon)$ going to zero as ϵ does so. A discussion about this property can be also found in [47]. It must be noted that (4.65) is valid also in the cases in which one of the two intersecting set is not a random fractal, but a deterministic fractal or an homogeneous set. The only essential point is that the sets S_1 and S_2 considered are independent.

4.11 Morphology and Voids

The concept of *lacunarity* has been introduced to further characterize fractal structures and in particular their voids [151]. However the definition of a "void" in space with $d > 1$ can be subtle and arbitrary (while in $d = 1$ its definition is evident an unique). For example one may consider spherical voids, but sometimes cubical or more anisotropic definitions can be more appropriate. Let us suppose we generate a fractal set with a lower cut-off $\Lambda_0 > 0$ (i.e. a so-called pre-factor). For example, we can consider a random Cantor set in $d = 1$ (see Appendix B) obtained by a finite number of iterations of the fragmentation algorithm. In this case it is possible to measure a void in units of λ_0 and to define unambiguously the lacunarity coefficient F as

$$Nr(\lambda > \Lambda) = F\Lambda^{-D} \tag{4.66}$$

where $Nr(\lambda > \Lambda)$ is the number of voids with size $\lambda > \Lambda$ and D is the fractal dimension of the set. The scaling behavior is completely determined by the fractal dimension, as shown by (4.66). However the lacunarity coefficient F, takes different value for different fractals with the same dimension. An example is shown in Fig. 4.18.

4.12 Angular and Orthogonal Projection of Fractal Sets

Let us now define the angular correlation function (ACF) which is very useful for the study of a fractal distribution of points. The definition we give holds both for fractal and homogeneous sets. This quantity is closely related to the two-point conditional density, i.e. to the three-point correlation function: in fact, we are looking for the probability that, given an occupied point (the "observer"), we find a pair of structure points with a certain angular separation θ. The ACF is defined in the following way: given a spherical sample of radius R_s (with $R_s \to \infty$) with the observer at the center, we look for the probability of finding a pair of structure points with a certain angular separation θ, and then we consider an average over different observers. That is

$$\Gamma(\theta) = A(R_s) \int_{C(R_s)} \int_{C(R_s)} \langle \hat{\rho}(\mathbf{r_1})\hat{\rho}(\mathbf{r_2}) \rangle_p \, \delta(\text{ang}(\mathbf{r}_1, \mathbf{r}_2) - \theta) d^3 r_1 dr^3 r_2 , \tag{4.67}$$

where the integral is performed in the spherical sample of radius R_s and $\text{ang}(\mathbf{r}_1, \mathbf{r}_2)$ is the angle between the two vectors. The coefficient $A(R_s)$ is an appropriate normalization factor chosen in order to have, for example,

$$\int_0^\pi d\theta \Gamma(\theta) = 1 .$$

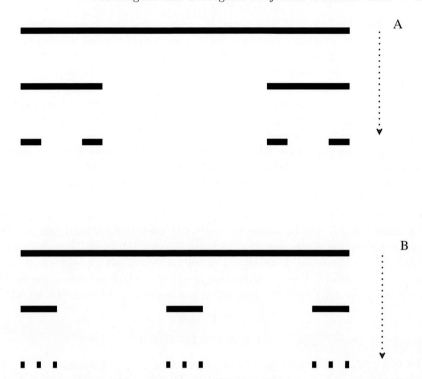

Fig. 4.18. Construction of two different deterministic Cantor sets with the same fractal dimension. The generator of the upper one consists of two segments of size $1/4$ while the generator of the lower one consists of three segments of size $1/9$. The fractal dimension is the same and $D = 1/2$: The two sets are distinguished by their different lacunarity

We note two important limitations for the case of real samples (for which we have in mind the galaxy catalogs discussed in Chap. 8): (i) The volume of integration $C(R_s)$ and the radius R_s are finite, and may change from one sample to another. (ii) It is often not possible to perform the average given by (4.67) because only angular coordinates are given. In this case we may only compute the non-average quantity, i.e. $\Gamma_i(\theta)$ from a single point i. Such a quantity in a fractal set of points can be strongly affected by finite size effects and intrinsic fluctuations as already shown for $N_i(R)$ from a single observer. Therefore the estimator $\Gamma_i(\theta)$ of the ACF from a single observer i in the case of fractal set can be substantially different from its ensemble value. On the other hand, in the case of a homogeneous set of points, we expect instead that for a scale R_s much larger than the homogeneity scale, we obtain a good estimator of the ACF even from a single observer. However an important difference of the ACF with respect to three-dimensional correlation functions such as the two-point conditional density must be pointed out: even in measurements from only one point, in order to find the angular correlation at a given angle, one has to perform an average considering all pairs of points (galaxies) which lie at a certain angular separation θ. This gives an additional source of smoothening for the signal. In this respect the angular dependence of the ACF is well estimated also by $\Gamma_i(\theta)$, even though a trace of the fractal self-similar fluctuations remains at any scale R_s [47, 223]. For what concerns the problem of finite size effects on the projection of a fractal, while the angular dependence can be recovered also with the measurement from a single point (although some important finite size effects must be taken into account – see below), the amplitude[4] of $\Gamma_i(\theta)$ is strongly perturbed by the intrinsic oscillations. This point will be discussed in more detail in Chap. 10.

Let us consider the angular conditional density in more detail. First of all we introduce a theorem about the properties of the *orthogonal* projection of a fractal which will be useful to study the more complicated case of the angular projection. Orthogonal projections preserve the sizes of objects in the perpendicular direction. If an object of fractal dimension D, embedded in a space of dimension d, is projected orthogonally on a plane (of dimension $d-1$) it is possible to show that the projection has dimension D' with [47, 83, 151]

$$D' = D \text{ if } D < d-1 ; \qquad (4.68)$$
$$D' = d-1 \text{ if } d-1 \leq D \leq d .$$

This explains, for example, why clouds which have fractal dimension $D \approx 2.5$, give rise to a compact shadow of dimension $D' = 2$. The angular projection is a more subtle problem because it mixes different length scales. Nevertheless the theorem given by (4.68) can be extended to the case of angular projections in the limit of small angles ($\theta \ll 1$) for which the related portion of the surface of the sphere can be considered flat. Therefore considering small angles θ we can write that θ scales as

$$N(\theta) = B_a \theta^{D_a} \qquad (4.69)$$

where, according to (4.68) $D_a = D'$ and B_a is related to the angular lower cut-off of the distribution. Equation (4.69) holds from every occupied point, and in the case of a homogeneous distribution in three dimensions we have $D_a = 2$. We define [47] the conditional average density as

$$\Gamma(\theta) = \frac{1}{S(\theta)} \frac{dN(\theta)}{d\theta} = \frac{BD_a}{2\pi} \theta^{-\gamma_a} \qquad (4.70)$$

where $S(\theta)$ is the differential solid angle element ($S(\theta) \approx 2\pi\theta$ for $\theta \ll 1$) and from (4.69) $\gamma_a = 2 - D_a$ is the angular correlation exponent (angular co-dimension). The last equality holds in the limit $\theta < 1$ where the angular problem can be treated as a two-dimensional Euclidean projection. We refer to [164] for a discussion about the measurement of the ACF in galaxy catalogs.

---

[4] In the case of galaxies the amplitude of the angular correlation function is fixed by the number counts as a function of apparent magnitude (see Chap. 10).

4.12.1 On the Uniformity of the Angular Projection

We discuss in this section important finite size effects which can occur in the angular projection of a fractal set, and in particular we make an explicit reference to the case of galaxies [70]. The main result is that, if one takes finite size effects properly into account, *a set of dimension 2 (or larger) in 3-dimensional space can have a non-fractal and quite homogeneous projection onto the celestial globe.* Clearly if $D > 2$ (or more generally $D > d-1$, d being the spatial dimension) one may obtain quite uniform angular distributions. We will explain this result below and put it into perspective with known mathematical results. Depending on how finite size effects are taken into account, sometimes the projection will be homogeneous and sometimes it will be fractal. In particular, we will argue that *there is no contradiction between observing a fractal of dimension 2 in the three-dimensional space and a uniform projection of equal size points onto the celestial globe.* We discuss this problem in relation to the projection of galaxies and to the uniformity of angular galaxy catalogs (see Chap. 8).

In order to explain the "projection paradox" and to put it into context, we first explain some mathematical aspects of the problem. As already mentioned, a set A in $\mathbb{R}^3$ is said to have Hausdorff dimension D if it can be covered by sets $S_{i,\delta}$ of diameter less than δ in such a way that the "Hausdorff measure"

$$H^s(A) = \liminf_{\delta \to 0} \sum_i (\text{diam} S_{i,\delta})^s \qquad (4.71)$$

is zero for $s > D$ and infinite for $s < D$ and D is the Hausdorff dimension. For $s = D$ one can have $H^D(A) = \infty$ or $H^D(A) < \infty$. Thus, if the Hausdorff dimension is D, then $H^D(A)$ can be finite or infinite. When D is an integer and $H^D(A)$ is finite, one can decompose A into a regular part (consisting of piecewise rectifiable sets of dimension D, such as lines or sheets) and a singular part consisting of "dust" (see [83] Sect. 6.2 and [159] Sect. 9). For non-integer D the regular part is absent. Almost every projection of the regular part onto a D-dimensional plane is of positive measure, while all projections of the singular part (which is the interesting case for the study of galaxies) have measure 0, i.e., they are very small. Finally, if $H^D(A)$ is infinite, then a clever method, called the "Venetian blind construction" shows that the projection onto *any* subspace can be prescribed, and can be essentially anything we want. For example, one can construct a set in the 3-dimensional space whose projection onto all two-dimensional planes is a "sundial" in the sense that it shows the hour (minutes, and seconds) of the current time in Roman numerals (see [83] Sect. 6.3).

Thus, in the light of a strict mathematical definition the projection has positive measure (and hence relatively smooth) *only* in the cases when $D > 2$ or when $D = 2$ and either $H^2 = \infty$ or A has a regular part. In this last case, the set A must contain rectifiable "lines" or "sheets." We disregard this situation as physically irrelevant because galaxies are considered as points or tiny disks in this analysis and thus form a "dust-like," i.e., irregular set. This is not the whole story, because some finite size effects come into play in a subtle way. In order to clarify this point, we first work with 1-dimensional sets in a 2-dimensional space and then illustrate the extension to 2-dimensional sets in 3-space.

We say that a set (of points, which in this context we think of as galaxies) has an effective dimension 1 inside a physical space of dimension 2 at length scale r if the mean number $n(r)dr$ of galaxies between distance r and $r+dr$ from an occupied point goes like Cdr, where C is a constant. Integrating from 0 to r we find for the total number of points, N:

$$N(r) = \int_0^r dr' \, n(r') \sim Cr \, , \qquad (4.72)$$

and this is characteristic of the distribution of a set of dimension 1.

We next study what happens when we project such a set onto the celestial globe (the unit circle in the case of a 2-dimensional universe). If the galaxies are considered just as a countable set of points, the question of the measure of their projection makes no sense, since it is equal to 0 by definition. However this is not what one means by the dimension of an experimentally measured set, in which there is in any case only a finite number of points [73]. In fact, dimension is experimentally a notion which holds only over a certain range of scales. Even taking this into account, the mathematical theorem cited above tells us that the projected density has *zero* measure when H^1 is finite and the set is singular.

We may now analyze in more detail various aspects which come into play when one considers the projection of galaxies onto the celestial globe. The issues we discuss here are *lacunarity*, the role of taking into account *apparent sizes*, and the influence of *opacity*. We first define these quantities. As already mentioned, the lacunarity describes the size distribution of voids in a fractal set. These voids can be large in a set of small dimension, but we shall show that they can be arbitrarily small in the projection. The apparent size problem has to do with whether we represent galaxies as points of equal size, or apparent size. The first representation will be called *pixel projection* and the second *apparent size projection*. Finally, opacity is related to the following observational problem: We assume there is a limit to how close together two galaxies can be observed, and this means that small opaque disks are drawn around each observed galaxy.

We illustrate all these phenomena for a set of dimension 1 in 2-space. We then generalize to the case of a set of dimension 2 in 3-space. Our example is constructed as follows: Divide the unit square into 25 equal small squares and fill the central square and the four corners (see Appendix B for more details). We call this "pattern 0." Similarly, we can fill 5 among 25 squares by patterns 1, 2, and 3 as shown in Fig. 4.19.

4.12 Angular and Orthogonal Projection of Fractal Sets

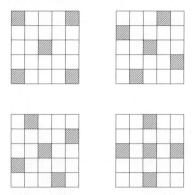

Fig. 4.19. The patterns 0 to 3 used to subdivide a five by five square. (From [70])

The fractal of Fig. 4.20 is then obtained by dividing the square recursively, choosing at each level randomly one of the 4 patterns. (An even more homogeneous projection is obtained by choosing at level n the pattern $(n \bmod 4)$ of Fig. 4.19.) These fractals have dimension 1, finite Hausdorff measure H^1, and are of the singular type described above. Of course, space becomes quite empty far away from the origin, but such fractals "block" the skylight in almost all directions for an observer at the center, and hence he will see an almost uniformly black sky as in Fig. 4.20. If the observer is not at a center, but still on a "galaxy," as we are, this argument continues to hold after a sufficient number of iterations of the inverse cascade [213]. To understand why the lacunarity decreases, we scale each point from the center to a fixed distance: $(x, y) \mapsto (x/r, y/r)$, where $r = (x^2 + y^2)^{1/2}$. In Fig. 4.21, we blow each "inner" point up to a position on the outer set of squares.

Since there is always one of 4 squares colored in each of the patterns 0–3, we see that the apparent open space can be at most about 3/5th of what it

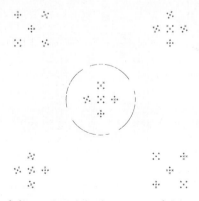

Fig. 4.20. A fractal of dimension 1 in 2-space, and its projection onto a circle. There are 7 levels of recursion. (From [70])

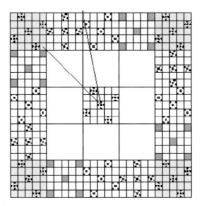

Fig. 4.21. Five levels of the same fractal as in Fig. 4.20. Each "inner" feature is scaled outward as shown by the lines for one feature. The corners contain the same feature as the central square. The outermost features of the central square are mapped as indicated by the radial lines. Going deeper into the recursion, each successive feature is projected clockwise to the boundary until one reaches again the corner squares in light gray which are mappings of the innermost feature. (From [70])

was at the previous level. Thus we get an estimate that *the maximal angular void scales like* $(3/5)^n$ *as the number n of levels grows*. There need not be any sizeable voids in the projection of a set of dimension 1 in 2-space (or for that matter, a set of dimension 2 in 3-space).

We next discuss how the projection onto the celestial globe can vary, depending on whether we show apparent size (as in Fig. 4.21) or just a pixel (as in the circle around Fig. 4.22). To simplify the discussion, we assume that all galaxies are small spheres of fixed diameter ϵ. The projection of a

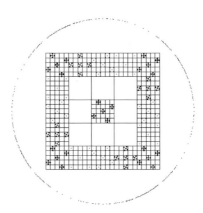

Fig. 4.22. The same as the previous figure, but now with "opaque" galaxies. Note that the projected density on the unit circle becomes quite uniform. (From [70])

galaxy at distance r has apparent size $\sim \epsilon/r$. One can view this in one of two ways: Either the remote galaxies have very small projections (somehow less than a pixel), or the close-by ones have very large projections. Note that we discuss here for simplicity a universe of points of *equal* size. How big is the area covered by these projections? Since

$$H^1(A) = \lim_{\delta \to 0} \sum_i \text{diam}(S_{i,\delta}) , \qquad (4.73)$$

and there are $\sim C$ galaxies at a distance r and $\text{diam}(S_{i,\delta}) = \delta$, we find that the number of galaxies ($\propto$ area of pixel projection) in a ring extending from $r_{\min}$ to $r_{\max}$ is given by

$$N(r_{\min}, r_{\max}) \sim \int_{r_{\min}}^{r_{\max}} dr' \, C \sim C(r_{\max} - r_{\min}) , \qquad (4.74)$$

while the *projected area* is

$$\int_{r_{\min}}^{r_{\max}} dr' \, C/r' = C \log(r_{\max}/r_{\min}) . \qquad (4.75)$$

(The area is smaller when the projected galaxies start to overlap, i.e., for very fat rings.) These equations explain why projected galaxies drawn as points (4.74) look more homogeneous after increasing the size of the annulus $(r_{\max}, r_{\min}) \to \lambda(r_{\max}, r_{\min})$; whereas a projection of the apparent area occupied by galaxies (4.75) is invariant under scaling transformations. In our simple fractal model, homogenization occurs for shells with $r_{\max} > 5^4 r_{\min}$[5]. At small radii, finite size effects dominate, as in Fig. 1 of [57]. Using smaller sub-divisions in the construction of the example, one can lower the necessary shell-width to $r_{\max} \sim 2 r_{\min}$. In the area projection, most of the area is covered by close-by disks, because they are larger.

4.13 Summary and Discussion

In this chapter we have presented the class of spatial distributions of point particles with unitary mass known as *fractals*. These mass distributions are characterized by being asymptotically empty and by displaying strongly correlated self-similar fluctuations at any length scale. This implies that they are characterized by voids and structures of any length scale without an intrinsic upper cut-off. The main effect of these strong correlations consists in the fact that the "mass" seen by a point belonging to the density field on average increases as R^D, where $D < d$ is called *fractal dimension* and d is the

---

[5] Note that the homogenization resulting in a point projection (4.74) has been used as proof for the homogeneity of the universe [57]. We thus see that the reasoning in [57] is not conclusive.

spatial dimension, while in ordinary uniform stochastic mass distributions the same mass increases as R^d. Moreover, because of the strong internal correlations, the relative fluctuations of mass with respect to the average is constant at any spatial scale. All these "critical" behaviors are very different from those observed in uniform mass distributions created by ordinary SSP or SPP, such as random particle distributions with or without correlations. For these reasons, many of the tools introduced in Chap. 2 for ordinary SSP and SPP cannot be used to study fractal mass distributions, and new ones must be considered: in particular, given that a fractal is asymptotically empty, only conditional statistical quantities are well-defined.

The introduction of fractal analysis is very important also for another task: let us suppose we have a sufficiently large but finite sample of a point-particle distribution. A common approach consists in applying, for the statistical analysis, directly the framework introduced in Chap. 2 for ordinary SPP. However, as shown throughout this chapter, in the case in which the particle distribution has fractal features, this kind of analysis will give spurious results. In fact the applicability of such a framework is based on the hypothesis that the mass distribution becomes sufficiently uniform at a spatial scale well within the sample size (implying in particular the existence of a well defined average density $\rho_0 > 0$). Therefore the fractal analysis can also be seen as the first step for the classification of particle distributions. In fact the study of the sample estimator of the conditional density $\langle \rho(\boldsymbol{r}) \rangle_p \equiv \Gamma(r)$ can discriminate between a fractal behavior up to the sample size or a crossover toward an uniform distribution at a scale (the so-called homogeneity scale) smaller than the sample size. Even in the case of a crossover to homogeneity at large scales the description of fractal properties up to a certain scale requires the appropriate statistical tools for these strongly fluctuating scale-invariant structures both at the level of data analysis and for their theoretical description. Only when fluctuations become very small can one use again the more refined statistical analysis presented in Chaps. 2 and 3 based on connected correlation functions and the power spectrum, without introducing strong spurious distortions depending on the sample size. Finally, we would like to stress again that the fractal analysis presented in this chapter represents a powerful set of statistical tools, even if fractality is limited to a finite range of scales smaller than the sample size presenting a crossover to a homogeneous (uniform) behavior. In fact if we want to analyze morphological properties of strongly correlated structures we have to use the set of tools presented in this chapter. For these reasons we have discussed many features of a fractal distribution ranging from the simple conditional density, to the angular and three-point properties, from the statistical properties of the intersection of two fractals to a characterization of the intrinsic self-similarity by studying spatial fluctuations of conditional integrated quantities such as the "mass" in a finite volume. For all these features a comparison with ordinary sufficiently uniform SSP and SPP has been given.

5 Multifractals and Mass Distributions

5.1 Introduction

In the previous chapter we have seen how to detect and quantify strong irregularities in point distributions through the introduction of *fractal analysis*. However sometimes large fluctuations appear as intrinsic features of more general mathematical measures than simple distributions of equal mass point-particles. There are different ways to associate a measure to a given physical phenomenon, and usually one has to focus on some relevant quantities defining the phenomenon itself, such as density of charges, mass or potentials depending on the different physical processes considered. For example, in the case of galaxy distributions one definition of the measure is given by galaxy masses [190, 220]. As discussed below, it is possible to use a mathematical framework to study in a unified approach the joint space and mass distributions: the so-called multifractal (MF) formalism (see Fig. 5.1).

The MF analysis is a refinement and a generalization in the sense of measure theory of fractal analysis for simple sets of points which arises naturally in the case of self-similar measures. In the more complex case of MF measure densities, simple fractal scaling properties can be different for points embedded in regions with different levels of the measure density, and one has to introduce a continuous set of exponents to characterize the system (the MF spectrum) [102, 171]. MF measures have a rich scaling structure and MF analysis deals with the description of these complex structures. Multifractality has been shown to exist in a wide variety of natural phenomena such as: dynamical systems [29, 102, 106, 110], fully developed turbulence [86, 161, 171], random resistor networks [6, 200] and wave functions of disordered systems [41, 172, 192]. For a comprehensive discussion of the subject see [171].

In this basic introduction to the field we firstly discuss the MF formalism and illustrate it with simple examples of deterministic and random MF measures [151, 201, 213]. We consider in more detail the properties of the measure distribution and its relation to the galaxy-like "mass-function". In Chap. 11 we discuss an application of the MF formalism to the case of galaxy and luminosity distributions, while in Chap. 12 we treat the case of galaxy clusters.

144   5 Multifractals and Mass Distributions

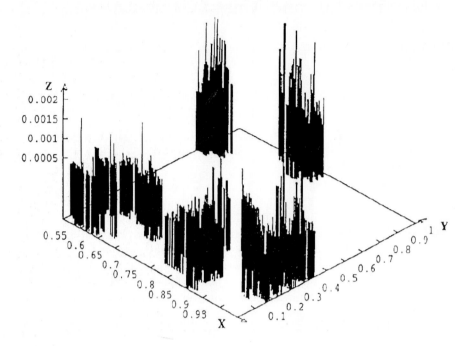

Fig. 5.1. A random multifractal obtained by the generalized Cantor set algorithm (see text) in the two dimensional Euclidean space. (From [223])

5.2 Basic Definitions

The main characteristic of MF measures is that the scaling properties vary from point to point depending on the value of their measure. Let us consider a measure $\mu_x(\epsilon)$ defined on the points x of a given space (e.g. $I\!R^3$). The quantity $\tilde{\mu}_x(\epsilon)$ is defined as the measure (the *mass*) of a small volume of linear size ϵ around the point x. If we define the point-wise dimension or singularity exponent α as

$$\lim_{\epsilon \to 0} \mu_x(\epsilon) \sim \epsilon^{\alpha(x)} \;, \tag{5.1}$$

then the characteristic feature of a MF measure is that α is fluctuating greatly as a function of the position x.

A way to characterize a MF measure is by means of the generalized or Renyi dimensions [102, 110, 171]. In order to define these dimensions let us partition the space into a grid of cells of lattice constant ϵ. Let us also suppose that the measure of the whole space is normalized to 1, as in usual probabilistic measures. We then introduce a q-partition function of such a discretized measure as

$$Z(q,\epsilon) \equiv \sum_i \mu_i(\epsilon)^q \,, \tag{5.2}$$

where $\mu_i(\epsilon)$ is the measure in the ith box of the grid centered around the point x_i. The exponent q is called the *structure parameter*. The set of generalized q-dimensions $D(q)$ is then defined by the following limit relation

$$(q-1)D(q) \equiv \tau(q) \equiv \lim_{\epsilon \to 0} \frac{\log Z(q,\epsilon)}{\log \epsilon} \tag{5.3}$$

implying a scaling behavior of the partition function $Z(q,\epsilon)$ for small ϵ

$$Z(q,\epsilon) \sim \epsilon^{\tau(q)} \,. \tag{5.4}$$

Note that for $q \to 0$ the partition function reduces to the number of ϵ-boxes (i.e. cells) containing a non-zero measure. This is called the *support* of the measure. Consequently $D(q=0)$ is nothing other than the box-counting fractal dimension of the support of the total measure. On the other hand, taking properly the limit $q \to 1$ (i.e. before of taking $\epsilon \to 0^+$ of (5.3)), we can find $D(1)$:

$$D(1) = \lim_{\epsilon \to 0} \frac{\sum_i \mu_i(\epsilon) \log \mu_i(\epsilon)}{\log \epsilon} \,. \tag{5.5}$$

Since (5.5) is strictly related to the information entropy of the measure (see Chap. 2) $D(1)$ is usually called the *information dimension*. It is possible to show that this gives the fractal dimension of the points on which the measure is mostly concentrated [191].

5.3 Deterministic Multifractals

Let us consider the simplest example of a deterministic MF, which in many respects can be considered the paradigm of any MF: the *multiplicative deterministic Cantor set* (MDCR). In particular we present its binomial version which is also called the *binomial MF measure*. We can define it by generalizing the simple algorithm defining the simple fractal deterministic Cantor set (DCR, see Appendix B) by a procedure called *curdling* [151]. The construction is illustrated in Fig. 5.2. The domain of definition of the measure is the one-dimensional segment $[0,1]$. We start by defining on it a constant measure density $\rho = 1$ in order to have, as usual, a unitary global measure. At this point a recursive deterministic algorithm of modification of the measure is defined as follows. The first iteration consists in dividing the unit interval into the two pieces $[0,1/2]$ and $[1/2,1]$ of equal length, and then redistributing the total measure to have a measure $\mu_1 > 0$ on the first segment $[0,1/2)$ (i.e. with a constant density $2\mu_1$) and a measure $\mu_2 > 0$ on the second one $[1/2,1]$ (i.e. with a constant density $2\mu_2$) in such a way that $\mu_1 + \mu_2 = 1$ in order to keep the right normalization of the total measure. We can suppose

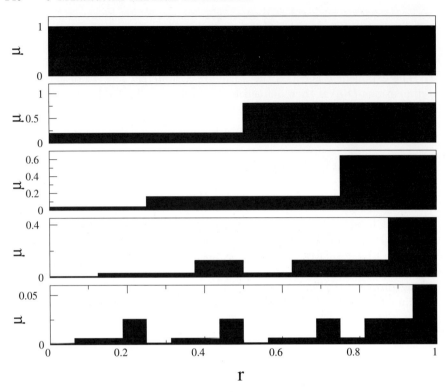

Fig. 5.2. First four iterations of the construction of a MF binomial measure with $\mu_1 = 1/5$ and $\mu_2 = 4/5$. The one dimensional spatial coordinate is plotted on the x-axis, while the y-axis shows the microscopic value of the measure

$\mu_2 > \mu_1$ without loss of generality. At the next iteration we apply the same procedure to both the segments $[0, 1/2)$ and $[1/2, 1]$ independently of each other. For example the segment $[0, 1/2)$ is divided into the two sub-segments $[0, 1/4)$ and $[1/4, 1/2)$ and the measure is redefined to be μ_1^2 in the first sub-segment (i.e. with a constant density $4\mu_1^2$) and $\mu_1\mu_2$ in the second one (i.e. with a constant density $4\mu_1\mu_2$). Analogously the second segment of the first iteration will be divided into two sub-segments $[1/2, 3/4)$ and $[3/4, 1]$ with respective measures $\mu_1\mu_2$ and μ_2^2. Note that the total measure of the system is conserved. Henceforth after the second iteration, we have 4 intervals of length $1/4 = (1/2)^2$, the left one with weight μ_1^2, the two central ones each one with weight $\mu_1\mu_2$, and the right one with weight μ_2^2. Now it is simple to repeat the algorithm for an arbitrary number of iterations. After an infinite number of iterations we are left with a well defined *binomial MF measure* [152, 191]. It is important to note that this algorithm defines a multiplicative process. This is a common feature of many deterministic or stochastic fractal and MF algorithms. In Fig. 5.3 we show the result obtained by iterating the

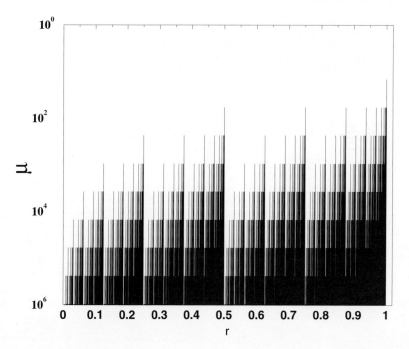

Fig. 5.3. The same as in Fig. 5.2 but after 10 iterations (where $\mu_1 = 1/5$). Note that the y-axis has now a log-scale

algorithm ten times. At the end we are left with $N = 2^{10}$ intervals of length $l = 2^{-10}$, $\binom{10}{n}$ of which are of measure $\mu_1^n \mu_2^{10-n}$ with $n = 0, 1, \ldots, 10$. Note that, after an arbitrary finite number of iterations the support of the measure is compact (i.e. the whole segment $[0,1]$) and hence $D(0) = 1$ which corresponds to an homogeneous distribution of points in $d = 1$.

We easily find for the partition function after k iterations

$$Z(q, \epsilon = 2^{-k}) = (\mu_1^q + \mu_2^q)^k = \chi(q)^k \qquad (5.6)$$

where $\chi(q)$ is defined to be the partition function after the first iteration and is called the *generator* of the MF. Using (5.3) and (5.6) we find for $D(q)$

$$(q-1)D(q) = -\log_2 \chi(q) \qquad (5.7)$$

(see Fig. 5.4). Note that the generalized dimensions $D(q)$ can be expressed completely in terms of the generator alone, reflecting the fact that the whole construction of the binomial measure is determined by the first iteration through a multiplicative process. This property is generally shared by all deterministic MF measures.

We can extend naturally the definition of the generalized Cantor set both in the number of partitions of the first iteration and in the dimensionality

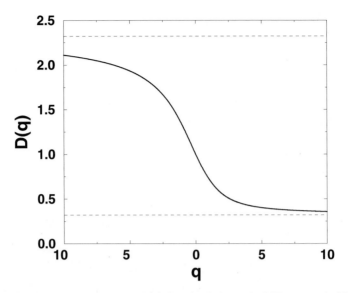

Fig. 5.4. Generalized dimension $D(q)$ for the binomial MF measure. Note that $D(q=0) = 1 = d$, hence the distribution has compact support

of the embedding space. In particular we can consider starting with the d-dimensional cube $[0,1]^{\otimes d}$ with a total measure 1 and uniform density $\rho = 1$. Then we can partition the cube in b^d smaller cubes of equal linear size $1/b$ and we can redefine the measure in each small cube in such a way to have the value μ_i in the ith small cube (i.e. with a uniform density $b^d \mu_i$), with the constraint of conserving the total measure (i.e. mass):

$$\sum_{i=1}^{b^d} \mu_i = 1 \ .$$

At this point the following iterations of the algorithm are analogous to those presented for the one-dimensional binomial Cantor set. The q-partition function $Z(q, \epsilon = b^{-k})$ after k iterations will be

$$Z(q, \epsilon = b^{-k}) = X(q)^k \ ,$$

where the *generator* $X(q)$ of this generalized version of the MF Cantor set is simply given by

$$\chi(q) = \sum_{i=1}^{b^d} \mu_i^q \ . \tag{5.8}$$

For the generalized dimensions we then find

$$\tau(q) \equiv (q-1)D(q) = -\log_b \chi(q) \ . \tag{5.9}$$

5.4 The Multifractal Spectrum

In this section we show how it is possible to extract information about the distribution of singularity exponents α (5.1) directly from the generalized dimensions $D(q)$ [86, 106]. Consider a partitioning of a measure S with boxes of size ϵ. Let $N_\alpha(\epsilon)$ be the number of boxes in which the measure is scaling as $\mu(\epsilon) \sim \epsilon^\alpha$ (5.1) with α in the range $[\alpha, \alpha + d\alpha]$. We can associate a dimension $f(\alpha)$ with this subset which we call S_α

$$f(\alpha) \equiv \lim_{\epsilon \to 0} \frac{\log N_\alpha(\epsilon)}{-\log \epsilon} \tag{5.10}$$

implying a scaling behavior of $N_\alpha(\epsilon)$ in the limit of small ϵ

$$N_\alpha(\epsilon) \sim \epsilon^{-f(\alpha)}. \tag{5.11}$$

The partition function $Z(q, \epsilon)$ can be then rewritten in terms of α and $f(\alpha)$ by splitting the sum

$$\sum_i \mu_i(\epsilon)^q$$

into a sum over α of the sum over the sub-set S_α of all the boxes carrying this singularity exponent

$$Z(q, \epsilon) = \sum_\alpha \sum_{S_\alpha} \mu_i(\epsilon)^q \sim \int \epsilon^{q\alpha - f(\alpha)} d\alpha. \tag{5.12}$$

For small ϵ we can apply the saddle point method to evaluate the integral in (5.12). Doing so we find that such an integral is dominated by the minimum of the exponent $q\alpha - f(\alpha)$. We can then write, using (5.3),

$$\tau(q) = q\alpha(q) - f(\alpha(q)) \quad \text{with} \quad f'(\alpha) = q$$
$$\frac{d\tau(q)}{dq} = \alpha(q). \tag{5.13}$$

Thus if we know the generalized dimensions $D(q)$ we can find the $f(\alpha)$-spectrum simply by performing the Legendre transformation given by the first of (5.13). The structure parameter q can be regarded as a "hunter" of different singularities α. For q positive, high density boxes with strong singularities are selected while for negative q weak singularities present in the low density boxes are magnified. In this way a MF measure S consisting of intertwined fractal subsets S_α can be decomposed.

In Fig. 5.5 we show the $f(\alpha)$-curve of the binomial MF measure as illustrated in Fig. 5.2. For this case it is easy to derive an analytical expression by applying the Legendre transformation (5.13)

$$f(\alpha) = -c(\alpha) \log_2 c(\alpha) - [1 - c(\alpha)] \log_2 [1 - c(\alpha)] \tag{5.14}$$

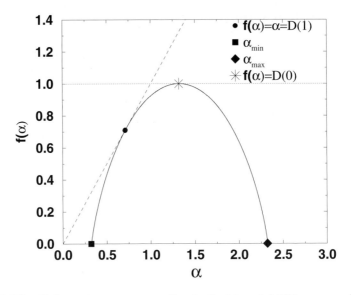

Fig. 5.5. The $f(\alpha)$ spectrum corresponding to the binomial MF measure shown in Figs. 5.2–5.4

with

$$c(\alpha) = \frac{\alpha - \alpha_{min}}{\alpha_{max} - \alpha_{min}}$$

$$\alpha_{min} = -\log_2 \mu_1$$

$$\alpha_{max} = -\log_2 \mu_2 \ .$$

One can in general relate some points of the $f(\alpha)$-spectrum directly to $D(q)$. The maximum of the $f(\alpha)$ is given by $f(\alpha(q=0)) = D(0)$. Indeed, in the limit of $\epsilon \to 0$ a box of size ϵ contains a singularity $\alpha(q=0)$ with probability 1. By using $\tau(1) = 0$ (normalization), we find directly for the information dimension $D(1) = f(\alpha(q=1)) = \alpha(q=1)$. This subset $S_{\alpha(q=1)}$ carries all the measure. Further, for the strongest and the weakest singularities one has in general that $\alpha_{min} = D(\infty)$ and $\alpha_{max} = D(-\infty)$. In some cases, as for instance the binomial MF measure, one can also calculate the $f(\alpha)$-spectrum directly, i.e. without calculating first the partition function and then performing a Legendre transformation. This histogram method [152] consists of a direct counting of singularities given a particular partition. For the above mentioned binomial measure we have after k iterations

$$N(\alpha_i) = \binom{k}{i} \quad i = 0, 1, \ldots, k \tag{5.15}$$

boxes of size $\epsilon = 2^{-k}$ with a singularity

$$\alpha_i = \frac{1}{k} \log_2 \left(\mu_1^i \mu_2^{k-i} \right) \quad i = 0, 1, \ldots, k \,. \tag{5.16}$$

Substituting (5.15) in the definition of $f(\alpha)$ (5.10) and making the Stirling approximation we get the same result as in (5.14).

Note that even in the case that the dimension of the support is compact $D(0) = d$ one has that the mass correlation dimension $D(2) < D(0)$ when the distribution is MF: the mass distribution is then characterized by non trivial correlation even if the space distribution is homogeneous.

5.5 Random Multifractals

We now study the general properties of MF measures obtained by stochastic algorithms or processes by studying directly the most famous example: the generalized multiplicative random Cantor set (MRCS) [214]. One can define a MRCS in d-dimensions as follows. Consider the usual d-dimensional cube $[0,1]^{\otimes d}$ with unitary total measure and uniform density $\rho = 1$. As in the deterministic version, we fragment the original cube into b^d smaller cubes of size $1/b$. In the binomial MF measure the way of assigning measures to the small fragmentation cube was unique. Instead in this case let us now define m different ways of redistributing the measure in the smaller cubes $\{\mu_{1,\gamma}, \mu_{2,\gamma}, \ldots, \mu_{b^d,\gamma}\}$ with $\gamma = 1, \ldots, m$ each one with the normalization constraint (total measure conservation)

$$\sum_{i=1}^{b^d} \mu_{i,\gamma} = 1 \,.$$

Let us give to each of these possibilities a probability p_γ with

$$\sum_{\gamma=1}^{m} p_\gamma = 1 \,.$$

Note that each of the m different ways of partitioning the measure defines a binomial MF measure whose generator is

$$\chi^{(\gamma)}(q) = \sum_{i=1}^{b^d} \mu_{i,\gamma}^q \quad \gamma = 1, 2 \ldots m \,. \tag{5.17}$$

At the first iteration, as we have said, the original cube $[0,1]^{\otimes d}$ is fragmented into the b^d smaller ones each of size b^{-1}, and one of the possible m redistributions of the measure, say the γth, is performed with the corresponding probability p_γ. Therefore, after the first iteration our measure consists of b^d cubes of size b^{-1} filling the domain $[0,1]^{\otimes d}$ and of respective measures $\{\mu_{1,\gamma}, \mu_{2,\gamma}, \ldots, \mu_{b^d,\gamma}\}$. At the second iteration we repeat the recipe of the first one, independently of each other, to each of the cubic cells i of size b^{-1} and measure $\mu_{i,\gamma}$ in which the original measure was redistributed at the first step, i.e., we fragment the ith cell of the first iteration into b^d sub-cells of size b^{-2}. Then we redistribute its measure $\mu_{i,\gamma}$ among these sub-cells by extracting once the possible m sets of weights, say the βth, with related probability p_β. In this way the jth sub-cell of the ith cell will have a measure $\mu_{i,\gamma}\mu_{j,\beta}$. This procedure is repeated for each cell of the first iteration with independent choices of the weights of redistribution of the measure. Note that in this way the total measure is conserved. Therefore, once completed the second iteration for each cell of the first step, the original measure consists of b^d groups of b^d sub-cells of size b^{-2} with the measure fragmented and redistributed in this multiplicative way. Further iterations are an obvious repetition of this algorithm to each sub-cell of the last step.

As an example consider the simple one-dimensional case in its binomial version, $b = 2$, where we have only three possible partitions of the measures: (1) $\{1/2, 1/2\}$ with probability $0 < p < 1$, (2) $\{0, 1\}$ with probability $\frac{1-p}{2}$, and (3) $\{1, 0\}$ with probability $\frac{1-p}{2}$. Note that the partition (3) is the specular version of (2), introduced to have a statistically isotropic algorithm. Therefore (2) and (3) have the same generator of the related binomial MF measure. Consequently this binomial algorithm has the two generators

$$\chi^{\gamma=1}(q) = 2\left(\frac{1}{2}\right)^q \text{ with } p_1 = p$$
$$\chi^{\gamma=2}(q) = 1 \text{ with } p_2 = 1 - p$$
(5.18)

as shown in Fig. 5.6. In addition to that, permutations of each realization of a generator appear with equal probability as indicated in the same figure in order to make the stochastic algorithm statistically isotropic. Note that this set is a generalized version of the random Cantor set discussed in Appendix B. The generators are the same, with the difference that now there is a measure associated to them.

Let us now compute $\tau(q)$ for the generalized MRCS as defined by (5.18). Consider the partition function for a given realization of the kth iteration of the fragmentation process

$$Z(q, \epsilon_k = b^{-k}) \equiv Z(q, \epsilon_k) = \sum_i \mu_i(\epsilon_k)^q \tag{5.19}$$

where $\mu_i(\epsilon_k)$ denotes the measure in the ith box of size ϵ_k. If we take a fragmentation one step further, every box will fine-grain according to a generator $\chi^\gamma(q)$ with a probability p_γ independently of each other. The partition function of the $(k+1)$th iteration can then be written down in a convenient way by grouping together into subsets $\{i\}_\gamma$ all the terms which fine-grain with the same generator

$$Z(q, \epsilon_{k+1}) = \sum_\gamma \sum_{\{i\}_\gamma} \mu_i(\epsilon_k)^q \chi^{(\gamma)}(q). \tag{5.20}$$

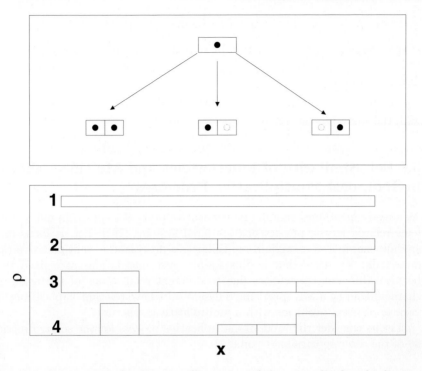

Fig. 5.6. *Upper panel*: A schematic illustration of the generalized multiplicative random Cantor set as given by (5.18). Note that the two permutations of χ^2 appear with equal probability. *Lower panel*: we show the first three iterations of a possible realization of the set

It can be easily shown that for those values of q for which $f(\alpha(q)) > 0$ we have [191]
$$\sum_{\{i\}_\gamma} \mu_i(\epsilon_k)^q = p_\gamma Z(q, \epsilon_k) . \qquad (5.21)$$
Substituting (5.20) gives
$$Z(q, \epsilon_{k+1}) = Z(q, \epsilon_k) \sum_\gamma p_\gamma \chi^{(\gamma)}(q) \qquad (5.22)$$
which leads finally, using the definition of $\tau(q)$, to
$$\tau(q) = \frac{\log\left(\sum_\gamma p_\gamma \chi^{(\gamma)}(q)\right)}{\log(b^{-1})} = \frac{\log\langle\chi(q)\rangle}{\log(b^{-1})} \quad \{\forall q \in \mathbb{R}^3 \,|\, f(\alpha(q)) > 0\} \qquad (5.23)$$

where $\langle \ldots \rangle$ denotes the average over disorder. The result of (5.23) implies that for a MRCS almost every realization gives the same $\tau(q)$, namely that which would be obtained by averaging the partition functions over all different realizations. One can easily show the scaling relation

$$\langle Z(q,\epsilon_{k+1})\rangle = \langle Z(q,\epsilon_k)\rangle\langle\chi(q)\rangle \tag{5.24}$$

of the averaged partition function by making use of the Markovian property that the fragmentation of a box does not depend on its history. Equation 5.24 directly implies that

$$\langle Z(q,\epsilon_k)\rangle = \epsilon_k^{\tau(q)} \tag{5.25}$$

with the same $\tau(q)$ of (5.23).

5.6 Self-Similarity of Fluctuations and Multifractality in Temporal Multiplicative Processes

As already mentioned random multiplicative processes appear in many problems related to the physics of disordered systems [191]. Let us focus on a specific example of a random multiplicative process: the multiplicative random walk. We show that in this simple case, fluctuations at a fixed time between different trajectories due to different realizations of the noise are characterized by a MF spectrum. These results have relevant implications for various physical problems with a multiplicative structure.

Let us consider the random multiplicative process for the scalar function ψ_t of the discrete variable t (time):

$$\psi_{t+1} = e^{\Delta_t}\psi_t = \exp\left(\sum_{t'=0}^{t}\Delta_{t'}\right)\psi_0 \tag{5.26}$$

where Δ_t is a dicotomic random variable whose PDF is

$$p(\Delta_t) = \frac{1}{2}\delta(\Delta_t - \Delta) + \frac{1}{2}\delta(\Delta_t + \Delta), \tag{5.27}$$

and no correlation is supposed between the values at different times t. Therefore each sequence of t values $\Delta_0, \Delta_1, \ldots, \Delta_{t-1}$ defines one and only one trajectory of the multiplicative walker ψ up to the time t. The generalization to more complex $p(\Delta_t)$ is quite simple [191]. We now describe the fluctuations of this process in relation to the properties of self-similarity. If we fix as initial condition for all the trajectories $\psi_0 = 1$, then the probability distribution $P_t(\psi_t)$ for the values of ψ_t at time t is given by the log-binomial distribution

$$P_t(\psi_{t,k}) = \left(\frac{1}{2}\right)^t \binom{t}{k}, \tag{5.28}$$

where

$$\psi_{t,k} = e^{(-t+2k)\Delta}, \quad k = 0, 1, 2, \ldots, t, \tag{5.29}$$

are all the possible values that ψ_t can take. In other words (5.28) gives the statistical weight of the whole set of trajectories such that $\psi_t = \psi_{t,k}$, with the condition $\psi_0 = 1$.

5.6 Self-Similarity of Multiplicative Processes

The problem we consider now concerns the possible self-similar nature of the fluctuations of this process. There are two kinds of fluctuations that can be considered: (i) *Noise fluctuations* which refer to the different possible values of ψ_t at a fixed time t for different sequences of the random process, i.e. different trajectories. (ii) *Time fluctuations* which refer to the sequence of values of ψ_t at different times t along a single trajectory.

Let us study the first case. In order to study the possible self-similar character of noise fluctuations, we will map all the possible trajectories of ψ_t onto the points x of the real interval $[0, 1)$ in such a way that the process described by (5.26) will correspond one to one to the multiplicative iterations of the fragmentation and redistribution of a measure on this interval identical to the binomial MF process. For this purpose it is useful to look at the tree of all the possible trajectories as shown in Fig. 5.7.

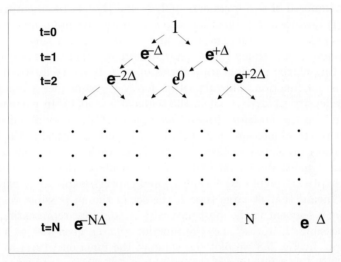

Fig. 5.7. Schematic view of the tree of all the possible trajectories (i.e. realization of the disorder) for the multiplicative process defined by (5.26)–(5.27) up to a time $t = N$. A trajectory is defined by the complete sequence of arrows connecting the point at $t = 0$ to any other one at $t = N$

Our precise goal is to map an entire trajectory from $t = 0$ to $t = N$ (with $N \to \infty$) onto a fragment of the whole interval $x \in [0, 1)$ with a certain length and measure (i.e. weight) in such a way that: (i) the length of the fragment represents the statistical weight of the trajectory, (ii) the assigned measure, uniformly distributed in the fragment, represents the value of ψ_N, and (iii) the spatial disposition of the different fragments at time $t = N$ represents the level $t = N$ of the tree of the trajectories. Since the probability of the noise at each time step is simply given by (5.27), all the trajectories of the same length

N have the same statistical weight $(1/2)^N$. On the other hand the measure, i.e. the value of ψ_N, fluctuates strongly from trajectory to trajectory.

The recipe to introduce such a representation is the following: let us suppose that at time t a trajectory is represented by a fragment $[a,b) \subset [0,1)$ with length $b - a = (1/2)^t$, and uniformly distributed measure ψ_t (i.e. with a constant measure density $\frac{\psi_t}{b-a}$). At the next time step this trajectory splits into two other trajectories depending on whether $\Delta_t = \Delta$ (with related probability $1/2$) or $\Delta_t = -\Delta$ (with the same related probability $1/2$). Therefore the segment $[a,b)$ is fragmented into two parts $\left[a, \frac{a+b}{2}\right)$ and $\left[\frac{a+b}{2}, b\right)$ of common length $(1/2)^{t+1}$ and respective uniformly distributed measures $\exp(-\Delta)\psi_t$ and $\exp(\Delta)\psi_t$. Therefore for $N = 0$, since there is only one trajectory with probability 1 and measure $\psi_0 = 1$, the segment $x \in [0,1)$ will be composed of only one fragment of length 1 (i.e. the whole segment itself) and uniformly distributed measure $\psi_0 = 1$. For $N = 1$ the segment $x \in [0,1)$ will be composed of two fragments $[0, 1/2)$ and $[1/2, 1)$ representing the two possible trajectories after one time step with respective uniformly distributed measures $\exp(-\Delta)\psi_0$ and $\exp(\Delta)\psi_0$, and so on at further time steps.

Therefore at any time t the *measure density function* $\psi(x;t)$ for $x \in [0,1)$, representing all the possible trajectories up to t, is a very fluctuating piecewise constant function. This algorithm for $N \to \infty$ defines a binomial MF measure density $\psi(x; N \to \infty)$ in the segment $x \in [0,1)$. In particular, if at any time step the *measure density function* $\psi(x;t)$ is normalized in such a way that the total measure in $[0,1)$ is 1 it is exactly identical to the binomial MF measure presented in Sect. 5.3 where at each step the measure of a fragment is redistributed into two sub-fragments of the same length with weights $\exp(-\Delta)/[\exp(-\Delta) + \exp(\Delta)]$ and $\exp(\Delta)/[\exp(-\Delta) + \exp(\Delta)]$.

This means that at large time t, the set of the all possible values of ψ_t settled on a segment in the most natural way, and best representing the tree of the possible "histories", i.e. the function $\psi(x;t)$, gives rise to a complex MF shape. In Fig. 5.8 we plot the value of the function $\psi(x;t)$ at different $t = 0, 1, 2, 3, 4$ as a function of the variable $x \in [0,1)$.

Up to now we have seen how to map the multiplicative random walk in a MF fragmentation process, and in particular in a binomial MF measure. An important point consists in finding an explicit formula to associate any trajectory to the left extreme of the fragment representing it. To this aim we introduce the characteristic function:

$$C_t = \begin{cases} 1 & \text{if } \Delta_t = \Delta \\ 0 & \text{if } \Delta_t = -\Delta \end{cases} . \tag{5.30}$$

Therefore given the trajectory $\mathcal{C}_t = \{\psi_0, \psi_1, \ldots, \psi_t\}$, the left extreme of the related fragment will be:

$$x(\mathcal{C}_t) = \sum_{t'=1}^{t} C_{t'} \left(\frac{1}{2}\right)^{t'}, \tag{5.31}$$

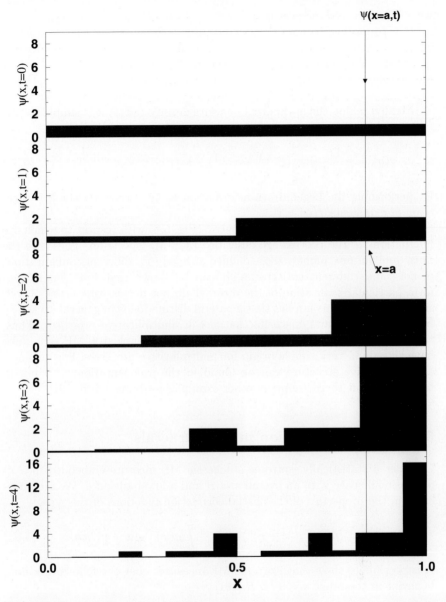

Fig. 5.8. Representation of the multiplicative process with $\Delta_0 = \log_e(2)$ as a series of non-normalized self-similar iterations for a distribution in the unit interval. A crucial point for this representation is the mapping of all possible trajectories (i.e. realizations of the disorder) onto the unit interval. This representation makes clear the MF nature of the disorder fluctuations

where the sum is done over the C'_t related to the chosen trajectory.

It is important to note that at any time t the measure density function $\psi(x;t)$ contains much more information that the simple log-binomial probability distribution given by (5.28) and (5.29). In fact it explains also how fluctuations between different trajectories of the same temporal length are built.

Finally, taking the normalized measure density $\psi(x;t)$ it is simple to perform the MF analysis giving that

$$\tau(q) = \frac{1}{\log_e 2} \left[\log_e(e^{-q\Delta} + e^{-q\bar{\Delta}}) - \log_e(e^{-q\Delta} - e^{-q\bar{\Delta}}) \right] . \tag{5.32}$$

By performing the Legendre transformation we may then get the MF spectrum of exponents $f(\alpha)$.

We have therefore identified a metric (Fig. 5.8) with respect to which the fluctuations of the process are described by a MF spectrum. These are the *noise fluctuations*, namely those defined at fixed t for different configurations of noise, i.e. different trajectories. It can be shown that *time fluctuations* for a particular realization of the disorder do not possess properties of self-similarity [191]. This implies that a careful distinction is in general necessary between disorder and space fluctuations in multiplicative processes. Thus, this example shows that, despite their apparent complexity, MF structures have actually a very simple origin for multiplicative processes. For instance, a multiplicative structure can be found in the fragmentation of eddies in turbulence, but there are many other examples as discussed in [191].

5.7 Spatial Correlation in Multifractals

Consider a statistically isotropic stochastic MF measure embedded in a d-dimensional space with an overall size R and a lower cut-off a. We are interested in the properties of the spatial correlation functions of the type

$$C_{mn}(r) = \langle \mu(x)^m \mu(x+r)^n \rangle \equiv R^{-d} \int \mu(x)^m \mu(x+r)^n dx . \tag{5.33}$$

It is possible to show that isotropic MF measures obey the following scaling behavior in the range $a \ll r \ll R$ [191]

$$C_{mn}(r) \sim \left(\frac{R}{a}\right)^y \left(\frac{r}{a}\right)^z$$

$$y = -\tau(m+n) - d \tag{5.34}$$

$$z = \tau(m+n) - \tau(m) - \tau(n) - d ,$$

where $\tau(m) \equiv (m-1)D(m)$ as defined in Sect. 5.2. As a direct result we find that the integrated correlation function scales with r as

$$I_{mn}(r) \equiv \int_0^r C_{mn}(y) y^{d-1} dy \sim r^w$$
$$w = z + d = \tau(m+n) - \tau(m) - \tau(n) . \tag{5.35}$$

So, if the scaling behavior of the moments represented by the $\tau(q)$ function is known, then one can calculate easily the scaling behavior of the correlation functions by using (5.35).

Let us consider two special cases. First if we set $m = n = 0$ then the correlations of the geometry of the support of the measure are probed and (5.35) reduces to the mass-length relation used to measure the fractal dimension $D(0)$ of ramified structures

$$I_{00}(r) = \left\langle \int_0^r \rho(\boldsymbol{x})\rho(\boldsymbol{x}+\boldsymbol{y}) y^{d-1} dy \right\rangle \propto r^{D(0)} , \tag{5.36}$$

where $\rho(\boldsymbol{x})$ is the microscopic density of points of the support of the measure. Secondly if we choose $m = n = 1$ then the correlation of the measure itself is probed and (5.35) gives

$$I_{11}(r) = \left\langle \int_0^r \mu(\boldsymbol{x})\mu(\boldsymbol{x}+\boldsymbol{y}) y^{d-1} dy \right\rangle \propto r^{D(2)} . \tag{5.37}$$

with $D(2)$ being the correlation dimension [191].

5.8 Multifractals and "Mass" Distributions

In this section we show how to apply the MF formalism and analysis to distributions of point-particles carrying different masses (measures). The application we have in mind, which is discussed in Chap. 11, is to the distribution of galaxies and their observed luminosity (related to the mass). As explained in Chap. 4, if one takes the masses of the particles all equal (say unitary), one has a simple set of points that we call the *support* of the measure distribution. The question of the self-similarity versus homogeneity of a set of points with equal mass can be exhaustively discussed in terms of the single correlation exponent which corresponds to the fractal dimension of the support of the measure distribution. Questions about multifractality and a complex spectrum of dimensions instead become interesting and physically relevant when the different masses are included and the entire matter distribution considered [47, 190]. In the case of galaxies the measure density is defined (see Chap. 11) by assigning to each galaxy a weight proportional to its mass.

Suppose that the total volume of the sample consists of a three-dimensional cube of size L. The microscopic mass density can be written as usual

$$\rho(\boldsymbol{r}) = \sum_{i=1}^{N} m_i \delta(\boldsymbol{r} - \boldsymbol{r}_i) , \tag{5.38}$$

where m_i is the mass of the ith galaxy and N is the number of points in the sample of volume $V = L^3$. We assume that this distribution corresponds to a measure defined on a set of points (the support) characterized by the average conditional density $\Gamma(r) \sim r^{D-3}$ (Chap. 4). It is possible to define the dimension-less normalized density function

$$\mu(\boldsymbol{r}) = \sum_{i=1}^{N} \mu_i \delta(\boldsymbol{r} - \boldsymbol{r}_i) \tag{5.39}$$

with $\mu_i = m_i/M_T$ and $M_T = \sum_{i=1}^{N} m_i$, the total mass in the sample. We divide this volume into boxes of linear size l. We label each box by the index i and construct for each box the function

$$\mu_i(\epsilon) = \int_{ith\,box} \mu(\boldsymbol{r}) d^3 r \tag{5.40}$$

where $\epsilon = l/L$ and $0 < \mu_i < 1$.

Suppose at this point that we have a MF measure distribution in the volume V, and we want to study the behavior of the number of boxes with measure in the range μ to $\mu + d\mu$, having fixed the partitioning of the measure with boxes of size ϵ. Changing variables and using (5.1), the measure distribution (5.11) becomes

$$N_\epsilon(\mu) d\mu \sim \epsilon^{-f(\alpha(\mu))} \frac{1}{|\log(\epsilon)|} \frac{d\mu}{\mu}. \tag{5.41}$$

From this equation it follows that the distribution of the measure, at fixed resolution ϵ, does not scale as a power-law in μ, because the exponent $f(\alpha(\mu))$ is a complex function of μ. The self-similarity of the distribution is recovered by looking at the measure distribution as a function of the scale ϵ.

Suppose we fix the box size at the scale ϵ: in the application to galaxies this can be taken to be the galactic scale, or the galaxy cluster scale. The function $N_\epsilon(\mu)$ is bell-shaped and convex with a maximum corresponding to the point at which

$$\frac{\partial N_\epsilon(\alpha)}{\partial \mu} = \frac{\partial N_\epsilon(\alpha)}{\partial \alpha} \frac{\partial \alpha}{\partial \mu} = -(f'(\alpha) + 1) N_\epsilon(\alpha) \frac{1}{\mu} = 0, \tag{5.42}$$

which corresponds to

$$\left(\frac{\partial f(\alpha)}{\partial \alpha} \right)_{\alpha_c} = -1. \tag{5.43}$$

The maximum of $N_\epsilon(\mu)$ fixes the most probable value of μ. Beyond this maximum the function can be well fitted by a power-law[1]. For still higher

---

[1] In practice this is the only observable part of the measure distribution in the case of galaxies because the higher values of α correspond to the smallest galaxies: in any galaxy sample there is a lower cut-off on the intrinsic luminosity of the objects.

values of μ the function shows an exponential-like decay. The tail is fixed by the point at which the derivative (5.43) has a maximum. This happens for $\alpha = \alpha_{min}$, namely at the value corresponding to the box which contains the maximum measure (i.e. the strongest singularity)

$$\mu^* \sim \epsilon^{\alpha_{min}} . \tag{5.44}$$

In order to compute the exponent characterizing the leading power-law behavior we study the derivative of $\log(N_\epsilon(\mu))$ with respect to $\log(\mu)$. We obtain

$$\frac{\partial \log(N_\epsilon(\mu))}{\partial \log(\mu)} = -\left(\frac{\partial f(\alpha)}{\partial \alpha} + 1\right) . \tag{5.45}$$

We can try to fit (5.41) with a power-law function of μ, plus an exponential tail. From (5.45) we can define an effective exponent δ, which depends explicitly on μ. This implies that the power-law approximation can be considered as a local fit

$$\delta = -\left(\frac{\partial f(\alpha)}{\partial \alpha} + 1\right) . \tag{5.46}$$

This leads to $\delta = 0$ for $\alpha = \alpha_c$ and $\delta = -1$ for α_0 such that $f(\alpha_0) = D(0)$. Locally we can expand $f'(\alpha)$ in a power series of μ, so that the measure distribution of (5.42) in a certain range of μ is well fitted by a power-law function with a cut-off

$$N(\mu) \sim \mu^\delta e^{-\frac{\mu}{\mu^*}} . \tag{5.47}$$

The exponent δ depends on the shape of the derivative of $f(\alpha)$, as well as on the value of α around which one expands $f(\alpha)$. In various cases the value of δ is in the range $[-2, -1]$ [220].

5.9 Summary and Discussion

In conclusion we can distinguish between three different scenarios for a measure distribution as illustrated by Figs. 5.9–5.11.

- *A non-fractal measure.* Any analytic measure gives rise to a non-fractal measure. As a result one finds $D(q) = d$, i.e. generalized dimensions are equal to the embedding space dimension of the measure (see Fig. 5.9). The $f(\alpha)$-spectrum reduces to one point: each point of the measure has the same trivial singularity exponent. A simple example is the following: A uniform distribution on the unit interval. This corresponds to a generalized Cantor set (5.8) with a generator where all the boxes have the same weight, $\mu_i = b^{-1}$ for $i = 1, \ldots, b$.
- *An homogeneous fractal measure.* In this case an homogeneous measure is defined on a fractal set with dimension D (see Fig. 5.10). The measure is homogeneous in the sense that the scaling properties do not change as a

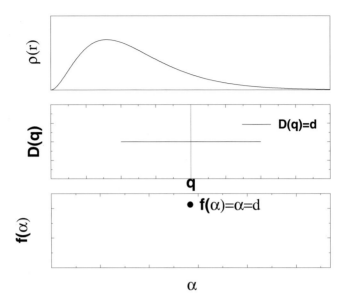

Fig. 5.9. Characteristics of a non-fractal homogeneous measure. In the *upper panel* is shown the spatial behavior of the measure (or density), in the *center panel* the $D(q)$ exponents and in the *lower panel* the $f(\alpha)$-spectrum

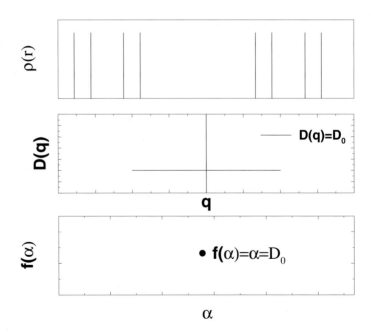

Fig. 5.10. As in the previous figure but for an homogeneous fractal measure

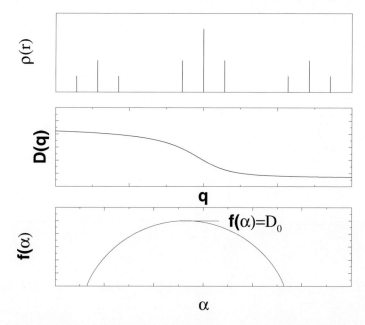

Fig. 5.11. As in the previous two figures, but for a MF measure

function of position. As a result $D(q) = D = D_0$ and $f(\alpha) = \alpha = D = D_0$. A simple example is a Cantor set (5.8) with a generator where the measure of $n(< b)$ boxes take the same non-zero value and where the remaining $b^d - n$ boxes are omitted, $\mu_i = 1/n$ for $i = 1, \ldots, n$.
- *A multifractal measure.* Here we have an inhomogeneous measure in the sense that the scaling properties of the measure vary as a function of position (see Fig. 5.11). As a result one typically finds a $D(q)$-curve that is varying with q and a $f(\alpha)$-curve with a parabolic shape. A simple example is given by a generalized Cantor (or random Cantor) set (5.8) with a generator where the boxes have different weights μ_i.

Part II

Applications to Cosmology

6 Fluctuations in Standard Cosmological Models: A Real Space View

6.1 Introduction

We start our foray into cosmology with a look at which instruments are used to describe fluctuations in cosmology, and some of their properties. As we will see the formalism used is essentially that described in the first two chapters of part I of this book: stationary stochastic processes with fluctuations about a well defined and positive mean density. We focus in particular on the fact that at large scales the correlations described in standard cosmological models fall into the very special category of super-homogeneous systems, which we discussed in Chap. 3 where we gave a general classification of all stationary stochastic mass density fields. This leads us in particular to emphasize the description of these perturbations in real space, which gives a different perspective to that usual in cosmology, where the k-space description is very dominant. In the latter part of the chapter we briefly describe also the real space features of the "acoustic" oscillations observed in recent years in the cosmic microwave background radiation (CMBR).

We do not enter here into a detailed discussion of the physical theories explaining the origin and evolution of the correlations studied, only very briefly sketching, for example, the origin of the "scale-invariant" spectrum of primordial fluctuations. A detailed treatment of these aspects can be found in many standard textbooks on cosmology (see e.g., [170, 185]). Here our focus is on the basic characteristics of these density fields from a statistical point of view.

6.2 Basic Properties of Cosmological Density Fields

In standard theories of structure formation in cosmology the density field in the early universe is described as a homogeneous (spatially uniform) and isotropic matter distribution, with fluctuations characterized by specific correlation properties (e.g., [185]). We have discussed in Chap. 2 the main characteristics of this kind of stationary stochastic process. These fluctuations are believed to be the initial seeds from which, through a complex dynamical evolution, galaxies and galaxy structures have emerged. In particular the initial fluctuations are taken to have Gaussian statistics and a spectrum which is exactly, or very close to, the so-called *Harrison-Zeldovich* (HZ) [109, 252] or "scale-invariant"[1] power spectrum (PS). Since the fluctuations are Gaussian, the knowledge of the PS, or its Fourier transform (FT), the real space two-point correlation function, gives a complete statistical description of the fluctuation field. The HZ type spectrum was first given a special importance in cosmology with arguments for its "naturalness" as an initial condition for fluctuations in the framework of the expanding universe cosmology. It is in this context that the use of the term "scale-invariant" to designate it can be understood. It subsequently gained in importance with the advent of inflationary models in the eighties, and the demonstration that such models quite generically predict a spectrum of fluctuations of this type. Since the early nineties, when the COBE experiment [27] measured for the first time fluctuations in the temperature in the CMBR at large angular separations, and found results consistent with the predictions of models with a HZ spectrum at such scales, the HZ type spectra have become a central pillar of standard models of structure formation in the universe.

The HZ spectrum arises in cosmology through a particular condition applied to perturbations of Friedmann-Robertson-Walker (FRW) models, which describe a uniform universe in expansion [91]. This condition – commonly referred to in cosmology as "scale-invariance" of the perturbations – gives rise to a spectrum (commonly called the "scale-invariant" perturbation spectrum) with $P(k) \sim k$ at small k. All current standard cosmological models of structure formation in the universe assume a spectrum exactly like this, or close to it, as initial condition for perturbations in the universe. In such models there is, at any time, a finite scale corresponding to the causal horizon, which increases with time, and below which causal physics can act to modify the spectrum. This causal physics depends, in general, on the details of the model, i.e., on the nature of its content in matter and radiation (or other forms of energy), until a characteristic time (the time when matter and radiation have comparable densities), after which purely gravitational evolution takes over. There are many variants on standard cosmological models, e.g., cold dark matter (CDM), the currently favored one with a non zero cosmological constant (ΛCDM) or the mixed dark matter (MDM), each of them leading to a different form for the spectrum at smaller scales (i.e., large k) which can be calculated. In CDM models (in which the predominant massive component driving collapse under gravity is cold dark matter, and where "cold" refers to the fact that the particles have little initial velocity dispersion) the PS decays at small scales (large k) as a power-law in k with negative exponent, while in hot dark matter (HDM) models (for which the prototype is a universe dominated by light neutrinos) there is an exponential cut-off in the spectrum (due essentially to the fact that the "hot" neutrinos wipe out structures at these

---

[1] We use "scale-invariance" following the terminology used in the cosmological literature. As we discuss below the term as used in this way has no relation to the concept of scale-invariance in statistical physics (and as described in Chap. 2).

6.2 Basic Properties of Cosmological Density Fields

scales with their large velocity dispersion). All of these models, however, have the same "primordial" HZ spectrum $P(k) \sim k$ (or very close to it) on large scales (i.e., small k), that is at scales which are large compared to the causal scale at the time of matter-radiation equality. This latter scale is of course much smaller than our present causal horizon (i.e., than the part of the universe we can probe today). This means, in particular, that these primordial density correlations should be imprinted in the distribution of matter at very large scales, and should in principle be detectable in the distribution of galaxies at very large scales, inside the present horizon. Until now the only probe of the behavior of density fluctuations on such scales is through the temperature variations in angle of the CMBR, as the angular correlations in temperature fluctuations are coupled directly to the three dimensional density fluctuations (see discussion in Sect. 6.6.1) [170]. From the COBE measurements [27] the amplitude of the relative fluctuations is inferred to be $\sim 10^{-5}$ at these scales (see also [26, 58, 107]).

Discussions of real space properties of the density fluctuations encountered in cosmology are extremely sparse in the literature on the subject. In [185] it is noted that a very particular characteristic of HZ models is that "on large scales the fluctuations have to be anti-correlated to suppress the root mean square mass contrast on the scale of the Hubble length". Indeed, we will emphasize below the fact that these models are characterized at large scales by a correlation function $\xi(r)$ which has a negative power-law tail: detecting it would be the real space equivalent of finding the turnover to HZ behavior $P(k) \sim k$. The preference for a k-space description in the cosmological literature is probably rooted in the fact that the linear dynamics, which are used to describe many problems in cosmology, are most naturally treated in this space. While it is true of course that this description in k-space is complete, this by no means implies that the complementary real space view is redundant, as is well known in many contexts in physics. One of the points of this chapter is to show that this complementary view of these models is at the very least interesting and, as we see in the next chapter, potentially useful.

A basic question we try to answer is the following: What "kind" of two-point correlation function is the one corresponding to the HZ behavior in cosmological models ? We compare it to some different statistical homogeneous and isotropic systems (see Chap. 3): (i) substantially Poisson distributions, (ii) super-Poisson i.e., systems with a power-law and mainly positive correlation function as in thermodynamical critical phenomena [148], and (iii) both stochastic and deterministic particle distributions characterized by a sort of long-range order (e.g., lattice or glass-like) [91, 92]. Through this comparison we can classify HZ models in the third category. As discussed in Chap. 3, we introduce the term super-homogeneous (or hyper-uniform [231]) to refer to this kind of distribution, as their primary characteristic is that mass density fluctuations decay at large scales faster than in a completely uncorrelated

(Poisson) stochastic mass density field. For critical systems one has instead a decay of the normalized mass variance which is slower than Poisson.

Formally the definition of this class of super-homogeneous distributions is given by having a PS which satisfies the condition $P(0) = 0$, or equivalently in real space that the integral of the two-point correlation function $\tilde{\xi}(r)$ over all space is zero. In the cosmological literature the latter property of cosmological models is often noted, but its meaning (as a strong *non-local* ordering condition on a stochastic process) is not completely appreciated, or worse misunderstood as a trivial boundary condition applying to any correlated system. In the textbook [170], for example, it is "proved" that the integral over all space of the correlation function $\tilde{\xi}(r)$ vanishes independently of its functional behavior. The error is in an implicit assumption made that the number of particles in a large volume in a single realization converges *exactly* to the ensemble average. This is not true because, in general, extensive quantities, such as particle number in a volume, have fluctuations which are increasing functions of the volume (e.g., Poisson, for which the integral is not zero). A slightly different, but common, kind of misunderstanding of the meaning of the vanishing of the integral of the correlation function is evidenced in [130]. There it is affirmed to be "...just a statement of mass conservation: if galaxies are clustered on small scales, then on large scale they must be "anti-clustered" to conserve the total amount of mass (number of galaxies)". This is a too deterministic interpretation of the condition of super-homogeneity. In general, if the distribution of mass is described by a stochastic process, there is no finite volume for which the mass is constant: in fact, as shown in Chap. 3, fluctuations of extensive quantities like the mass diverge at least as fast as the surface of the volume in which they are calculated[2]. Only fluctuations of *intensive* quantities like the mass density vanish at large scales.

The source of confusion in the cosmological literature about this point seems to be the so-called "integral constraint" in data analysis (see Sect. 6.5) which imposes such a condition on the *estimator* of the correlation function in a *finite* sample, due to the fact that the (unknown) average number of points in such a sample is estimated by the (exactly known) number of points in the actual sample. Despite their apparent similarity, these are different conditions: the first (infinite volume) integral constraint provides non-trivial physical information about the intrinsic probabilistic nature of fluctuations, while the second is just an artifact of the boundary conditions which holds, for a statistical estimator, in a finite sample independently of the nature of the underlying correlations and independently of the size of the sample. We will discuss this point in Sect. 6.5 below.

---

[2] Note that it is true that a stochastic fluctuation field constructed starting from a completely uniform mass density, by a dynamics which conserves mass "locally", is necessarily super-homogeneous. This does not imply that the distribution of galaxies, for example, must be such a distribution.

6.3 The Cosmological Origin of the HZ Spectrum

We describe firstly the physical argument[3] that singles out the HZ spectrum in cosmology, and why the term "scale-invariant" is applied to it. In a homogeneous FRW cosmology there is only one fundamental characteristic length scale, the horizon scale $R_H(t)$. It is simply the distance light can travel from the Big Bang singularity at $t = 0$ until any given time t in the evolution of the universe. The HZ criterion can be written

$$\sigma_M^2(R = R_H(t)) = \text{constant}, \tag{6.1}$$

i.e, it requires that *the relative normalized mass variance at the horizon scale be constant*. Equivalently, given the proportionality of gravitational potential to mass, it can be stated as the constraint that the normalized variance in the gravitational potential be constant at the horizon scale. If we take any other prescription than (6.1) (e.g., a Poisson distribution), such a description will always break down in the past or future, as the amplitude of the perturbations become arbitrarily large or small. It is in this specific sense that the resulting PS is said to be "scale-invariant": there is no characteristic scale at which fluctuations at the horizon scale become large (or small), or put another way, they have the same amplitude as a function of the only scale in the model. As discussed in Chaps. 2–3, it has nothing to do with the same term as understood in statistical physics. There scale-invariance is a characterization not of the amplitude of fluctuations, but rather is associated to a particular range of power-law behaviors in the correlation function.

More precisely the form of the HZ spectrum is arrived at from the condition (6.1) in the following way. We move necessarily to a k-space description, as we need to include the dynamical evolution of the density field to infer the PS inside the horizon today. Let $\delta_k(t)$ be the amplitude of the Fourier component of the density contrast

$$\delta_\rho(\mathbf{r}) = \frac{\rho(\mathbf{r}) - \rho_0}{\rho_0}$$

(where ρ_0 is the positive average density) as a function of time. To every such mode k we can associate a time t_c at which it "enters the horizon", i.e., at which the wavelength $2\pi k^{-1}$ is equal in size to the horizon. Here we work (as almost always in cosmology) with a k which is the FT with respect to the spatial coordinates which do not change with the expansion, the so-called "comoving" coordinates. In these coordinates the time at which the mode enters the horizon is given by $k\eta = 1$ where η is the so-called "conformal" time

---

[3] We choose here a particular (but commonly used) way of describing the HZ spectrum which allows us to avoid too much extra formalism [91]. For a commonly used formulation preferred by many cosmologists, in terms of a constant "gauge independent" potential, see for example [144].

given by $\eta = \int dt/a(t)$, with $a(t)$ the scale factor describing the expansion of all physical scales in the universe. (The horizon scale is simply $R_H(t) = a(t)\eta$, corresponding to horizon crossing criterion $(k/a)R_H(t) = 1$.) The PS today (at $t = t_o$, say) given by $|\delta_k(t_o)|^2$ can be written in term of the amplitude of each mode k when it entered the horizon. In linear perturbation theory, in the matter dominated universe (i.e., recent epochs), the mode evolves as

$$\delta_k(t_o) = \left(\frac{a(t_o)}{a(t)}\right) \delta_k(t) . \tag{6.2}$$

In the matter dominated FRW cosmology we have $a \propto t^{2/3}$ and thus $\eta \propto t^{1/3}$, so that the time $t_c(k)$ when the mode k crosses the horizon follows $t_c(k) \propto 1/k^3$ and therefore

$$\delta_k(t_o) \propto k^2 \delta_k(t_c) . \tag{6.3}$$

The HZ choice for the primordial PS $|\delta_k(t_o)|^2 \propto k$ is then singled out by imposing the criterion

$$k^3 |\delta_k(t_c)|^2 = \text{constant} , \tag{6.4}$$

which is identified as the normalized mass variance at the horizon scale $\eta = k^{-1}$ [170]. We note immediately, following the discussion in Chap. 3, that the latter identification between $k^3|\delta_k(t_c)|^2$ and the normalized mass variance is in fact valid only for power spectra k^n with $n < 1$. Strictly speaking therefore it is impossible to satisfy the HZ criterion as it is understood naively; or, to put it another way, the HZ spectrum, that which satisfies (6.4), does not satisfy the condition of exact "scale-invariance" since the mass variance at the horizon scale ($\propto \eta$) is dominated in this case by the power at the cut-off scale, not by the modes $k \sim \eta^{-1}$. Taking a spectrum $k^{1-\epsilon}$ ($\epsilon > 0$) one can get arbitrarily close to satisfying the HZ criterion, but the condition of "scale-invariance" (in the sense just explained) is not physically satisfiable (for $n = 1$ there are logarithmic corrections). To avoid this conclusion the criterion could be refined to be that the mass variance in Gaussian spheres of radius of the horizon size be constant in time. While it does allow a mathematically coherent formulation, from a physical point of view it is an artificial way of avoiding the problem, which is that the variance at a given real space scale has nothing to do in principle with the amplitude of the PS at the inverse scale for $n \geq 1$. This is, as we have discussed in Chap. 3, a real physical property of such systems, not a mathematical artifact.

The HZ spectrum can equivalently be characterized in term of fluctuations in the gravitational potential, $\delta\phi(\mathbf{r})$, which are linked to the density fluctuations $\delta\rho(\mathbf{r})$ via the gravitational Poisson equation [4]:

---

[4] We simplify here to Newtonian gravity, which becomes a good approximation on sub-horizon scales. The comments given below can however be generalized to a rigorous formulation of perturbations in a FRW model.

$$\nabla^2 \delta\phi(\boldsymbol{r}) = -4\pi G \delta\rho(\boldsymbol{r}) \ . \tag{6.5}$$

From this, transformed to Fourier space, it follows that the PS of the potential $P_\phi(k) = \left\langle |\delta\hat{\phi}(\boldsymbol{k})|^2 \right\rangle$ is related to the density PS $P(k)$ as:

$$P_\phi(k) \sim \frac{P(k)}{k^4} \ .$$

The HZ spectrum corresponds therefore to $P_\phi(k) \propto k^{-3}$; or, considering the variance in real space spheres of the gravitational potential fluctuations, which as for the density fluctuations is related to the PS, one finds that this variance is constant as a function of R. This is the alternative form in which the HZ condition is often formulated. Note that the Wiener-Khinchin theorem (see Chaps. 2–3) requires that a well defined SSP $\delta\phi(\boldsymbol{r})$ has $P_\phi(k) \sim k^a$ with $a > -3$ for $k \to 0$, so that the HZ condition corresponds to the limiting (disallowed) behavior. Equivalently the exact constancy of the variance of the gravitational potential is in contradiction with (2.30) which requires that the asymptotic variance be zero (in order to have a well defined mean about which fluctuations are defined). The HZ spectrum can thus be seen as the limiting behavior for the potential fluctuations to be treatable as an SSP. That such a treatment is applicable to the potential fluctuations is however not a physical requirement for what concerns the gravitational field. The work of Chandrasekhar [43] (see Chap. 14) treats the gravitational force probability distribution in a point set and, in particular, shows it to be well defined even in the Poisson case, for which the potential fluctuations are not a well defined SSP ($n = 0$). To treat the force field as an SSP requires only the weaker condition $P(k) \sim k^n$ with $n > -1$ for the mass density fluctuation field [91].

6.4 The Real Space Correlation Function of CDM/HDM Models

All of the current "viable" standard type cosmological models have a "primordial" PS which is the HZ one (or very close to it) down to some arbitrarily small scale (i.e., large k). During cosmological evolution causal physics modifies this spectrum at large k, corresponding roughly to the scales within the causal horizon at that time. Around the time at which the matter in the universe (with density scaling as $1/a^3$) begins to dominate over the radiation (with density scaling as $1/a^4$), the evolution becomes purely gravitational at all but the very smallest scales, while prior to this time it depends strongly on the details of the particular model. As a result all such models are HZ for $k < k_{eq}$, but "turn-over" at this scale to a PS decreasing as a function of k. The form of the spectrum in this region depends on the details of the particular model. Since the scale k_{eq}^{-1}, being the size of the causal horizon at the time of matter-radiation equality, is much smaller than the causal horizon today, the primordial HZ PS is in principle detectable today. Indirect evidence for its reality come from the measurements of temperature fluctuations in the CMBR, which show a dependence on angular scale quite consistent with the HZ spectrum, with the power-law spectrum $P(k) \sim k^n$ giving a fit in the range $n = 1.1 \pm 0.5$ [27]. The search to observe this "turn-over" to HZ behavior directly in three dimensions in the distribution of matter at large scales – a central prediction and check on such models – has so far proved elusive, because of weak statistics at large scales in observations of the distribution of galaxies. It is anticipated that forthcoming surveys, now being made [250] or close to completion [180], will sample a range of scale large enough to include the theoretically predicted scale of the turn-over (see the discussion in Chaps. 8–13). Let us now look at the characteristic real space features which should be found in these galaxy surveys if the underlying behavior is HZ.

We consider first the connected two-point correlation function. In general the FT of the PS of standard cosmological models must be done numerically. Before doing so for some standard models we consider a simple PS which can be Fourier transformed analytically, a HZ spectrum with a simple exponential cut-off:

$$P(k) = Ak e^{-\frac{k}{k_c}}, \qquad (6.6)$$

where A is the amplitude and k_c^{-1} the cut-off scale (in the cosmological context $k_c \approx k_{eq}$). The correlation function is exactly given by

$$\tilde{\xi}(r) = \frac{A}{\pi^2} \frac{\left(\frac{3}{k_c^2} - r^2\right)}{\left(\frac{1}{k_c^2} + r^2\right)^3}. \qquad (6.7)$$

For $r \ll r_c \equiv k_c^{-1}$ we have

$$\tilde{\xi}(r) \simeq \tilde{\xi}(0) \simeq \frac{3A}{\pi} k_c^4 > 0,$$

changing at $r \sim r_c$ to an asymptotic behavior $\xi(r) \sim -r^{-4}$. Note that the correlation function does not oscillate, its only zero crossing being at $r = \sqrt{3} r_c$. Simply because of the condition $P(0) = 0$, which implies that the integral of the correlation function must be zero, the correlation function must change sign and in this case it only does so once and thus it remains negative at large scales.

The normalized mass variance $\sigma^2(R)$ shows a corresponding change in behavior from being approximately constant at small scales $R < r_c$ to a $\ln R / R^4$ decay at large scales, as was shown in Chap. 3. Note that, unlike for the normalized mass variance in spheres, there is no limit to the rapidity of the decay of the correlation function. Despite the weakness of this correlation at large scales, however, the variance in spheres does not behave like that of a Poisson system, because of the balance between positive correlations at

6.4 The Real Space Correlation Function of CDM/HDM Models

small scales and negative correlations at large scales imposed by the non-local condition $P(0) = 0$ (see Chap. 3).

In cosmological HDM models the form of the PS is very similar to that we have just considered with an exponential cut-off of the form [170]

$$P(k) \sim k \exp(-k/k_c)^{3/2} . \tag{6.8}$$

A numerical integration verifies that the main features of the two-point correlation function are essentially unchanged.

For CDM models, the class by far favored in the last few years, the form of the PS at k larger than the turn-over from HZ behavior is considerably more complicated. In a linear analysis the PS of the CDM matter density field decays below the turn-over with a power-law $\sim k^{-3/2}$ until a larger k above which it is more rapidly cut-off. Numerical studies of these models designed to include the non-linear evolution bring further modifications, increasing the exponent in the negative power-law regime of the PS at large k. We take here an analytic approximation to the final PS given by [76], and computed numerically the FT to obtain the two-point connected correlation function. We also compute directly the normalized variance $\sigma^2(r)$ in spheres of radius r as defined in Chap. 3. This form of the PS is given in terms of the various cosmological parameters. Here we consider for simplicity the case with the small baryon density set to zero ($\Omega_b = 0$), which gives a PS without the oscillations reportedly detected in recent observations of the CMBR [26, 58, 107]. This structure is not of primary interest to us here because it can modify the correlation function only at small scales (it arises from causal physics at early times) and it will be treated in more detail in Sect. 6.6.2. In Figs. 6.1–6.2 we show respectively the behavior of the PS and of the correlation function for two quite different values of the total matter density of the model $\Omega = 1, 0.2$. Minor differences will result in the case that there is a cosmological constant $\Omega_\Lambda \neq 0$ [76].

Let us now return to the discussion of the mass variance (see Chap. 3 and particularly Sect. 3.4) and take a PS

$$P(k) = A k^n e^{-k/k_c}$$

(where A and k_c are two constants). We consider $n > -3$ and take the cut-off to satisfy the convergence properties of the Wiener-Khinchin theorem. It is easy to check subsequently that the results we derive are not sensitive to the form of this cut-off at large k. As discussed in Chap. 3 we may have three different behaviors. To summarize clearly: For a power-law

$$P(k) \sim k^n \text{ for } k \to 0$$

(with an appropriate cut-off around the wavenumber k_c) the normalized mass variance in real spheres with radius $R \gg k_c$ is given by

1. For $n < 1$, $\sigma^2(R) \sim 1/R^{3+n}$ and the dominant contribution comes from the PS modes at $k \sim R^{-1}$.

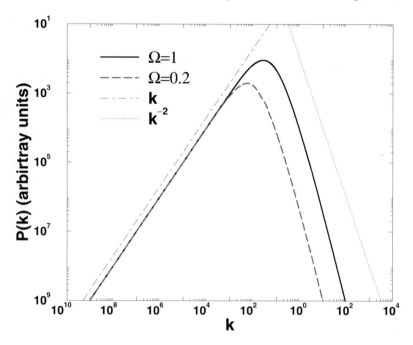

Fig. 6.1. Behavior of the power spectrum for a CDM model with $\Omega = 1, 0.2$ respectively. The two reference lines have exponents k, k^{-2}. (From [91])

2. For $n > 1$, $\sigma^2(R) \sim 1/R^4$ and the dominant contribution comes from the PS modes at k_c^{-1}.
3. For $n = 1$, we have the limiting logarithmic divergence with $\sigma^2(R) \sim (\ln R)/R^4$.

In Fig. 6.3 we show the behavior of the normalized mass variance, computed in real-space spheres. We see again a clear convergence in both models to the predicted $1/R^4$ (up to logarithmic corrections) behavior beyond the scale characterizing the "turn-over" (i.e., at $R \gg k_{eq}^{-1}$).

The two simple real space characteristics of the distribution of matter coming from the primordial HZ PS are thus a *negative non-oscillating power-law tail in the two-point correlation function $\xi(r) \sim -r^{-4}$, and a $(\ln R)/R^4$ decay in the variance of mass in spheres of large radius R*. These are the distinctive features of HZ type spectra in real space [91]. The relation to the galaxy distribution involves however an additional sampling which can change qualitatively these behaviors (see Chap. 13).

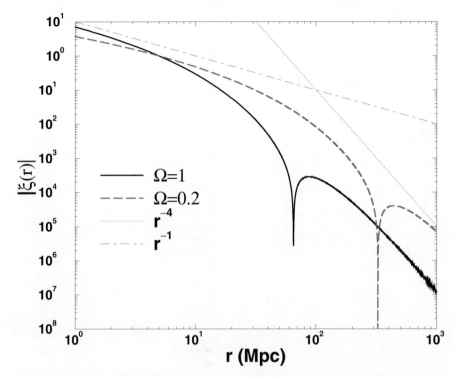

Fig. 6.2. Behavior of absolute value of the reduced two-point correlation function for the two CDM models with $\Omega = 1, 0.2$ and $h = 0.5$. The two reference lines are r^{-4} and r^{-1}. Note that at small scale $\tilde{\xi}(r) > 0$, with a zero crossing at a scale depending on the location of the peak or "turn-over" in the PS, after which it remains negative ($\tilde{\xi}(r) \sim -r^{-4}$) at larger separations. The correlation function has been normalized so that $\tilde{\xi}(r_0) = 1$ for $r_0 = 5$ Mpc. (From [91])

6.5 $P(0) = 0$ and Constraints in a Finite Sample

As we have mentioned in the introduction the physical meaning of the constraint $P(0) = 0$ is often missed in the cosmological literature because of a confusion with the so-called "integral constraint", which is another apparently similar, but actually completely different constraint. Let us clarify this point (see Chap. 9 and Appendix F).

The "integral constraint" refers in this context to a constraint which appears in the estimation of the correlation function in a finite sample (S, say). It can take the form

$$\int_S d^3r \; \tilde{\xi}_E(r) = 0 \tag{6.9}$$

where the subscript indicates that the integral is over the finite sample volume, and $\tilde{\xi}_E(r)$ is the value of the *estimator* of the correlation function

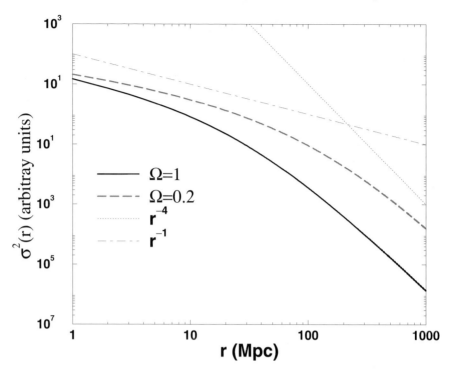

Fig. 6.3. Behavior of the normalized mass variance in spheres for the two CDM models $\Omega = 1, 0.2$ and $h = 0.5$. The two vertical lines show the transition to the $\xi(r) \sim r^{-4}$ behavior for the two models. The r^{-4} behavior is a clear and distinctive feature corresponding to the $P(k) \sim k$ behavior. (From [91])

obtained by the statistical analysis of S. This is in general a quantity calculable from the sample whose ensemble average converges to the real correlation function at any finite scale when the boundaries of the sample go to infinity.

While a condition like (6.9) resembles the super-homogeneity condition $P(0) = 0$ (which imposes that the integral *over all space* of the (real) correlation function is zero), the two are in fact in no way related. This is clear from the fact that (6.9) will be true independently of what kind of underlying distribution the sample is taken from. Its origin is simple: the mean density, relative to which fluctuations are estimated, is estimated in the sample itself, for instance as the ratio of the number of objects in S and its volume. Therefore, roughly speaking, the positive correlations measured relative to this estimated density are constrained to be balanced by anti-correlations at larger scales, giving rise to a constraint like (6.9). More specifically, in the case of a stochastic distribution of identical particles, the (non diagonal part of the) reduced two-point correlation function can be written exactly as

$$\xi(r) = \frac{\langle \rho(\mathbf{r}) \rangle_p}{\rho_0} - 1 \,, \tag{6.10}$$

where $\rho(\mathbf{r})$ is the microscopic density given by (2.1), $\langle \rho(\mathbf{r}) \rangle_p$ is the conditional average density at distance r from an occupied point and $\rho_0 > 0$ the ensemble average unconditional mean density. Integrating this expression over the volume of the sample S gives the relation

$$\langle N_S \rangle_p - \langle N_S \rangle = \rho_0 \int_S d^3r \; \xi(r) \tag{6.11}$$

where $\langle N_S \rangle_p$ is the average number of points (e.g., galaxies) in the sample volume, with a point at the origin by construction, while $\langle N_S \rangle$ is the average number of points in the same volume, but without the condition that there is a point at the location of the observer. Now in a finite and spherical sample the best estimate of $\langle N_S \rangle_p$ and $\langle N_S \rangle$ is simply given, for many choices of the estimators (see Chap. 9 and Appendix F), in both cases by the actual number of points N_S in the given sample. This implies that, independently of the shape of $\xi(r)$, its estimate $\xi_E(r)$ in the same finite sample has to satisfy

$$\int_S d^3r \; \langle \xi_E(r) \rangle = 0. \tag{6.12}$$

In summary there are necessarily constraints on the estimator of the two-point correlation function $\tilde{\xi}_E(r)$ measured in a finite sample, which may take a form similar to the condition $P(0) = 0$ defining super-homogeneous distributions, but over a finite integration volume. These two kinds of constraint have a completely different origin and meaning, one ($P(0) = 0$) describing an intrinsic property of the fluctuation field in a well-defined class of distributions, the other (equation (6.9)) a property of the estimated correlation function of any distribution as measured in a finite sample. Their formal resemblance however is not completely without meaning and can be understood as follows: in a super-homogeneous distribution the fluctuations between samples are extremely suppressed, being smaller than Poisson fluctuations; in a finite sample a similar behavior is artificially imposed since one suppresses fluctuations at the scale of the sample by construction by measuring fluctuations only with respect to the "local" sample average density.

6.6 CMBR Anisotropies in Direct Space

One of the primary ways in which the spectrum we have just discussed is probed is through the associated anisotropies in the CMBR. In this section we briefly discuss these anisotropies, explaining how the "scale-invariant" spectrum we have been examining is linked to these observations. Further, in the spirit of our emphasis on a real-space view of the correlations in these models complementary to the usual reciprocal space one, we discuss how the "acoustic" peak structure in the CMBR anisotropies gives rise to a very particular feature of the correlations in angular space.

6.6.1 CMBR Anisotropies and the Matter Power Spectrum

The anisotropies of the CMBR represent one of the most important observations in modern cosmology. Since they were first detected by the COBE [27] experiment in 1992, dramatic progress has been made in this domain with a whole host of experiments probing them from large to small angular scales (for a review see [115]). In 2003 the WMAP experiment [26] – the first satellite experiment since COBE, giving full sky coverage with a calibration accurate better than 0.5% and with a signal-to-noise ratio larger than one up to $l = 658$ (i.e., about ~ 0.3 degrees) – confirmed the picture which had emerged since COBE of these anisotropies.

In standard cosmological theories the CMBR represents a bridge between the very early universe and the universe as we observe it today (and in particular the galaxy structures which we will focus on in some of the subsequent chapters). On the one hand the CMBR probes the very early hot universe at extreme energies and the theories proposed – notably "inflation" – to explain the origin of these perturbations. On the other hand the anisotropies reflect the local very small amplitude perturbations which give the initial conditions for the gravitational dynamics which should subsequently generate the galaxy structures observed today.

In the CMBR one measures fluctuations in temperature on the sky i.e., on the celestial sphere. We will not enter here into the detail of the physical theory in standard models which link these temperature fluctuations to the mass density field [170]. It is useful however for what follows to give the precise relation between the two quantities (we follow the notation of [170], where one can find a more detailed description of its derivation). The temperature fluctuation field

$$\frac{T(\theta,\phi) - \langle T \rangle}{\langle T \rangle} \equiv \frac{\delta T}{T}(\theta,\phi) ,$$

where θ, ϕ are the two angular coordinates, is conventionally decomposed in spherical harmonics on the sphere:

$$\frac{\delta T}{T}(\theta,\phi) = \sum_{l=0}^{\infty} \sum_{m=-l}^{+l} a_{lm} Y_{lm}(\theta,\phi) . \qquad (6.13)$$

The variance of these coefficients a_{lm} is then related to the matter PS through

$$C_l \equiv \langle |a_{lm}|^2 \rangle = \frac{H_0^4}{2\pi} \int_0^\infty dk \frac{P(k)}{k^2} |j_l(k\eta)|^2 \qquad (6.14)$$

where j_l is the spherical Bessel function and $\eta \simeq 2H_0^{-1}$ is a constant at fixed time (H_0 is the Hubble constant today). Note that the ensemble average contains no dependence on m because of the assumption of statistical isotropy.

With these definitions it is possible to show that the angular space correlation function is given in terms of the C_l by

$$C(\theta) \equiv \left\langle \frac{\delta T}{T}(\theta_1, \phi_1) \frac{\delta T}{T}(\theta_2, \phi_2) \right\rangle = \frac{1}{4\pi} \sum_{l=0}^{\infty} (2l+1) C_l P_l(\cos(\theta)) \quad (6.15)$$

where $P_l(\cos(\theta))$ are the Legendre polynomials, and $\theta = \theta_2 - \theta_1$.

Taking $P(k) = Ak^n$ in (6.14) we get for $n < 3$,

$$C_l = \frac{A H_0^{n+3}}{16} \frac{\Gamma(3-n)}{\Gamma^2[(4-n)/2]} \frac{\Gamma(2l+n-1)}{\Gamma[(2l+5-n)/2]} . \quad (6.16)$$

For $n = 1, l = 2$ one obtains

$$C_2 = \frac{A H_0^4}{24} \quad (6.17)$$

which is just the quadrupole moment. By measuring this one can probe directly the amplitude of the matter power spectrum A. The other multi-poles are given by

$$C_l = \frac{6C_2}{l(l+1)} , \quad (6.18)$$

so that the scale-invariant $n = 1$ spectrum corresponds to a constant value of the quantity $l(l+1)C_l$. For this reason it is usually in terms of this combination of l and C_l that the data from the CMBR are represented.

In terms of mean square normalized temperature fluctuations one has that[5]

$$\left(\frac{\Delta T}{T}\right)_{rms}^2 \equiv \left\langle \left(\frac{\delta T}{T}(\theta,\phi)\right)^2 \right\rangle \equiv \frac{1}{4\pi} \sum_{l=2}^{\infty} (2l+1) C_l , \quad (6.19)$$

and

$$\left(\frac{\Delta T}{T}\right)_Q^2 \equiv \frac{5}{4\pi} C_2 , \quad (6.20)$$

where the quadrupole moment Q includes only the contribution for $l = 2$ in (6.15). For $n = 1$, using the expressions above, one has

---

[5] This expression is formally divergent but is always regularized by a small scale cut-off. Note that the term corresponding to $l = 1$ has been assumed to be removed as it is associated to the Earth's motion with respect to the CMBR. For more detailed formulas, which take into account also the angular resolution of the insrtument, see [170].

$$\frac{(\Delta T)_{rms}}{(\Delta T)_Q} \simeq 2.3 \tag{6.21}$$

which can be compared with observations of CMBR anisotropies.

The COBE satellite gave the first measurement of the root mean square fluctuations in the range of angular scale 15°–165°, and found [27]

$$\left(\frac{\Delta T}{T}\right)_{rms} = (1.1 \pm 0.2) \times 10^{-5} . \tag{6.22}$$

Further the quadrupole contribution to anisotropies was determined to be:

$$\left(\frac{\Delta T}{T}\right)_Q = (0.48 \pm 0.15) \times 10^{-5} . \tag{6.23}$$

Using the relations above these are equivalent to $n = 1.1 \pm 0.5$, $A = (24\,\mathrm{Mpc/h})^4$.

It is interesting to comment on how the normalization obtained from the CMBR is related to the amplitudes inferred from galaxy catalogs, or other observations probing the fluctuations in matter today. One can roughly parameterize a CDM-type PS *today* in the form

$$P(k) = \frac{Ak^n}{1 + (k/k_c)^m} \tag{6.24}$$

assuming that $n = 1$. The index m controls the large k behavior, above the characteristic "turn-over" at $k = k_c$. These parameters (or similar ones in other more complicated parameterizations of the PS today) are in principle determined in any given theory, and one of the goals of numerical simulations is to calculate them.

In any case the overall normalization from the CMBR ends up as an overall normalization of the PS today, as in linear theory – valid for small fluctuations at large scales – the spectrum undergoes a simple renormalization between the time probed by the CMBR and today. In principle then one can calculate, for example, the normalized mass variance in spheres using (6.24) and determine whether this is consistent with observations today. In practice such a direct matching does not work: the reason for this is ascribed to a real difference between the correlation properties of what is probed through galaxy observations (hot baryons) and dark matter, which dominates the gravitational dynamics (and fluctuations) in these theories. This is what is known as "biasing". Phenomenologically such an effect is usually described by the introduction of a single parameter ("linear bias") giving the relative amplitudes of the dark matter PS $P(k)$ and that of visible matter $P_b(k)$:

$$P_b(k) = b^2 P(k)$$

and analogously for the mass variance. We will return to discuss at length in later chapters the problematic aspects of simple models justifying such a parameterization.

6.6.2 The Origin of Oscillations in the Power Spectrum

The main result of the COBE observations was that they permitted a measurement of the amplitude of the primordial fluctuations and the large scale exponent of perturbations. The latter in particular was in line with the HZ spectrum predicted by inflation. In the last few years the exploration of the anisotropies at small angular scales has led to the discovery of another predicted feature of these models, but now at smaller ("sub-horizon") scales. These are the so-called "acoustic" oscillations, or *acoustic peaks*. We first describe briefly their physical origin, and then turn to an analysis of their real space counterpart. We show that this consists in a very localized feature in the reduced two-point correlation function (see also [20, 136, 165]).

The physical description which gives rise to the oscillations is based on fluid mechanics and gravity: when the temperature of the CMBR was greater than $1000K$, photons were hot enough to ionize hydrogen. The strong coupling between the baryons and photons which results, means that they can be described as a single fluid. Gravity attracts and compresses this fluid into the potential wells associated with the local density fluctuations. Photon pressure resists this compression and sets up acoustic oscillations in the fluid. Regions that have reached maximal compression by recombination become hotter and hence are now visible as local positive anisotropies in the CMBR [115]. Such oscillations have been recently observed by [26, 58, 107].

The essential qualitative features of the oscillations in the photon-baryon fluid can in fact be derived by neglecting the dynamical effects of gravity and the (inertia of the) baryons. Perturbations in this fluid can be simply described by a continuity equation and an Euler equation (see [115] for a recent review on the subject). The description is given in Fourier space because perturbations are very small, and in linear perturbation theory different Fourier modes evolve independently.

The appropriate continuity equation (which includes the effects of the cosmological expanding background), at linear order in perturbations, and in Fourier space, gives

$$\dot{\Theta} = -\frac{1}{3}kv_\gamma \qquad (6.25)$$

where $\Theta = \Theta(\boldsymbol{k},t)$ is the FT of the temperature fluctuation field

$$\frac{\Delta T}{T}(\theta)$$

(where we have assumed isotropy) and v_γ is the photon fluid velocity. The Euler equation, which is an expression of momentum conservation, gives

$$\dot{v}_\gamma = k\,\Theta\,. \qquad (6.26)$$

Differentiating (6.25) and inserting in the Euler equation yields the simple oscillator equation

$$\ddot{\Theta} + c_s^2 k^2 \Theta = 0 \ . \tag{6.27}$$

Thus the pressure gradients act as a restoring force to any initial perturbation in the system which thereafter oscillates at the speed of sound. Physically these temperature oscillations represent the heating and cooling of a fluid that is compressed and rarefied by a standing acoustic wave. The solution of (6.27) is given by

$$\Theta(\boldsymbol{k},t) = \Theta(0)\cos(k\,s_* + \phi(\boldsymbol{k})) \tag{6.28}$$

where $s_* = c_s t$ is the distance sound can travel up to the time t and $\phi(\boldsymbol{k})$ is a relative phase. For large scales $k\,s_* \ll 1$ the perturbation is frozen. On small scales the amplitude of Fourier modes exhibit temporal oscillations. This picture can be refined by considering the gravitational forcing, the inertia of baryons and other smaller effects, but it does not change qualitatively. (These other effects modify the spread and amplitude of the "peaks").

The observed acoustic peak structure is obtained with one further ingredient: coherence of the phases of the different modes, which means that these temporal oscillations are associated to oscillation in k-space (at a fixed time). Given such coherence, modes which arrive at the maximum or minimum of oscillation at recombination (when the photons decouple and begin to travel to us on straight lines) are seen as k-oscillations in the PS. In the cosmological context – and in particular in the context of inflation – such coherence between the modes is predicted generically in the initial conditions. This is because the evolution of the modes outside the horizon (i.e., prior to the phase we have described, which is the causal "sub-horizon" evolution) have solutions which are a linear combination of a growing solution and a decaying solution. For generic initial conditions imposed in the very early universe, the former solution is dominant, and gives the temporal coherence of the modes.

6.6.3 A Simple Example of k-Oscillations

Using this coherence of the modes imposed in the initial conditions, it follows from (6.28) that the PS (both in 2 and 3 dimensions) has oscillations, being modulated by a term which goes as $\cos^2(ak)$ with appropriate a. To illustrate the effect of these oscillations let us consider the following simple PS:

$$P(k) = k\,e^{-k/k_c} \tag{6.29}$$

and the corresponding one with k-oscillations

$$P_o(k) = k\,e^{-k/k_c}\,\cos^2(ak) \tag{6.30}$$

(see Fig. 6.4 where we set $a = k_c = 1$ for simplicity). For both PS the real space reduced two-point correlation function can be computed analytically. For (6.29) we obtain

$$\tilde{\xi}(r) = 2\frac{3 - r^2}{1 + 3\,r^2 + 3\,r^4 + r^6} \tag{6.31}$$

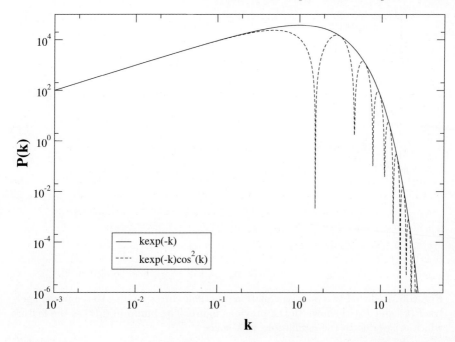

Fig. 6.4. Behavior of a typical HZ power spectrum $P(k) = ke^{-k}$ and of the corresponding one with k-oscillations $P_o(k) = ke^{-k}\cos^2(k)$

while for (6.30) we get (see Fig. 6.5)

$$\tilde{\xi}_o(r) = -2\frac{r^{14} - 23175 - 9r^{12} + 25871r^2 - 3r^{10} - 10635r^4 - 805r^8 + 5043r^6}{(5 + r^2 + 4r)^3(5 + r^2 - 4r)^3(1 + r^2)^3}. \tag{6.32}$$

The principal point to note is that while k-oscillations are de-localized, in real space the correlation function shows a localized "bump". This is not really surprising: indeed the FT of an oscillating function is typically a localized function.

6.6.4 Oscillations in the CDM PS

To see in further detail the relation between the PS in a CDM model and the two-point correlation function in this case, we consider the approximation of the CDM PS as given by [76]. The PS is in general a function of various parameters – in particular, limiting our scope, of the dark matter density Ω_c and of the baryon density Ω_b. In this case the amplitude of the k-oscillations in the PS is controlled by Ω_b; when its value is increased, the amplitude of the oscillating term increases (see Fig. 6.6). The reduced two-point correlation function $\tilde{\xi}(r)$ can be obtained by a numerical FT of the PS (see Fig. 6.7). We

186   6 Fluctuations in Standard Cosmological Models: A Real Space View

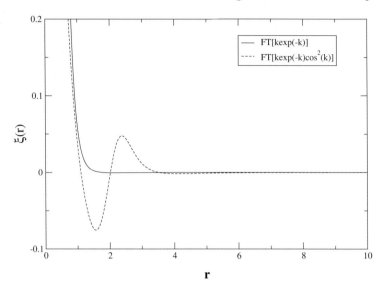

Fig. 6.5. Reduced two-point correlation functions corresponding to the power spectra $P(k)$ and $P_o(k)$ shown in the previous figure. The bump in $\xi_o(r)$ is the feature which arises due to the oscillations in k space

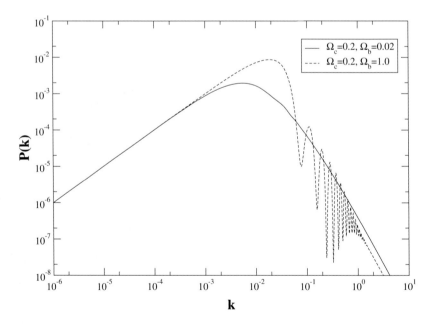

Fig. 6.6. Power spectrum for a CDM model with $\Omega_c = 0.2$ and Ω_b varying in the range indicated

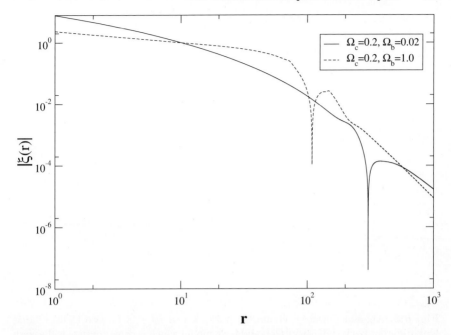

Fig. 6.7. Correlation function $\tilde{\xi}(r)$ for a CDM model with $\Omega_c = 0.2$ and Ω_b varying in the range indicated

see that, just as in the previous simple example, the k-oscillations correspond in real space to a well-localized bump-like feature in the correlation function.

6.6.5 Oscillations in the CMBR Anisotropies

We now return to the CMBR anisotropies and the oscillations as they are manifested in these. For any given standard-type model, the C_l spectrum can be calculated. In practice there are now several standardized numerical codes [249] available which allow one to simply give the parameters of the model and generate the spectrum. The angular correlation function of the temperature fluctuations can then be simply reconstructed using (6.15).

As in the 3-d case, we restrict ourselves to the the two parameters Ω_c and Ω_b. In particular we fix $\Omega_c = 0.2$ and then vary Ω_b in the range $[0.02, 0.2]$ (and the amplitude of the k-oscillations). In Fig. 6.8 we show the behavior of $C_l(l(l+1))$ as a function of the angular wave-number l. In Fig. 6.9 we show the same data in a log-log plot. We have included also, for comparison, an example where the C_l are almost a power-law which decays as $\sim l^{-2}$ without k-oscillations (more exactly we take $C_l(l(l+1)) \sim$ const., plus a smooth cut-off at very large l). In Fig. 6.10 we show the corresponding angular correlation function $C(\theta)$. We see that, just as in the 3-d examples, the effect of the

188   6 Fluctuations in Standard Cosmological Models: A Real Space View

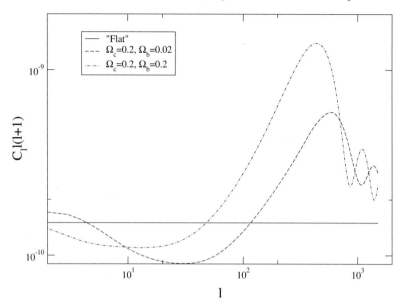

Fig. 6.8. Angular power spectrum C_l multiplied by $l(l+1)$ for two CDM models with different baryon density (Ω_b), and for a model which is constant in this diagram (i.e., $C(l) \sim 1/l(l+1)$) and without k-oscillations for comparison (in the label shown by "Flat")

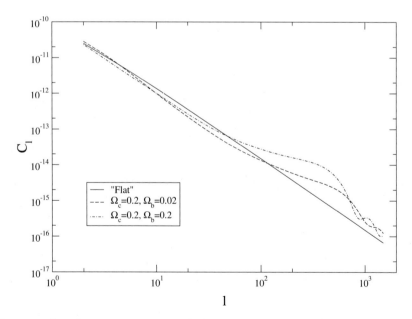

Fig. 6.9. Angular power spectrum C_l for the same cases shown in the previous figure

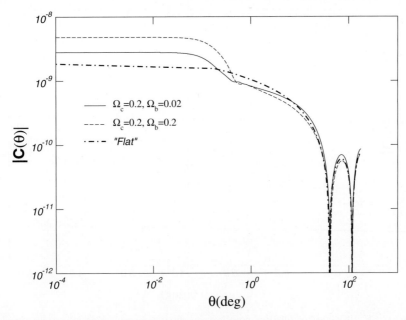

Fig. 6.10. Two-point angular correlation function $C(\theta)$ (absolute value) for two values of Ω_b, as well as that corresponding to $C(l) \sim 1/l(l+1)$

oscillations is to introduce a localized "bump" structure. Figure 6.11 shows in more detail the region around this bump.

6.7 Summary and Discussion

We have described in Sect. 6.3 the use of the term "scale-invariance" with respect to fluctuations in cosmology: it refers to the fact that the variance of the mass has an amplitude at the horizon scale which does not depend on time. The PS associated with this behavior is that of a correlated system with surface (sub-Poisson) fluctuations: a so-called super-homogeneous distribution. This use of the term "scale-invariance" is therefore very different to its (original) use in statistical physics. In this context it is associated with a distinctly different class of distributions which have special properties with respect to scale transformations: typically critical systems, like a liquid-gas coexistence phase at the critical point, which have a well defined homogeneity scale and a reduced two-point correlation function which decays as a non-integrable power-law: $\xi(r) \sim r^{-\gamma}$ with $0 < \gamma < 3$. In particular the term scale-invariance does not in this usage have anything to do with the *amplitudes* of fluctuations being independent of scale: the amplitudes of fluctuations vary with scale, while the system is correlated at all scales.

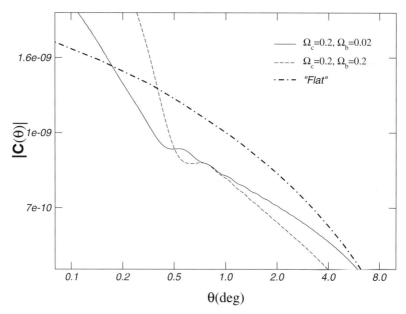

Fig. 6.11. Behavior of $C(\theta)$ at small angles for the cases reported in the previous figure

We have highlighted the fact that all current cosmological models will share at large scales the characteristic behavior in real space of the HZ spectrum. Specifically we note primarily the very characteristic lattice-like behavior of the variance in spheres $\sigma^2(R) \sim R^{-4}$ (up to a small correction which is formally logarithmic for the case of exact HZ), as well as the characteristic negative (non-oscillating) power-law tail in the two-point correlation function $\xi(r) \sim -r^{-4}$.

We have not at all discussed here so far observations of the distribution of matter today, and in particular the degree to which they are compatible with the models which we have described. It is a question we will consider at length in Chaps. 8–12 and Chap. 13. The central point of our analysis will in fact be to broaden the class of distributions used to describe galaxy structures to include fractals, in which there is no well defined (sample independent) mean density. Further we will discuss how observations of galaxy catalogs actually show, up to the scales which can currently be robustly probed without a priori assumptions of homogeneity, that there is no clear convergence to a well defined density. Thus at these scales the models we have described here cannot be used even at zeroth order to describe these observed structures. This does not mean, however, that these models cannot describe successfully galaxy structures: but to establish whether they can, it must first be shown from observations that there is a clear crossover toward homogeneity i.e., a scale beyond which the average density becomes a well-defined (i.e., sample-independent) positive quantity. These models then predict that, on much larger scales (beyond the turn-over scale in the PS), galaxy structures should present the super-homogeneous character of the HZ type PS. Indeed this should in principle be a critical test of the paradigm linking the measurements of CMBR on large scales [26, 27, 58, 107] to the distribution of matter. Observationally a crucial question is the feasibility of measuring the transition between these regimes directly in galaxy distributions. With large forthcoming galaxy surveys it may be possible to do so, but this is a question which must address exactly the statistics of these surveys and the exact nature of the signal in any given model. This is an issue we will address only partially in this book. In particular we will look at one of the important elements in it: the galaxy distribution is a discrete set of objects whose properties are related in a non-trivial way to the ones of the underlying continuous field. To understand the relation between the two, one has to consider the additional effects related to sampling the continuous field. This is intimately related to the problem of "biasing" between the distribution of visible and dark matter, which we mentioned in Sect. 6.6.1 above. We will discuss a simple (but canonical) model for this in Chap. 13, addressing in particular how selection modifies the super-homogeneous properties of the theoretical mass distribution at large scales.

7 Discrete Representation of Fluctuations in Cosmological Models

7.1 Introduction

Having discussed the properties of standard cosmological models of mass fluctuations, we turn to a different question in this chapter: how to produce spatial distributions of point-particles which give a representation of these correlations. This is interesting firstly from a purely heuristic point of view: it answers the question as to what such mass density fields "look like". It is also of interest from a practical point of view: the primary instrument for studying the problem of structure formation in cosmology is gravitational N-body simulations, and in this context it is necessary to produce point processes approximating the correlation properties of cosmological models of primordial perturbations.

Dealing with this question leads us to two different approaches. On the one hand we identify a system which is well known and much studied in statistical physics, showing the same kind of correlations as in the cosmological case: the one component plasma (OCP) which we have already briefly described in our discussion in Chap. 3 of super-homogeneous systems. Here we discuss a modification of such systems which may produce at thermal equilibrium precisely the spectra of cosmological models at sufficiently large scales (rather than simply the qualitative super-homogeneous feature). In the second approach we analyze with complete generality the method currently used to produce initial conditions for N-body simulation: the displacement of points in an initial highly uniform – super-homogeneous – point process (lattice or glass-like spatial distribution of point-particles) from their initial positions in a way prescribed by the cosmological power spectrum. We derive exact expressions for the power spectra of the point process obtained in this way.

Before turning to the discussion of the two methods there is an important point we briefly discuss: the cosmological spectra being considered are, as we have emphasized, for continuous mass fields, and we are considering their representation with a discrete set of identical massive point-particles. We discuss first more precisely the sense in which a discrete point process represents the continuous one.

7.2 Discrete versus Continuous Density Fields

In order to take a spatial distribution of identical point-like masses (i.e., a point process) as giving a representation of a continuous mass density field, we need to specify explicitly how to relate the two. There is in fact no unique prescription to pass from a discrete distribution of identical particles to a continuous mass field [92].

A simple prescription is given by requiring that a physical smoothing of the point process gives the continuous one. This corresponds to taking a regularization of the Dirac delta function in the expression of the microscopic density function (2.1) with a function $W_L(r)$ with the property

$$W_L(r) = L^{-3} W_o\left(\frac{r}{L}\right), \qquad \int W_o(r) d^3r = 1 \qquad (7.1)$$

where L is the characteristic scale introduced by the regularization e.g., the Gaussian function in three dimensions

$$W_L(r) = \left(\frac{1}{\sqrt{2\pi}L}\right)^3 \exp\left(-\frac{r^2}{2L^2}\right). \qquad (7.2)$$

For any finite value of L we can define a continuous density field $\rho_L(r)$ as the convolution of this function with the density field $\rho(r) = \sum_i \delta(r - r_i)$ of the point process:

$$\rho_L(r) = \int d^3y\, W_L(|y - r|) \rho(r) = \sum_i W_L(|r_i - r|). \qquad (7.3)$$

The pair correlation function of the continuous field obtained in this way can then be written also as a double convolution integral of $W_L(r)$ and the correlation function for the point-particle density $\rho(r)$. The singularity in $r = 0$ in the latter i.e., the diagonal part (see Chap. 2) of the correlation function of the point process, is thus also removed by the smoothing implied by the regularization (7.3). The PS of this continuous field is then simply given as

$$P_L(k) = |\tilde{W}_L(k)|^2 P_D(k) \qquad (7.4)$$

where $\tilde{W}_L(k)$ is the FT of the regularization function $W_L(r)$, and $P_D(k)$ is the PS of the point-particles distribution. In particular for the Gaussian smoothing function (7.2) we have

$$P_L(k) = \exp(-k^2 L^2) P_D(k). \qquad (7.5)$$

As we have already mentioned, the context in which such discrete representations of continuous fields of mass fluctuations are of interest in cosmology is given by gravitational N-body simulations, where such a representation must be given to the continuous theoretical density field (usually CDM). In particular the initial configuration of the N points must represent the initial conditions of the CDM field over some appropriate range of scales. The way this is done in practice [12, 112, 246] is the following: one starts from a "pre-initial" configuration, which is usually a perfect simple lattice, or sometimes a "glassy" configuration of identical particles obtained by running the N-body code with the sign of gravity reversed[1]. In either case this pre-initial configuration is understood to be a "sufficiently uniform" discretization of a constant density background. Most importantly they are configurations which are unstable equilibria – if unperturbed they evolve negligibly on the timescales simulated. This configuration is then perturbed by applying to all particles a displacement field which is prescribed by the PS of density perturbations of the continuous model one wishes to represent. The prescription for the displacements is given by the so-called "Zeldovich approximation" (ZA), which is a perturbative solution to the equations describing the evolution of a self-gravitating fluid in the Lagrangian formalism (see [38] for details). The displacement field in the ZA is directly determined by the perturbed density field, which one takes to be that one wishes to represent. We will explain below in Sect. 7.7 that this is equivalent to taking a displacement field with a PS for the displacement field given by $P(k)/k^2$, which should generate a PS $P(k)$ of the final particle distribution. This is only true neglecting all the effects associated to the discreteness – and fluctuations – in the "pre-initial" distribution. The very general results we derive in the second part of this chapter will allow us to include these effects and determine their importance.

It is interesting to note that in the case we have just described – generation of initial conditions for N-body simulations by displacement of point-particles off a lattice or glass – the continuous system is not necessarily represented by the discrete distribution in the sense we defined above. One can seek a function $\tilde{W}(k)$ which satisfies

$$|\tilde{W}(\boldsymbol{k})|^2 = \frac{P_D(\boldsymbol{k})}{P(k)} \tag{7.6}$$

where $P(k)$ is the theoretical model one wishes to represent and $P_D(\boldsymbol{k})$ is the PS of the discrete particle distribution obtained by the displacement algorithm applied to the pre-initial particle configuration. However it is not evident that a function $W(\boldsymbol{r})$ (the FT^{-1} of $\tilde{W}(\boldsymbol{k})$) can be defined which corresponds to a localized function, describing a physical smoothing of the massive point-particles.

The sense in which the discretization represents the continuous model is in fact thus a much weaker one in this context. It is simply the requirement that $P(k) \approx P_D(\boldsymbol{k})$ for some range of k i.e., that the theoretical PS be represented well in k space in some range. For the case of the perfect lattice, which has no power for k below the Nyquist frequency, taken as "pre-initial" configuration,

---

[1] The latter is in fact essentially just a limiting case of the OCP, with a PS proportional to k^2 at small k.

we will see in Sects. 7.7 and 7.8 that the displacement algorithm has a PS which approximates extremely well the theoretical model below the Nyquist frequency. This does not imply, as we have noted, that a smoothing of the massive point-particles of the perturbed lattice in real space approximate well the density field in real space of the theoretical field.

7.3 Super-Homogeneous Systems in Statistical Physics

We have emphasized in the previous chapter that the mass distributions described by standard cosmological models are super-homogeneous, according to the definition of this term given in Chap. 3. As discussed in Chap. 3, a simple cubic lattice of identical point-particles is the simplest example of a discrete set of points which shows this super-homogeneous behavior ($\sigma^2 \propto 1/R^4$) of the normalized mass variance [125]. A better (for what concerns stochasticity and statistical stationarity) example of the same kind of distribution is the so-called *shuffled lattice* [91]; this is a lattice whose sites are independently randomly displaced by a distance x in all directions from their initial position according to some PDF $p(x)$ of the displacement which has a finite second moment. In this case, which we discuss below in Sect. 7.7.3, we find $P(k) \sim k^2$ at small k and, consequently, again $\sigma^2(R) \propto 1/R^4$ at large R. The simple lattice, however, is not a *stochastic* SPP, and even the shuffled lattice, though it is stochastic, is not spatially stationary and isotropic because the underlying lattice structure in general is not completely erased by the shuffling [96].

To construct a statistically isotropic and homogeneous particle distribution with the same behavior of $\sigma^2(R)$ is non-trivial. A particular example, in two dimensions, is the so-called "pinwheel" tiling of the plane [198, 199]. It is defined by a *deterministic* generation algorithm consisting in taking a right angled triangle with sides of respective length one and two (and hypotenuse $\sqrt{5}$) and, at the first step, forming five similar square triangle of sides $1/\sqrt{5}$ and $2/\sqrt{5}$ respectively as shown in Fig. 7.1. At the second step we expand these new triangles, to the size of the original triangle, and repeat the procedure *ad infinitum*, so that they cover the plane completely. Finally, placing a point-particle randomly inside each elementary triangle will give the super-homogeneous point process which is statistically isotropic (with a continuous PS) [92], and whose PS is $P(k) \sim k^2$ at small k.

As discussed in Chap. 3 one context in which such isotropic distributions are produced by a physical mechanism is in the study of the OCP, which is simply a system of charged point-particles interacting through a repulsive $1/r$ potential, in a uniform background which gives overall charge neutrality [22]. At thermal equilibrium, at sufficiently high temperature, this system has a PS for the charge or mass fluctuations which has the behavior $P(k) \sim k^2$ as $k \to 0$. With respect to the problem of representation of cosmological models this suggests an interesting possibility: to seek a system whose equilibrium

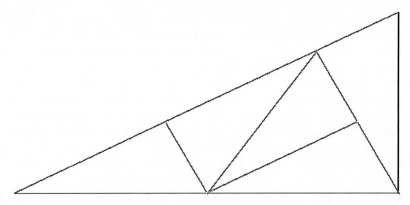

Fig. 7.1. Fragmentation step for the "pinwheel" tiling of the plane. (From [92])

two-point correlations are not simply in this same class of super-homogeneous distributions, but which represents (in the sense described in Sect. 7.2 above) the PS of the theoretical model. This is what we describe in the following section. In principle once such a system is given, this gives a method (alternative to the standard one we have briefly described above) of producing appropriate initial conditions for N-body simulations: one simply needs to simulate the thermal equilibrium of the N particles interacting through the prescribed potential and take a typical equilibrium configuration.

7.4 HZ as Equilibrium of a Modified OCP

We have seen in Chap. 3 that the three-dimensional OCP equilibrium correlations give surface fluctuations ($\langle \Delta N^2 \rangle \sim R^2$), but with a PS at small k which goes like k^2. By considering instead a repulsive $1/r^2$ potential, whose FT is $2\pi^2/k$, we obtain, by following the screening argument shown in (3.46)–(3.50), that

$$P(k) \simeq \frac{k}{2\pi^2 n^2 \beta}, \text{ for } k \sim 0 \qquad (7.7)$$

and

$$\xi(r) \simeq -\frac{1}{2\pi^4 \beta n^2} \frac{1}{r^4} \qquad (7.8)$$

for $r \gg (2\pi^2 \beta m)^{-1}$. The change from the exponential decay of $\xi(r)$ to a power-law decay is a result of the different analyticity properties of the two power-spectra: the k^2 behavior is analytic at the origin, guaranteeing a rapidly decaying behavior of its FT, while the k spectrum is not.

In the context of cosmological N-body simulations what one needs is not simply the primordial HZ PS with some appropriate small scale cut-off: what is simulated is only a part of the cosmological evolution, starting from a time at which the initial spectrum of fluctuations is already significantly modified from its primordial form at large k. As we discussed in the previous chapter, while purely gravitational evolution at these early times does not modify the HZ spectrum, non-gravitational effects, present until the time when the universe becomes dominated by matter, do so significantly. The nature of these modifications depends on the details of the cosmological model, but in all cases it affects only k larger than a characteristic k_{eq} corresponding to the causal horizon at the "time of equality" (when matter starts to dominate over radiation). One then has a PS of the form

$$P(k) = k^n f(k) \tag{7.9}$$

with n exactly or very close to unity and $f(k)$ such that $f(0) = a > 0$ (see Fig. 6.1) and going to zero for $k \to \infty$ in a model dependent way.

To produce such a spectrum as the equilibrium one of an OCP like system requires further modification of the form of the interaction potential [92]. Just like the standard OCP, an unmodified $1/r^2$ potential will give, in the weak-coupling limit, a spectrum which becomes flat (i.e., Poisson like) at large k. A crude guess of what potential would produce the behavior of a typical cosmological model can be obtained by supposing that at small scales

$$1 + \xi(r) \approx e^{-\beta V(r)} \tag{7.10}$$

which corresponds to completely neglecting collective effects ($1+\xi$ represents the relative probability, compared to completely random, of finding a particle at distance r from a given one). Given that the desired fluctuations always have small amplitudes ($|\xi| \ll 1$), we would then need to be in the regime of temperature and scales such that $|\beta V(r)| \ll 1$, so that $\xi(r) \simeq -\beta V(r)$. The potential should thus be attractive at smaller scales, as the system is positively correlated at those scales. A k^{-2} behavior at large k, which is often used (see Fig. 6.1) as an initial condition approximating cosmological models (CDM type) in this regime (beyond the "turn-over"), may be obtained from an attractive $1/r$ potential. For instance we can take

$$V(r) = \frac{1}{r^2} - \eta \frac{e^{-\mu r}}{r} \ . \tag{7.11}$$

By modifying the parameters μ and η, as well as the temperature, both the amplitude of the $P(k)$ and the location of a change from a $P(k) \sim k^{-2}$ to a behavior $P(k) \sim k$ for small k can be controlled. The potential (7.11) is repulsive at short distances, but it may be necessary to make it more strongly repulsive in order to ensure that the system is not unstable to collapse [205].

An assumed property of the primordial fluctuation density field in standard cosmological models is that it is Gaussian distributed. Since the density field is inherently positive this assumption of Gaussianity is more properly attributed to fluctuations (expected to be small) around the uniform density. Small fluctuations in the discrete OCP are in fact also Gaussian to a good approximation [156].

7.5 A First Approximation to the Effect of Displacement Fields

We now move on to a very wide-ranging study of the effect of superimposing a generic displacement field, specified itself as a stationary stochastic process with given correlation properties, on a generic spatial distribution of point-particles. The motivation for this study comes from the method we briefly described in Sect. 7.2 to generate initial configurations for N-body simulations. We will recover from the results below the justification for the standard method of building initial conditions as a special case. It corresponds in particular to the limit of small displacements, and to the neglect of the power associated with the "pre-initial" configuration. Our results give analytical forms for the more general case, allowing one to determine what precisely are the initial conditions generated in this way. This can be important, for example, in trying to understand the role played by effects which arise from the discretization of the density field in these numerical simulations, an issue which is unresolved and evidently important e.g. see [160, 216, 132, 12, 13, 14].

In Chap. 2 we have introduced the general framework for the study of statistically stationary stochastic discrete or continuous processes. In particular, in order to characterize properly the internal spatial correlations of the system, we have defined the two-point correlation function $C_2(\boldsymbol{r})$ (or equivalently $\tilde{\xi}(\boldsymbol{r})$ if $\rho_0 > 0$) and the PS $S(\boldsymbol{k})$ (or $P(\boldsymbol{k})$ if again $\rho_0 > 0$) which is the Fourier counterpart. We now specialize the discussion to discrete SSPs of particles with the same unitary mass, studying the problem of how a stationary stochastic displacement field affects the correlation properties of a given point-particle distribution.

Before entering into a more detailed and rigorous discussion, we give a simple argument which roughly describes the effect of a displacement field on a "sufficiently uniform" distribution of point-particles. It is based on the fact that a displacement process conserves the mass. Hence it must satisfy a form of *continuity equation*. If we call $\rho_{in}(\boldsymbol{r})$ the initial microscopic density of particles and $\rho(\boldsymbol{r})$ the final one after the application of the displacement field $\boldsymbol{u}(\boldsymbol{r})$, we can write the continuity relation

$$\rho(\boldsymbol{r}) - \rho_{in}(\boldsymbol{r}) + \boldsymbol{\nabla} \cdot [\rho_{in}(\boldsymbol{r})\boldsymbol{u}(\boldsymbol{r})] = 0 . \tag{7.12}$$

Let the average density (which is not modified by the displacement field) be $\rho_0 > 0$. If $\rho_{in}(\boldsymbol{r})$ is "sufficiently uniform" with respect to $\rho(\boldsymbol{r})$, we can approximate it with a truly uniform density field ρ_0. With this approximation we can rewrite (7.12) as

$$\rho(\boldsymbol{r}) - \rho_0 + \rho_0 \boldsymbol{\nabla} \cdot \boldsymbol{u}(\boldsymbol{r}) = 0 . \tag{7.13}$$

Taking as usual,

$$\delta_\rho(\boldsymbol{k}) = FT\left[\frac{\rho(\boldsymbol{r}) - \rho_0}{\rho_0}\right] , \tag{7.14}$$

and $v(k) = FT[u(r)]$, we have:

$$|\delta_\rho(k)|^2 = |k \cdot v(k)|^2 \,. \tag{7.15}$$

If we suppose that the stochastic displacement field is statistically isotropic, from (7.15) and (3.7), we can say that the PS of the particle distribution is roughly proportional to k^2 times the PS of the displacement field [12].

This argument is in fact precisely equivalent to that used in the setting up the initial conditions of N-body simulations through the ZA as the displacement field in this case obeys the continuity equation and the same two approximations made in obtaining the result (7.15). The first consists in taking the initial microscopic density to be an exactly uniform continuous mass field i.e., $\rho_{in}(r) = \rho_0$. The second is that the displacements are small (i.e., that the k considered is less than the inverse of the typical displacement). In the exact equations we will derive in this chapter describing the effect of the displacement field on an initial point-process, we thus expect to find additional terms coming from the "granularity" of the initial particle distribution, as well as corrections at large k (see [94, 95] for more details).

7.6 Displacement Fields: Formulation of the Problem

We start by considering a SPP in a volume $V \to \infty$ with microscopic density

$$\rho_{in}(r) = \sum_i \delta(r - r_i) \,,$$

where r_i is the position of the ith particle of the distribution. We assume that the distribution has a finite *homogeneity scale* (i.e., that it is sufficiently uniform at large scales) having a well defined positive average:

$$\langle \rho_{in}(r) \rangle = \rho_0 > 0 \,,$$

where $\langle \ldots \rangle$ is the usual ensemble average. In this case we can introduce the FT of the so-called *normalized density contrast field* (7.14) the PS of which is defined by (see Chap. 3)

$$P_{in}(k) = \lim_{V \to \infty} \frac{\langle |\delta_\rho(k)|^2 \rangle}{V} \,.$$

Since the particle distribution is assumed to be statistically stationary in space, the connected two-point correlation function is simply given by

$$\tilde{\xi}_{in}(r) = FT^{-1}[P_{in}(k)] \,,$$

where as usual FT^{-1} is the inverse Fourier transform as defined in Chap. 3.

7.6 Displacement Fields: Formulation of the Problem

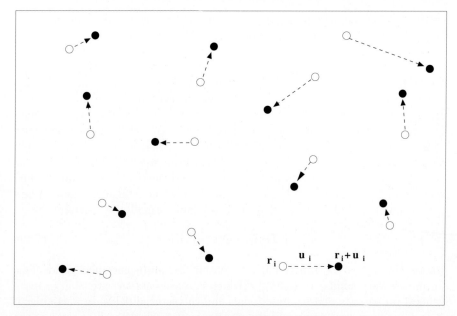

Fig. 7.2. The figure illustrates the superposition of a stochastic displacement field on a particle distribution. The particles pass, through the displacements (*arrows*), from the old positions (*white circles*) to the new ones (*black circles*), and, consequently, spatial correlations change

Let us now introduce a stationary stochastic displacement field which moves each particle from its initial position. In general this displacement process changes the correlation properties of the initial point-particle distribution from $P_{in}(\boldsymbol{k})$ (or $\tilde{\xi}_{in}(\boldsymbol{r})$) to a new $P(\boldsymbol{k})$ (or a new $\tilde{\xi}(\boldsymbol{r}) = FT[P(\boldsymbol{k})]$). If we apply to the particle i the displacement $\boldsymbol{u}_i$, its position changes from $\boldsymbol{r}_i$ to $\boldsymbol{r}_i + \boldsymbol{u}_i$ (see Fig. 7.2), therefore the final particle density field can be written as

$$\rho(\boldsymbol{r}) = \sum_i \delta(\boldsymbol{r} - \boldsymbol{r}_i - \boldsymbol{u}_i) \,. \tag{7.16}$$

A stochastic displacement field can be seen as a *vectorial* continuous stochastic process. We can think to "attach" a displacement vector $\boldsymbol{u}(\boldsymbol{r})$ to each spatial point $\boldsymbol{r}$, even though it acts on the particle distribution only if $\boldsymbol{r}$ is occupied by a point-particle. Therefore, in analogy with what has been discussed in Chap. 2 about scalar stochastic fields, we can say that the statistics of the displacement field is completely determined by a *probability density functional* $\mathcal{Q}[\boldsymbol{u}(\boldsymbol{r})]$ which defines the probability of each possible realization $\boldsymbol{u}(\boldsymbol{r})$ of the displacement field i.e., it defines the statistical ensemble of the possible realizations of the displacement field with relative weights. If the displacement field is statistically stationary in space the functional $\mathcal{Q}[\boldsymbol{u}(\boldsymbol{r})]$ is invariant under translations i.e., $\mathcal{Q}[\boldsymbol{u}(\boldsymbol{r} + \boldsymbol{r}_0)] = \mathcal{Q}[\boldsymbol{u}(\boldsymbol{r})]$ for any translation vector r_0. Analogously, if the displacement field is statistically isotropic, the functional $\mathcal{Q}[\boldsymbol{u}(\boldsymbol{r})]$ is invariant under any spatial rotation.

We consider here the general case in which displacement-displacement spatial correlations can be present in $\mathcal{Q}[\boldsymbol{u}(\boldsymbol{r})]$. However we assume that the probability of a single realization of the displacement field $\boldsymbol{u}(\boldsymbol{r})$ in *each* point of the space does not depend on the single realization of the initial configuration of the particle distribution. That is, we exclude the case in which $\mathcal{Q}[\boldsymbol{u}(\boldsymbol{r})]$ depends explicitly on the initial coordinates $\{\boldsymbol{r}_i\}$ of the particles.[2] This implies that the joint probability of having a specific initial microscopic particle density $\rho_{in}(\boldsymbol{r})$, and a specific realization $\boldsymbol{u}(\boldsymbol{r})$ of the displacement field, factorizes into the product of the two independent probabilities. The average of any functional $A[\boldsymbol{u}(\boldsymbol{r})]$ of the displacement field in one or more spatial points is indicated with $\overline{A}$, and can thus formally be written as:

$$\overline{A} = \int D[\boldsymbol{u}(\boldsymbol{r})] \mathcal{Q}[\boldsymbol{u}(\boldsymbol{r})] A[\boldsymbol{u}(\boldsymbol{r})] , \qquad (7.17)$$

where $D[\boldsymbol{u}(\boldsymbol{r})]$ is the functional differential of the displacement field. This formula is very similar to (3.53), of which it represents an extension to more general probability density functionals and to vectorial fields. In particular if A is a function only of the displacements $\{\boldsymbol{u}_i\}$ applied to the points of the space occupied by the point-particles of the initial particle distribution, then the average can be limited to these displacements:

$$\overline{A} = \int .. \int \left[\prod_j d^d u_j \right] h(\{\boldsymbol{u}_i\}) A(\{\boldsymbol{u}_i\}) , \qquad (7.18)$$

where we have indicated with $h(\{\boldsymbol{u}_i\})$ the joint PDF of all the displacements applied to the particles of the initial distribution. In general it depends parametrically on the initial positions of the particles. Note, however, that this does not mean at all that the statistical weight $\mathcal{Q}[\boldsymbol{u}(\boldsymbol{r})]$ of a given displacement field $\boldsymbol{u}(\boldsymbol{r})$ depends on the single initial configuration of the particle distribution. This is a subtle but important point to appreciate. Even though we assume that the probability of having $\boldsymbol{u}(\boldsymbol{r})$ is independent of that of having $\{\boldsymbol{r}_i\}$, the parametric dependence on the initial coordinates of the particles in $h(\{\boldsymbol{u}_i\})$ comes directly by its definition which is

$$h(\{\boldsymbol{u}_i\}) = \int D[\boldsymbol{u}(\boldsymbol{r})] \mathcal{Q}[\boldsymbol{u}(\boldsymbol{r})] \prod_i \delta\left(\boldsymbol{u}(\boldsymbol{r}_i) - \boldsymbol{u}_i\right) , \qquad (7.19)$$

i.e., the parametric dependence on $\{\boldsymbol{r}_i\}$ of $h(\{\boldsymbol{u}_i\})$ comes only from the fact that we are limiting the analysis of the displacement field to those spatial points occupied by particles.

---

[2] At most the *ensemble* properties of the displacement field can be related to those of the *ensemble* of the initial point-particle distribution i.e., the correlation functions characterizing the probability density functional $\mathcal{Q}[\boldsymbol{u}(\boldsymbol{r})]$ can be related to the correlation functions of the initial SPP.

In the case of statistically stationary and isotropic displacement fields, $h(\{u_i\})$ depends parametrically on the vectorial distances between all the couples of points of the initial distribution and is invariant under spatial rotations.

Finally, if we have a function of the final microscopic (i.e., after displacements) density $\rho(r)$ of the point-particles of the system, then the ensemble average over all the possible final configurations of the particle distribution is done by averaging both over the initial configurations $\langle \ldots \rangle$ and over all the possible displacements as in (7.18). In fact the *ensemble* of all possible final particle configurations is found by considering all possible initial configurations, and for each of these all final configurations obtained by applying the *ensemble* of the displacement fields. Hence the correlation function of the "displaced" particle distribution will be:

$$\tilde{\xi}(r) = \frac{\left\langle \overline{\rho(r_0 + r)\rho(r_0)} \right\rangle}{\left\langle \overline{\rho(r_0)} \right\rangle^2} - 1 . \tag{7.20}$$

Given the assumption of statistical independence between the displacement field and the initial point process, the order in which the two averages $\langle \ldots \rangle$ and $\overline{(\ldots)}$ are performed is arbitrary.

7.7 Effects of Displacements on One and Two-Point Properties of the Particle Distribution

We can now study how two-point correlations change under the effect of the displacement field. The aim of this section is to relate the new correlation properties of the point-particle distribution to its old ones and to those of the applied displacement field by using (7.20) and other similar relations. We limit our discussion to the case of statistically stationary displacement fields and initial particle distributions (i.e., point processes). Thus also the final particle distribution will be statistically stationary.

Let us start from (7.16) and evaluate the average mass density. The next step will consist in finding the transformation equation for the PS $P(k)$ (or equivalently the connected two-point correlation function $\tilde{\xi}(r)$) of the final particle distribution.

Since the displacement process does not create or destroy any particle and is statistically stationary, the average mass density remains equal to the initial one ρ_0:

$$\left\langle \overline{\rho(r)} \right\rangle = \rho_0 .$$

This can be also proved by direct calculation using that $\langle \rho_{in}(r) \rangle = \rho_0$. First of all we note that (7.16) is a sum of terms depending only on a single displacement. Therefore, in order to evaluate the displacement average $\overline{\rho(r)}$

we need only the one displacement PDF $p(\mathbf{u})$. This is obtained using (7.18) with, for instance, $A(\{\mathbf{u}_i\}) = \delta(\mathbf{u}_1 - \mathbf{u})$ (i.e., by integrating $h(\mathbf{u}_i)$ over all but one of the displacements). Therefore we can write

$$\overline{\rho(\mathbf{r})} = \sum_i \int d^d u_i \, p(\mathbf{u}_i) \delta(\mathbf{r} - \mathbf{r}_i - \mathbf{u}_i) = \sum_i p(\mathbf{r} - \mathbf{r}_i) \, .$$

From this we obtain

$$\left\langle \overline{\rho(\mathbf{r})} \right\rangle = \left\langle \sum_i p(\mathbf{r} - \mathbf{r}_i) \right\rangle =$$

$$\left\langle \int d^d y \, p(\mathbf{y}) \sum_i \delta(\mathbf{y} - \mathbf{r} + \mathbf{r}_i) \right\rangle = \rho_0 \int d^d y \, p(\mathbf{y}) = \rho_0 \, ,$$

where we have used the statistical spatial stationarity of $\rho_{in}(\mathbf{r})$ (i.e., $\langle \rho_{in}(\mathbf{r}) \rangle = \langle \rho_{in}(\mathbf{y} - \mathbf{r}) \rangle = \rho_0$).

We now face the problem of calculating the new two-point correlation function $\tilde{\xi}(\mathbf{r})$ and the new PS $P(\mathbf{k})$. The key step is to calculate the average

$$\left\langle \overline{\rho(\mathbf{r})\rho(\mathbf{r}')} \right\rangle \, .$$

Since the product $\rho(\mathbf{r})\rho(\mathbf{r}')$ is a sum of terms containing at most two different displacements, we do not need to know the complete joint PDF $h(\{\mathbf{u}_i\})$, but only the two-displacement PDF $f(\mathbf{u}, \mathbf{v})$ which is obtained from $h(\{\mathbf{u}_i\})$ by integrating out all but two of the displacements, i.e., using (7.18) with

$$A(\{\mathbf{u}_i\}) = \delta(\mathbf{u}_1 - \mathbf{u})\delta(\mathbf{u}_2 - \mathbf{v}) \, .$$

In general $f(\mathbf{u}, \mathbf{v})$ will depend parametrically on the coordinates of the two points of application of the displacements. In the hypothesis of a statistically stationary displacement field, $f(\mathbf{u}, \mathbf{v})$ will only depend parametrically on the vector separation $\mathbf{r}$ between these two points. For this reason we write it $f(\mathbf{u}, \mathbf{v}) \equiv f(\mathbf{u}, \mathbf{v}; \mathbf{r})$ putting in explicit evidence this dependence[3]. Note that the function $f(\mathbf{u}, \mathbf{v}; \mathbf{r})$ carries much more information than the simple knowledge of the average displacement $\overline{\mathbf{u}} = \mathbf{u}_0$ and the two-displacement correlation matrix whose generic element is

$$G_{\mu\nu}(\mathbf{r} - \mathbf{r}') = \overline{\left(u^{(\mu)}(\mathbf{r}) - u_0^{(\mu)}\right)\left(u^{(\nu)}(\mathbf{r}') - u_0^{(\nu)}\right)} \quad \text{with } \mu, \nu = 1, \ldots, d \, ,$$
(7.21)

where $u^{(\mu)}$ is the μth component of the vector $\mathbf{u}$. In fact $G_{\mu\nu}(\mathbf{r} - \mathbf{r}')$ is only the average value of

---

[3] If the stochastic displacement field is also statistically isotropic, $f(\mathbf{u}, \mathbf{v}; \mathbf{r})$ will depend on $\mathbf{r}$ in such a way that it is invariant under a rigid rotation of the configuration formed by the two displacements $\mathbf{u}$ and $\mathbf{v}$ connected by the vector $\mathbf{r}$.

7.7 Displacements Field on Particle Distributions

$$\left(u^{(\mu)}(\boldsymbol{r}) - u_0^{(\mu)}\right)\left(u^{(\nu)}(\boldsymbol{r'}) - u_0^{(\nu)}\right)$$

calculated with the PDF $f(\boldsymbol{u},\boldsymbol{v};\boldsymbol{r})$ itself.

In general $f(\boldsymbol{u},\boldsymbol{v};\boldsymbol{r})$ will satisfy the following limit conditions at $r=0$:

$$f(\boldsymbol{u},\boldsymbol{v};0) = \delta(\boldsymbol{u}-\boldsymbol{v})p(\boldsymbol{u}) \tag{7.22}$$

$$\lim_{r\to\infty} f(\boldsymbol{u},\boldsymbol{v};\boldsymbol{r}) = p(\boldsymbol{u})p(\boldsymbol{v}) \ . \tag{7.23}$$

The former equation is trivial, while the latter says that the correlation between displacements must go to zero as the separation between the points of application goes to infinity. First of all let us evaluate the average of $\rho(\boldsymbol{r})\rho(\boldsymbol{r'})$ over the displacements:

$$\overline{\rho(\boldsymbol{r})\rho(\boldsymbol{r'})} = \sum_{i,j}\int\int d^d u_i d^d u_j f(\boldsymbol{u}_i,\boldsymbol{u}_j;\boldsymbol{r}_{ij})\delta(\boldsymbol{r}-\boldsymbol{r}_i-\boldsymbol{u}_i)\delta(\boldsymbol{r'}-\boldsymbol{r}_j-\boldsymbol{u}_j)$$

$$= \sum_{i,j} f(\boldsymbol{r}-\boldsymbol{r}_i,\boldsymbol{r'}-\boldsymbol{r}_j;\boldsymbol{r}_{ij}) \ , \tag{7.24}$$

where $\boldsymbol{r}_{ij} = \boldsymbol{r}_i - \boldsymbol{r}_j$. Note that the limit condition given by (7.22) permits the evaluation of the average without separating the diagonal contribution $i=j$ from the non-diagonal part $i \neq j$ of the double sum in (7.24) by averaging the former using the one-displacement PDF $p(\boldsymbol{u}_i)$ and the latter using the two-displacement PDF $f(\boldsymbol{u}_i,\boldsymbol{u}_j;\boldsymbol{r}_{ij})$ with $i \neq j$.

At this point let us evaluate the average $\langle\ldots\rangle$ on the ensemble of initial particle configurations. To this end we make the following consideration: if we have a function of the initial configuration that can be written in the form $\sum_{i,j}\phi(\boldsymbol{r}_i,\boldsymbol{r}_j)$, with $\phi(\boldsymbol{r},\boldsymbol{r'})$ a generic two-point function, we can write rigorously

$$\left\langle\sum_{i,j}\phi(\boldsymbol{r}_i,\boldsymbol{r}_j)\right\rangle \equiv \left\langle\int\int d^d r_a d^d r_b \phi(\boldsymbol{r}_a,\boldsymbol{r}_b) \sum_{i,j}\delta(\boldsymbol{r}_a-\boldsymbol{r}_i)\delta(\boldsymbol{r}_b-\boldsymbol{r}_j)\right\rangle$$

$$= \int\int d^d r_a d^d r_b \phi(\boldsymbol{r}_a,\boldsymbol{r}_b)\left\langle\sum_{i,j}\delta(\boldsymbol{r}_a-\boldsymbol{r}_i)\delta(\boldsymbol{r}_b-\boldsymbol{r}_j)\right\rangle$$

$$= \int\int d^d r_a d^d r_b \langle\rho_{in}(\boldsymbol{r}_a)\rho_{in}(\boldsymbol{r}_b)\rangle \phi(\boldsymbol{r}_a,\boldsymbol{r}_b) \ . \tag{7.25}$$

As shown in Chap. 2, we can write

$$\langle\rho_{in}(\boldsymbol{r}_a)\rho_{in}(\boldsymbol{r}_b)\rangle = \rho_0^2\left[1+\tilde{\xi}_{in}(\boldsymbol{r}_{ab})\right] \ , \tag{7.26}$$

where $\boldsymbol{r}_{ab} = \boldsymbol{r}_a - \boldsymbol{r}_b$. Note that the diagonal part $\delta(\boldsymbol{r})/\rho_0$ of the connected two-point correlation function $\tilde{\xi}_{in}(\boldsymbol{r})$ takes correctly into account the diagonal term $i=j$ of the sum over i,j of (7.25).

By applying (7.25) and (7.26) to (7.24), we obtain:

$$\left\langle \overline{\rho(\boldsymbol{r})\rho(\boldsymbol{r}')} \right\rangle = \rho_0^2 \int\int d^d r_a d^d r_b \left[1 + \tilde{\xi}_{in}(\boldsymbol{r}_{ab})\right] f(\boldsymbol{r}-\boldsymbol{r}_a, \boldsymbol{r}'-\boldsymbol{r}_b; \boldsymbol{r}_{ab}) \, . \tag{7.27}$$

It is convenient to rewrite (7.27) by separating the two terms coming respectively from the diagonal and the non-diagonal parts of $\tilde{\xi}_{in}(\boldsymbol{r})$, i.e. by writing

$$\tilde{\xi}_{in}(\boldsymbol{r}) = \delta(\boldsymbol{r})/\rho_0 + \xi_{in}(\boldsymbol{r}) \, ,$$

where $\xi_{in}(\boldsymbol{r})$, as usual, is the non-diagonal part of the initial connected two-point correlation function of the particle distribution. By applying this prescription, we obtain:

$$\left\langle \overline{\rho(\boldsymbol{r})\rho(\boldsymbol{r}')} \right\rangle = \rho_0 \delta(\boldsymbol{r}-\boldsymbol{r}') + \rho_0^2 \int\int d^d r_a d^d r_b \left[1 + \xi_{in}(\boldsymbol{r}_{ab})\right] f(\boldsymbol{r}-\boldsymbol{r}_a, \boldsymbol{r}'-\boldsymbol{r}_b; \boldsymbol{r}_{ab}) \, . \tag{7.28}$$

Finally, the new connected two-point correlation function $\tilde{\xi}(\boldsymbol{r})$ of the final particle distribution will be given by

$$\tilde{\xi}(\boldsymbol{r}) = \frac{\left\langle \overline{\rho(\boldsymbol{r}_0+\boldsymbol{r})\rho(\boldsymbol{r}_0)} \right\rangle}{\rho_0^2} - 1 \, , \tag{7.29}$$

with the numerator taken from (7.28)[4].

We have now all the ingredients to find the relation between the initial and the final PS of the particle distribution. Let us start this analysis from the simplest case of uncorrelated displacements; then we will come back to the general case for other considerations and some examples.

7.7.1 Uncorrelated Displacements

In this section we focus on the analysis of the case of uncorrelated displacements, that is the case in which the displacement applied to a given point of the distribution does not depend on the displacements applied to other points. Therefore the statistics of the stochastic displacement field is completely determined by the one-displacement PDF $p(\boldsymbol{u})$ of the displacement $\boldsymbol{u}$ applied to a generic point of the space. The joint PDF of n displacements $\boldsymbol{u}_1, \boldsymbol{u}_2, \ldots, \boldsymbol{u}_n$ in n different points of the space factorizes as follows:

$$h^{(n)}(\boldsymbol{u}_1, \boldsymbol{u}_2, \ldots, \boldsymbol{u}_n) = \prod_{i=1}^{n} p(\boldsymbol{u}_i) \, . \tag{7.30}$$

In particular for the two-displacement PDF, we can write

---

[4] Note that from (7.29), as expected, one has that the diagonal part of $\tilde{\xi}(\boldsymbol{r})$ is again $\delta(\boldsymbol{r})/\rho_0$ as it must be for any particle distribution with average density ρ_0.

7.7 Displacements Field on Particle Distributions

$$f(\boldsymbol{u},\boldsymbol{v};\boldsymbol{r}) = \begin{cases} \delta(\boldsymbol{u}-\boldsymbol{v})p(\boldsymbol{u}) & \text{for } r = 0 \\ p(\boldsymbol{u})p(\boldsymbol{v}) & \text{for } r \neq 0 \end{cases}. \quad (7.31)$$

Note that $f(\boldsymbol{u},\boldsymbol{v};\boldsymbol{r})$ is discontinuous at $r = 0$: as shown below this is not the case for truly continuous correlated displacement fields (i.e., belonging to the class of continuous SSP). We can now apply (7.31) to (7.28) in order to find the two-point correlation function of the final system:

$$\left\langle \overline{\rho(\boldsymbol{r})\rho(\boldsymbol{r}')} \right\rangle = \rho_0^2 + \rho_0 \delta(\boldsymbol{r}-\boldsymbol{r}') + \rho_0^2 \int\int d^d r_a d^d r_b \, p(\boldsymbol{r}-\boldsymbol{r}_a) \xi_{in}(\boldsymbol{r}_a-\boldsymbol{r}_b) p(\boldsymbol{r}'-\boldsymbol{r}_b) \,. \quad (7.32)$$

Note that, as, in this case, $f(\boldsymbol{u},\boldsymbol{v};\boldsymbol{r})$ is discontinuous at $r = 0$, it has been important to separate the contributions of the diagonal and the non-diagonal part of $\tilde{\xi}_{in}(\boldsymbol{r})$ in (7.28). In fact, in this case, any element of the connected two-displacement correlation matrix $G_{\mu\nu}(\boldsymbol{r})$, given by (7.21), vanishes for any $r > 0$ while

$$G_{\mu\nu}(\boldsymbol{0}) = \delta_{\mu\nu} U_\mu^2 > 0$$

where U_μ^2 is the positive variance of the μth component of the single displacement (i.e., $G_{\mu\nu}(\boldsymbol{r})$ is discontinuous at $r = 0$). This is a very particular case, as in truly correlated continuous SSP $G_{\mu\nu}(\boldsymbol{r})$ is continuous[5] everywhere [100].

Recalling now that the PS $P(\boldsymbol{k}) = FT[\tilde{\xi}(\boldsymbol{r})]$ (Chap. 3), with $\tilde{\xi}(\boldsymbol{r})$ given by (7.29), we can Fourier transform (7.32) to obtain

$$P(\boldsymbol{k}) = \frac{1 - |\hat{p}(\boldsymbol{k})|^2}{\rho_0} + |\hat{p}(\boldsymbol{k})|^2 P_{in}(\boldsymbol{k}) \,, \quad (7.33)$$

where $\hat{p}(\boldsymbol{k})$ is the characteristic function of the one-displacement PDF

$$\hat{p}(\boldsymbol{k}) = FT[p(\boldsymbol{u})] \,,$$

and we have used

$$FT[\xi_{in}(\boldsymbol{r})] = P_{in}(\boldsymbol{k}) - 1/\rho_0 \,.$$

By definition $\hat{p}(0) = 1$.

Equation (7.33) gives the relation between the PSs of the point-particle distribution before and after the application of the uncorrelated displacements field. First of all we note that, if the initial point process is the Poisson one, then, as shown in Chap. 3, $P_{in}(\boldsymbol{k}) = 1/\rho_0$. This implies $P(\boldsymbol{k}) = 1/\rho_0$ too, regardless of the form of $p(\boldsymbol{u})$, i.e., the particle distribution is still Poissonian after the application of the random displacement field. This is quite easy to understand: as the displacement field is without correlations, it tends to randomize the particle distribution, but the Poisson particle distribution is already the most random possible SPP. In general it is evident that this

---

[5] Including the case in which $G_{\mu\nu}(\boldsymbol{r})$ diverges continuously for $\boldsymbol{r} \to 0$.

kind of displacement field can neither increase the degree of correlation nor introduce ordering in the particle distribution.

Secondly, we note that the right hand side of (7.33) is the sum of two terms: the former (which we call the *granularity* term) is proportional to the inverse average mass density of particles $1/\rho_0$, and is independent of the initial PS of the particle distribution, while the latter depends on ρ_0, only through $P_{in}(\mathbf{k})$ which satisfies the condition $P_{in}(\mathbf{k} \to \infty) = 1/\rho_0$ because of the diagonal term of $\tilde{\xi}_{in}(\mathbf{r})$.

If the initial point process is statistically isotropic as well as stationary, then $\tilde{\xi}_{in}(\mathbf{r})$, as aforementioned, is a function only of $r = |\mathbf{r}|$ and $P_{in}(\mathbf{k})$ only of $k = |\mathbf{k}|$. Furthermore, if also the displacement field is statistically isotropic, then $\overline{u^{(\mu)}} = 0$ for each $\mu = 1, \ldots, d$, and $p(\mathbf{u})$ depends only on $u = |\mathbf{u}|$. This implies that also $P(\mathbf{k})$ depends only on k (and $\tilde{\xi}(\mathbf{r})$ on r).

7.7.2 Asymptotic Behavior of $P(k)$ for Small k

It is of interest to study the asymptotic behavior of (7.33) for $k \to 0$. We limit the discussion to the case in which both the point process generating the initial particle distribution and the stochastic displacement field are statistically isotropic. As noted above, with this hypothesis $P_{in}(\mathbf{k}) = P_{in}(k)$, $p(\mathbf{u}) = p(u)$, and $P(\mathbf{k}) = P(k)$.

The first step is to study the small k behavior of the characteristic function $\hat{p}(\mathbf{k})$. By definition we have

$$\hat{p}(\mathbf{k}) = \int d^d u \, e^{-i\mathbf{k} \cdot \mathbf{u}} p(\mathbf{u}) , \qquad (7.34)$$

therefore, as aforementioned $\hat{p}(0) = 1$, and, as $p(\mathbf{u}) = p(u)$, $\hat{p}(\mathbf{k}) = \hat{p}(k)$. Let us suppose that for large u we have

$$p(u) \simeq B u^{-\beta} ,$$

where $\beta > d$ because $p(u)$ must be by definition integrable over all space ($\beta \to +\infty$ includes any decay faster than any power, e.g., exponential). Using this property and the definition of $\hat{p}(k)$, it is simple to show (see Appendix A) that to the lowest order in k larger than zero, one has

$$\hat{p}(k) \simeq 1 - Ak^\alpha \text{ with } \begin{cases} \alpha = \beta - d & \text{if } \beta \leq d+2 \\ \alpha = 2 & \text{if } \beta > d+2 \end{cases} \qquad (7.35)$$

where $A > 0$. This means that if the probability density function $p(u)$ has a finite variance $\overline{u^2} < +\infty$, then at the lowest orders in k, we have

$$\hat{p}(k) \simeq 1 - Ak^2 \text{ with } A = \frac{\overline{u^2}}{2} .$$

If, instead, $\overline{u^2} = +\infty$, then

$$\hat{p}(k) \simeq 1 - A k^{\beta-d} \quad \text{with} \quad A > 0$$

characterizing the singular part of the Taylor expansion. In this case A is completely determined by the large u tail of $p(u)$ i.e., by β and B (see Appendix A).

Note that in all cases $0 < \alpha \le 2$. This implies that for $k \ll A^{1/\alpha}$

$$1 - |\hat{p}(k)|^2 \simeq 2A k^\alpha \;.$$

On the other hand, we have seen in Chap. 3 that for any homogeneous SPP the PS for $k \to 0$ behaves in general as $P_{in}(k) \sim k^b$ with $b > -d$. Therefore we can draw the following conclusions for the small k behavior of $P(k)$ in (7.33):

1. Since $\alpha > 0$ in all cases, if $-d < b \le 0$ or $0 < b < \alpha$, $P(k) \sim k^b$ as for the initial PS $P_{in}(k)$ in the initial particle distribution, and it is independent of the displacement field. This is due to the fact that an *uncorrelated* displacement field cannot destroy the strong correlations already present in the system, as it tends only to increase the random noise in the system.
2. On the contrary if $0 < \alpha \le b$ then, the small k behavior is completely determined by the displacement field, resulting in $P(k) \sim k^\alpha$. In fact, as shown in Chap. 3, an SPP having $b > 0$ can be considered to show a sort of long-range order which is partially destroyed by the noise injected into the system by the uncorrelated displacement field if $b \ge \alpha$. In this respect, note that if $\overline{u^2} = +\infty$ then $\alpha < 2$. Thus, in this case, the randomization of the system introduced by the uncorrelated displacement field is more effective.

7.7.3 The Shuffled Lattice with Uncorrelated Displacements

In this subsection we present a specific but important example of the application of the case just analyzed the random shuffling of a regular lattice of particles (see Fig. 7.3). Its interest lies in the fact that a *perturbed* lattice, of which this is the simplest case, often used as the initial condition for N-body cosmological simulations.

In Chap. 3 we have shown that for a distribution of particles of unitary mass occupying the sites of a regular cubic lattice, the PS is [253]

$$P_{in}(\boldsymbol{k}) = (2\pi)^d \sum_{\boldsymbol{H} \ne 0} \delta(\boldsymbol{k} - \boldsymbol{H}) \;, \tag{7.36}$$

where the sum runs over all the sites $\boldsymbol{H}$ of the *reciprocal* lattice [253] with the exception of the origin $\boldsymbol{0}$. We recall that, if the lattice spacing of the direct space lattice is l, each component of $\boldsymbol{H}$ is any integer multiple of $2\pi/l$ including zero.

We can now apply (7.33) in order to find the final PS $P(\boldsymbol{k})$ (see Fig. 7.4) after the random shuffling (i.e., the random displacement field):

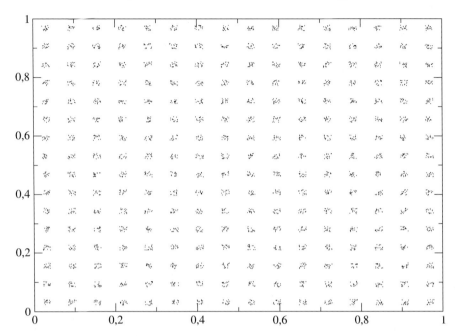

Fig. 7.3. A randomly shuffled lattice in 3 dimensions. This is a projection on the x-y plane along the z-axis. Each lattice site becomes a small square because of the shuffling of the lattice chain along the z-axis

$$P(\mathbf{k}) = \frac{1 - |\hat{p}(\mathbf{k})|^2}{\rho_0} + (2\pi)^d \sum_{\mathbf{H} \neq 0} |\hat{p}(\mathbf{H})|^2 \delta(\mathbf{k} - \mathbf{H}) . \quad (7.37)$$

Two important observations are the following: (1) the random shuffling in general does not erase completely the presence of the so-called *Bragg peaks* (i.e., the delta functions), but only modulates their amplitude and adds a continuous contribution typical of a truly stochastic point process[6]; (2) around $k = 0$ (more precisely in the so-called *first Brillouin zone* [8] of the reciprocal lattice) $P_{in}(\mathbf{k}) = 0$ at any order. Consequently, as clear from (7.37), in this region $P(\mathbf{k})$ is determined by only the behavior of the characteristic function $\hat{p}(\mathbf{k})$ of the displacement field. As shown above, if the displacement field is statistically isotropic, $\hat{p}(\mathbf{k}) \equiv \hat{p}(k)$. Therefore, even though the lattice is anisotropic, the shuffled one has isotropic mass fluctuations at large scales. This implies that, while in a cubic lattice, because of its internal symmetry, calculating $\langle \Delta N^2(R) \rangle$ in spheres of radius R or in cubes of the same size gives completely different scaling behavior in R (which can be considered a

---

[6] The complete cancellation of the Bragg peaks contribution to $P(\mathbf{k})$ is possible only in the very particular case in which $\hat{p}(\mathbf{H}) = 0$ for every reciprocal lattice vector.

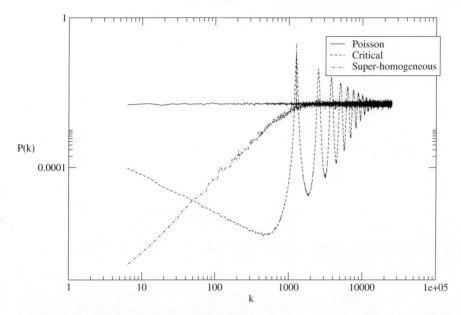

Fig. 7.4. Power spectrum of a shuffled lattice in $d=1$, compared with a Poisson distribution, and to a lattice with correlated critical displacements ($P(k) \sim k^{-0.5}$)

real *pathology* of regular lattices), after the shuffling, the scaling behavior in R of $\langle \Delta N^2(R) \rangle$ has the same exponent in the two cases.

Since in the first Brillouin zone $P(\boldsymbol{k})$ is completely determined by $\hat{p}(\boldsymbol{k})$, the asymptotic behavior at small k of $P(\boldsymbol{k})$ can be summarized in Sect. 7.7.2 and in particular by the relation (7.35). In particular if the variance $\overline{u^2}$ of the displacement field is finite, we find

$$P(\boldsymbol{k}) \sim k^2 \text{ for } k \to 0,$$

independently of the particular form of $p(u)$. This is a case of *universal behavior* for any displacement PDF $p(u) \sim u^{-\alpha}$ at sufficiently large u with $\alpha > d+2$. In the case in which $\overline{u^2}$ diverges (i.e., $d < \alpha \le d+2$), this universality is lost, and we have $P(\boldsymbol{k}) \sim k^a$ with $0 < a < 2$, with a one-to-one correspondence between the exponents α and a as shown in Sect. 7.7.2. A similar case of universality (see Sect. 2.10) is found in random walks with independent steps. In fact if the variance of the steps is finite (ordinary random walks) the average quadratic distance $\langle \Delta x^2(t) \rangle$ reached by the walker after a large number t of steps satisfies the scaling relation $\langle \Delta x^2(t) \rangle \sim t$ independently of the precise functional form of PDF of the single step. On the other hand, if the single step variance is infinite (Levy flights) this is no longer true, $\langle \Delta x^2(t) \rangle$ being infinite, and the PDF of $|\Delta x(t)|$ has a power-law tail with an exponent in a one-to-one correspondence with that characterizing the tail of the PDF of a single step.

7.8 Correlated Displacements

Let us now return to (7.28), and consider the general case of a stationary stochastic displacement field with correlations. In this case $f(\boldsymbol{u}, \boldsymbol{v}; \boldsymbol{r})$ cannot be factorized as in (7.31) for $r > 0$.

In order to write the equation of transformation of the PS let us recall that

$$\int\int d^d r\, d^d r'\, e^{-i(\boldsymbol{k}\cdot\boldsymbol{r} + \boldsymbol{k}'\cdot\boldsymbol{r}')} \tilde{\xi}(\boldsymbol{r} - \boldsymbol{r}') = (2\pi)^d \delta(\boldsymbol{k} + \boldsymbol{k}') P(\boldsymbol{k}) , \qquad (7.38)$$

and that, by definition,

$$\overline{\langle \rho(\boldsymbol{r})\rho(\boldsymbol{r}') \rangle} = \rho_0^2 (1 + \tilde{\xi}(\boldsymbol{r})) . \qquad (7.39)$$

Furthermore, we define the functions $\hat{f}(\boldsymbol{k}_1, \boldsymbol{k}_2; \boldsymbol{r})$ and $F(\boldsymbol{k}_1, \boldsymbol{k}_2; \boldsymbol{q})$ respectively by the following FT's:

$$\hat{f}(\boldsymbol{k}_1, \boldsymbol{k}_2; \boldsymbol{r}) = \int\int d^d u\, d^d v\, e^{-i(\boldsymbol{k}_1\cdot\boldsymbol{u} + \boldsymbol{k}_2\cdot\boldsymbol{v})} f(\boldsymbol{u}, \boldsymbol{v}; \boldsymbol{r}) \qquad (7.40)$$

$$F(\boldsymbol{k}_1, \boldsymbol{k}_2; \boldsymbol{q}) = \int d^d r\, e^{-i\boldsymbol{q}\cdot\boldsymbol{r}} \hat{f}(\boldsymbol{k}_1, \boldsymbol{k}_2; \boldsymbol{r}) . \qquad (7.41)$$

The function $\hat{f}(\boldsymbol{k}_1, \boldsymbol{k}_2; \boldsymbol{r})$ is the characteristic function of the joint two-displacement PDF. By definition $\hat{f}(\boldsymbol{k}_1, \boldsymbol{k}_2; \boldsymbol{r})$ satisfies the following limit conditions:

$$\hat{f}(0, 0; \boldsymbol{r}) = 1 \text{ for any } \boldsymbol{r} ,$$

and

$$\hat{f}(0, \boldsymbol{k}; \boldsymbol{r}) = \hat{f}(\boldsymbol{k}, 0; \boldsymbol{r}) = \hat{p}(\boldsymbol{k}) \text{ for any } r > 0 .$$

By using (7.38), (7.39), and (7.40), we can write:

$$P(\boldsymbol{k}) = \frac{1}{\rho_0}\left(1 - \frac{1}{(2\pi)^d}\int d^d q\, F(\boldsymbol{k}, -\boldsymbol{k}; \boldsymbol{q})\right)$$
$$+ \int d^d r\, e^{-i\boldsymbol{k}\cdot\boldsymbol{r}} \hat{f}(\boldsymbol{k}, -\boldsymbol{k}; \boldsymbol{r})\left(1 + \tilde{\xi}_{in}(\boldsymbol{r})\right) - (2\pi)^d \delta(\boldsymbol{k}) . \qquad (7.42)$$

Note that the term

$$\frac{1}{(2\pi)^d}\int d^d q\, F(\boldsymbol{k}, -\boldsymbol{k}; \boldsymbol{q})$$

must be treated with care: because of the properties of the inversion of the Fourier transform, it cannot be substituted directly with $\hat{f}(\boldsymbol{k}, -\boldsymbol{k}; 0)$ if $f(\boldsymbol{u}, \boldsymbol{v}; \boldsymbol{r})$ is discontinuous at $r = 0$ as in the case of uncorrelated displacements treated above. Instead it must be understood that

$$\frac{1}{(2\pi)^d}\int d^d q\, F(\boldsymbol{k}, -\boldsymbol{k}; \boldsymbol{q}) = \lim_{r\to 0} \hat{f}(\boldsymbol{k}, -\boldsymbol{k}; \boldsymbol{r}) .$$

More precisely, if $f(\boldsymbol{u},\boldsymbol{v};\boldsymbol{r})$ is continuous at $r=0$ i.e.,

$$\lim_{r\to 0} f(\boldsymbol{u},\boldsymbol{v};\boldsymbol{r}) = f(\boldsymbol{u},\boldsymbol{v};0) \equiv \delta(\boldsymbol{u}-\boldsymbol{v})p(\boldsymbol{u}),$$

then we have

$$\frac{1}{(2\pi)^d}\int d^d q\, F(\boldsymbol{k},-\boldsymbol{k};\boldsymbol{q}) = 1.$$

This condition is valid in all the cases in which the stochastic displacement field is a real continuous correlated stochastic process (see below the Gaussian example). In fact, as aforementioned in Chap. 2, in this case there is a theorem [100] saying that the two-displacement correlation function is continuous everywhere, being equal to the one-displacement variance at $r=0$. Thus, in the case of an uncorrelated stochastic displacement field this is no longer true (it is not a truly correlated stochastic process, but a form of white noise), and as already shown $f(\boldsymbol{u},\boldsymbol{v};\boldsymbol{r})$ is discontinuous at $r=0$, giving

$$\frac{1}{(2\pi)^d}\int d^d q\, F(\boldsymbol{k},-\boldsymbol{k};\boldsymbol{q}) = |\hat{p}(\boldsymbol{k})|^2.$$

With this prescription it is simple to recover (7.33) from (7.42) in the case of uncorrelated displacements.

Instead, in the case of a really correlated stationary stochastic displacement field, (7.42) can be rewritten:

$$P(\boldsymbol{k}) = \int d^d r\, e^{-i\boldsymbol{k}\cdot\boldsymbol{r}} \hat{f}(\boldsymbol{k},-\boldsymbol{k};\boldsymbol{r})\left(1+\tilde{\xi}(\boldsymbol{r})\right) - (2\pi)^d\delta(\boldsymbol{k}), \qquad (7.43)$$

or equivalently

$$P(\boldsymbol{k}) = F(\boldsymbol{k},-\boldsymbol{k};\boldsymbol{k}) + \frac{1}{(2\pi)^d}\int d^d q\, F(\boldsymbol{k},-\boldsymbol{k};\boldsymbol{q}) P_{in}(\boldsymbol{k}-\boldsymbol{q}) - (2\pi)^d\delta(\boldsymbol{k}). \qquad (7.44)$$

Equation (7.43) can be further simplified by noting that in the case of spatial statistical stationarity, the effect of the displacement field on the PS of the particle distribution does not depend separately on the couple of displacements $\boldsymbol{u}$ and $\boldsymbol{v}$ applied at two points separated by the distance vector $\boldsymbol{r}$, but only on the relative displacement $\boldsymbol{w}=\boldsymbol{u}-\boldsymbol{v}$. In fact let us call $s(\boldsymbol{w};\boldsymbol{r})$ the PDF that two points, separated by the distance vector $\boldsymbol{r}$, undergo a relative displacement $\boldsymbol{w}$; clearly, by definition

$$s(\boldsymbol{w};\boldsymbol{r}) = \int\int d^d u\, d^d v\, f(\boldsymbol{u},\boldsymbol{v};\boldsymbol{r})\delta(\boldsymbol{w}-\boldsymbol{u}+\boldsymbol{v}). \qquad (7.45)$$

If we take the FT of (7.45) with respect to $\boldsymbol{w}$, we have

$$\hat{s}(\boldsymbol{k};\boldsymbol{r}) = \hat{f}(\boldsymbol{k},-\boldsymbol{k};\boldsymbol{r}), \qquad (7.46)$$

where

$$\hat{s}(\boldsymbol{k};\boldsymbol{r}) = \int d^d u\, e^{-i\boldsymbol{k}\cdot\boldsymbol{w}} s(\boldsymbol{w};\boldsymbol{r})\,.$$

Therefore (7.43) can be rewritten

$$P(\boldsymbol{k}) = \int d^d r\, e^{-i\boldsymbol{k}\cdot\boldsymbol{r}} \hat{s}(\boldsymbol{k};\boldsymbol{r}) \left(1 + \tilde{\xi}_{in}(\boldsymbol{r})\right) - (2\pi)^d \delta(\boldsymbol{k})\,. \tag{7.47}$$

It is important to note that, while $\tilde{\xi}_{in}(\boldsymbol{r})$ depends on ρ_0 (average number density of the particle distribution)[7] at least through its diagonal part $\delta(\boldsymbol{r})/\rho_0$, the probability density function $f(\boldsymbol{u},\boldsymbol{v};\boldsymbol{r})$, in our hypothesis, is in general supposed not to. Therefore, differently from the case of uncorrelated displacements, both (7.43) and (7.44) can be divided into two parts, one dependent on ρ_0, and the other independent of it. Consequently, in (7.43) and (7.44) there is a part depending on the discretization process and another part independent of it. This is of importance in the context of gravitational N-body simulations, in which the continuous density field of matter is usually represented by a distribution of equal mass particles.

In the next subsection the general properties of a correlated displacement field on a given particle distribution will be further clarified through the discussion of a very important example: the correlated Gaussian displacement field. The mathematical treatment is simplified by the fact that (scalar) Gaussian fields have already been extensively analyzed in Sects. 2.7 and 3.6. This case is also of particular interest because of the application to initial condition of cosmological N-body simulations, where Gaussianity of perturbations (i.e., particle displacements) is assumed.

7.8.1 Correlated Gaussian Displacement Field

In this subsection we treat the case of a *one-dimensional* statistically stationary particle distributions $\rho_{in}(x)$ perturbed by a statistically stationary and isotropic stochastic displacement field $u(x)$. The probability density functional $\mathcal{Q}[u(x)]$ can be derived (see Sect. 3.6) to be

$$\mathcal{Q}[u(x)] \sim \exp\left[-\frac{1}{2}\int_{-\infty}^{+\infty}\int_{-\infty}^{+\infty} dx\, dy\, u(x) K(|x-y|) u(y)\right]\,, \tag{7.48}$$

where $K(x)$ (depending only on $|x|$ because of isotropy) is the positive definite correlation kernel of the Gaussian displacement field. We have $\bar{u}=0$, because of the supposed statistical isotropy.

---

[7] It is simple to verify that if all the particles of the distribution have the same *non-unitary* mass $m > 0$, the connected and normalized two-point correlation function $\tilde{\xi}_{in}(\boldsymbol{r})$ of the microscopic *mass* density is independent of m. It coincides with the connected and normalized two-point correlation function of the microscopic *number* density.

7.8 Correlated Displacements

By applying the recipe presented in Sect. 3.6, it is simple to find the connected two-displacement correlation function to be

$$g(x) \equiv \overline{u(x_0 + x)u(x_0)} = FT^{-1}\left[\frac{1}{FT[K(x)]}\right], \qquad (7.49)$$

which depends only on $|x|$, and is a continuous function as $u(x)$ is assumed to be a continuous stochastic process. The joint two-displacement PDF can be written as:

$$f(u,v;x) = \frac{1}{2\pi\sqrt{g^2(0) - g^2(x)}} \exp\left[-\frac{g(0)(u^2 + v^2) - 2g(x)uv}{2(g^2(0) - g^2(x))}\right], \qquad (7.50)$$

where $g(0) = \overline{u^2} \equiv \overline{v^2} < +\infty$. It is simple to verify that, in order to have the PDF (7.50) well defined at all x, the correlation function $g(x)$ must satisfy the following constraint:

$$|g(x)| \le g(0) \text{ for any } x \ne 0.$$

By taking the FT of (7.50) with respect to both u and v, we find the following simple relation for $\hat{f}(k, -k; x)$[8]:

$$\hat{f}(k, -k; x) = e^{-k^2(g(0) - g(x))}. \qquad (7.51)$$

Therefore the relation (7.43) between the PS of the particle distribution after the application of the displacement field, and its initial correlation function $\tilde{\xi}_{in}(x)$ will be

$$P(k) = e^{-k^2 g(0)} \int_{-\infty}^{+\infty} dx\, e^{-ikx + k^2 g(x)} \left(1 + \tilde{\xi}_{in}(x)\right) - 2\pi\delta(k). \qquad (7.52)$$

Note that since at large x both $g(x)$ and $\tilde{\xi}_{in}(x)$ go to zero, the FT in (7.52) is not well defined and contains a Dirac delta function contribution compensating the last delta function term. This can be made clearer in the following way. The Dirac delta function of (7.52) can be rewritten as:

$$2\pi\delta(k) = 2\pi e^{-k^2 g(0)}\delta(k) = e^{-k^2 g(0)}\int_{-\infty}^{+\infty} dx\, e^{-ikx}.$$

Using this relation, (7.52) can be recast in a form containing only well defined FTs and no artificial delta function contribution:

---
[8] Note that if the Gaussian displacement field is a truly continuous correlated SSP and, consequently, $\lim_{x \to 0} g(x) = g(0)$, then $\hat{f}(k, -k; 0) = 1$. If instead it is uncorrelated (i.e., $g(x) = 0$ for $x \ne 0$), then as shown in the previous section, $\hat{f}(k, -k; 0) = \exp[-k^2 g(0)] \equiv |\hat{p}(k)|^2$, as $\hat{p}(k) = \exp[-k^2 g(0)/2]$ is the characteristic function of the one-displacement PDF.

$$P(k) = e^{-k^2 g(0)} \left[\int_{-\infty}^{+\infty} dx\, e^{-ikx} \left(e^{k^2 g(x)} - 1 \right) + \int_{-\infty}^{+\infty} dx\, e^{-ikx + k^2 g(x)} \tilde{\xi}_{in}(x) \right] . \tag{7.53}$$

In cosmological applications the second FT contribution is neglected (see for instance [227]). It is interesting to study the small k behavior of (7.53). By considering that $g(x)$ is a bounded function and using the relation $FT[\tilde{\xi}_{in}(x)] = P_{in}(k)$, we can write for $k \to 0$:

$$P(k) \simeq [1 - k^2 g(0)][P_{in}(k) + k^2 \hat{g}(k)] + \frac{k^2}{2\pi} \int_{-\infty}^{+\infty} dq\, \hat{g}(q) P_{in}(k - q) \tag{7.54}$$

where $\hat{g}(k) = FT[g(x)]$ is the PS of the displacement. A similar relation can be found in $d > 1$ and also in the case of non-Gaussian displacement fields.

Equation (7.54) is more complex than the result obtained in a *naive* way in Sect. 7.7 by simply using the continuity equation for the conservation of mass which led to:

$$P(k) = k^2 \hat{g}(k) .$$

We see that with respect to this simple approximation, even in the small k limit (i.e., large scales), there are different corrections coming both from the internal correlations of the initial mass distribution and from the interaction between these and the internal correlation structure of the displacement field.

Before concluding this discussion about the Gaussian displacement field, it is useful to observe that in order to extend the discussion to $d > 1$ dimensions, even in the case of spatial statistical stationarity and isotropy, the more general form of $f(\boldsymbol{u}, \boldsymbol{v}; \boldsymbol{r})$ is much more complex. In fact in this case not only parallel components of $\boldsymbol{u}$ and $\boldsymbol{v}$ can be correlated, but also the perpendicular components. This leads to a displacement-displacement correlation *matrix* depending on $\boldsymbol{r}$ and invariant under spatial rotations. In particular for a d-dimensional spatially stationary Gaussian displacement field, we can write:

$$\hat{s}(\boldsymbol{k}; \boldsymbol{r}) \equiv \hat{f}(\boldsymbol{k}, -\boldsymbol{k}; \boldsymbol{r}) = \exp\left[-\sum_{\mu,\nu}^{1,d} k^{(\mu)} k^{(\nu)} [G_{\mu,\nu}(\boldsymbol{0}) - G_{\mu,\nu}(\boldsymbol{r})] \right] , \tag{7.55}$$

where $k^{(\mu)}$ is the μth component of $\boldsymbol{k}$, and the displacement-displacement correlation matrix $G_{\mu,\nu}(\boldsymbol{r})$ is defined by (7.21). If moreover the Gaussian displacement field is also isotropic then $\boldsymbol{u}_0 = \bar{\boldsymbol{u}} = 0$, and

$$G_{\mu,\nu}(\boldsymbol{r}) = g_\|(r) \text{ if } \mu = \nu$$
$$G_{\mu,\nu}(\boldsymbol{r}) = g_\perp(r) \text{ if } \mu \neq \nu$$

However the general features of the effect of a Gaussian displacement field are well described by the one-dimensional case we have analyzed.

7.9 Summary and Discussion

In this chapter we have considered various aspects of the problem of how to represent with a set of point-like masses the mass density fluctuations of standard cosmological models. This problem is interesting because numerical simulations for the formation of structure through the action of gravity require such a discrete representation of the initial fluctuations prescribed by these models. In a first approach we have briefly described one method, different from the standard one, which could in principle be used to generate these configurations: it involves determining the appropriate two-body interaction potential which will lead to such correlations at thermal equilibrium. In the second part of the chapter we have given a thorough generalized analysis of the procedure used in the standard method of generating such configurations by the superimposition of a displacement field on an initial point-particle configuration, usually taken to be a perfect lattice. We have studied rigorously the changes induced in the two-point correlation properties of the particle distribution by the displacements. We have seen explicitly that there are contributions to the PS (and to the two-point correlation function) additional to what is given by the usual *naive* result (and adapted in the standard method), which in particular involves the approximations of neglecting the correlation properties of the initial particle distribution and from the finite (i.e., non-infinitesimal) size of the applied displacements. We have distinguished the two cases of uncorrelated (i.e., random shuffling) and correlated displacement fields, giving for both cases the rigorous equations of transformation of the PS of the particle distributions. Moreover we have studied in particular the specific cases of (1) the random shuffling of a regular lattice array of particles, and (2) a correlated Gaussian displacement field.

8 Galaxy Surveys: An Introduction to Their Analysis

8.1 Introduction

We now turn to observations. This is where the wider framework offered by modern statistical physics to describe structures, which we have developed in the first part of the book, becomes useful. In this chapter we will give a brief basic introduction to the properties of galaxy surveys, and the fundamental features which must be taken into account in extracting information from them about the correlation properties of the distribution of visible matter. We will then discuss how galaxy catalogs probing the three dimensional distribution of matter, as they emerged in the eighties (see Fig. 1.3), made it clear that there is good cause to place in question the assumption that fluctuations become small at the scale of the surveys – recent catalogs (see Fig. 1.4) have revealed even larger structures. This is an assumption which is made in the standard analysis of this data, which uses the language of reduced correlation functions in which fluctuations are described with respect to a well defined positive mean density. This mean density is taken from the sample, which is thus implicitly assumed to have only very small fluctuations itself with respect to the assumed true (asymptotic) mean density.

To undertake an analysis without the a priori assumption of homogeneity in a given sample, one needs to consider the broader framework of statistically stationary and isotropic distributions without a well defined mean density in the finite sample, i.e. fractal-like distributions. It is the purpose of this chapter to provide the general background and then motivation for such an analysis. In the following two chapters we then describe the full details of this analysis applied to galaxy catalogs. In particular in Chap. 9 we describe the use of the average conditional density, which we use to probe the two-point properties, and to detect the presence (or absence) of a cross-over to homogeneity within the sample limit. In Chap. 10 we describe non-averaged statistics – number counts from the origin in redshift and magnitude space – which allow one to probe, again without the assumption of homogeneity (but with less accuracy), the nature of the underlying distribution of visible matter at larger scales.

We will discuss here both three-dimensional (redshift) surveys and two-dimensional (angular) catalogs. We describe in particular the importance of *selection effects and biases:* the galaxies included in a survey are strongly conditioned by the fact that the sample is selected by the very special observer who is located in our galaxy. The most important effect comes from the fact that surveys are limited by the flux of light received by the observer and not that emitted by the galaxies. A survey is thus a (complicated) convolution of galaxy positions *and* luminosities. There are also several other effects related to the particular position of the observer which we will briefly discuss. The construction of *volume-limited* (VL) subsamples, which we describe, is a way of correcting for these selection effects, and gives samples which are most appropriate for a real space statistical analysis of correlation properties without assumptions about the properties of the underlying distribution. There can also be numerous biases intrinsic to the method of galaxy selection (e.g. systematic loss of bright galaxies due to photometric limitations, or of low-surface brightness galaxies due to telescope sensitivity). Such effects, which are not taken into account in the construction of VL samples, must be considered in the interpretation of the results.

Having given this basic introduction to galaxy surveys and the construction of samples for analysis, we turn to considering their striking features: that they reveal the existence of structures on very large scales, up to the limits of current surveys (see Figs. 1.3–1.4). We describe why results obtained with the standard analysis – using instruments which build in the assumption of homogeneity at the scale of the sample – can lead to misleading results and we give the clear *prime facia* case for an analysis of galaxy correlations without this underlying assumption.

8.2 Basic Assumptions and Definitions

The elementary object of our discussion is a galaxy, which is characterized by, among other parameters, its space position and its luminosity. We may define the microscopic number density as in Chaps. 2 and 4:

$$n(\mathbf{r}) = \sum_i \delta(\mathbf{r} - \mathbf{r}_i) , \qquad (8.1)$$

where the sum extends to galaxies of any luminosity. We call this microscopic density $n(\mathbf{r})$ instead of the usual $\rho(\mathbf{r})$ in order to make clear that in this context we deal only with the number of objects and not with their mass (different galaxies can have very different masses).

In general, one should consider the joint conditional probability of finding a galaxy of luminosity L at distance $\mathbf{r}$ from another galaxy, i.e. the (ensemble) conditional average number of galaxies $\langle \nu(L,\mathbf{r}) \rangle_p d^3r dL$ with luminosity in the range $[L, L+dL]$ and in the volume element d^3r at distance r from an observer located on a galaxy. The function $\langle \nu(L,\mathbf{r}) \rangle_p$ can have a complex dependence on $\mathbf{r}$ and L. However here we make for the moment a greatly simplifying *assumption*, as a zero order approximation, that

$$\langle \nu(L,\mathbf{r}) \rangle_p = \phi(L) \times \langle n(\mathbf{r}) \rangle_p . \qquad (8.2)$$

The function $\langle n(\mathbf{r}) \rangle_p$ is the *average conditional density* which, as already discussed in Chaps. 2–4, when multiplied by d^3r gives the probability that, given a galaxy in the origin, there is another galaxy in the volume element d^3r around the position $\mathbf{r}$. The function $\phi(L)$ is the *luminosity function* (or probability density functional – PDF) such that $\phi(L)dL$ gives the probability that a randomly chosen galaxy has luminosity in the range $[L, L+dL]$.

By writing $\langle \nu(r,L) \rangle_p$ as a product of the conditional space density and the luminosity function, we have implicitly assumed that galaxy positions are independent of galaxy luminosity. Although there is clear evidence that there is a correlation between them (e.g. [32]), it has been tested that this is nevertheless a reasonable assumption in the galaxy catalogs available so far (e.g. [140, 223]). We will return subsequently in Chap. 11 to this approximation, and consider an analysis which tests its validity.

The galaxy luminosity function (see Chap. 11) has been found to have a shape characterized by a power-law behavior at faint luminosity, followed by an exponential cut-off at the bright end [32]:

$$\phi(L) = A L^\delta e^{-\frac{L}{L_*}} . \tag{8.3}$$

This is the so-called Schechter function [211] with parameters L_* (luminosity cut-off) and δ (power-law exponent) which can be determined experimentally (see Chap. 11). The pre-factor A is the normalization constant for the luminosity PDF. Note that in models in which $\delta < -1$, (8.3) is not integrable in $L = 0$. For this reason, in this case, an observationally justified lower cut-off for the luminosities $L_{min} > 0$ is introduced (as in a given survey one cannot measure arbitrarily faint objects) guaranteeing the normalization of $\phi(L)$. If $\delta > -1$ the constant A is automatically fixed by δ and L_*, while for $\delta \le -1$ the normalization A depends also on the luminosity lower cut-off L_{min}.

We note that in the literature about galaxy structures (e.g. [185] equation 5.128), instead of (8.2) one usually writes

$$\langle \nu(L,\mathbf{r}) \rangle_p dL = \phi_* \left(\frac{L}{L_*} \right)^\delta e^{-\frac{L}{L_*}} \frac{dL}{L_*} \tag{8.4}$$

for the mean number of galaxies per unit volume with luminosity in the range $[L, L+dL]$. In this case, ϕ_* (number of galaxies per unit volume) is taken to be a constant and hence a purely Poisson nature of the galaxy distribution (i.e. $\langle n(\mathbf{r}) \rangle$ is a constant independent of $\mathbf{r}$) is implicitly assumed. We will work, instead, with a formulation which takes explicitly into account the possibility that the spatial galaxy distribution (in a given sample) is not simply Poisson-like.

8.3 Galaxy Catalogs and Redshift

Galaxy catalogs providing information about the distribution of galaxies at large scales are divided in two classes: two-dimensional and three-dimensional.

In the first "angular catalogs" the angular position on the sky of the galaxies is measured, given by two angular coordinates, conventionally α (right ascension which is in the range $0° \leq \alpha \leq 360°$) and the declination δ (such that $-90° \leq \delta \leq 90°$). Beyond this the galaxy's apparent magnitude m (defined below) is measured, which is directly related to the flux received by the observer from the galaxy. As galaxies have a very wide range of intrinsic luminosities – varying over six orders of magnitude – these catalogs are a projection over a very wide range of distances, giving thus only an indirect probe of the full three dimensional distribution.

For what concerns the three dimensional distribution, much more information is given by so-called "redshift surveys". In addition to what is given in an angular catalog – from which the galaxies are in fact chosen – the redshift of the galaxies

$$z = \frac{\lambda_o - \lambda_e}{\lambda_e} \tag{8.5}$$

is also measured. This is the relative difference of the wavelength of spectral lines observed in the spectrum of the galaxy λ_o and in the laboratory λ_e (see e.g. [206]). The linear Hubble law [118] relates this redshift (for small redshifts $z \ll 1$) to the absolute distance of a given object, measured with other methods (for example "standard candles", i.e. objects of approximately constant intrinsic luminosity):

$$r = \frac{c}{H_0} z \tag{8.6}$$

where c is the speed of light, H_0 is an experimentally determined parameter and r is the distance. Using (8.6) one can infer the distance of the galaxy from its redshift, and thus obtain a three dimensional map. If H_0 is measured in km/sec/Mpc[1] then r is given in Mpc. Recent estimations of the Hubble constant give $55 \leq H_0 \leq 75$ km/sec/Mpc (e.g. [226]). It is common to define a dimensionless parameter h (with $0.55 \leq h \leq 0.75$) and hence to write $H_0 = 100h$ km/sec/Mpc.

The linear Hubble law (8.6) is valid only at low redshift. It is important to note that it is an experimental fact [17], which can be made use of without assuming a cosmological model. Its canonical interpretation is in terms of the framework of Friedmann-Robertson-Walker (FRW) models of the expanding universe (see [17, 18] for a good discussion of the assumptions in these models about the physical origin of redshift). In different cosmological models based on the FRW metric, the relation $r = r(z)$ becomes non-linear when the redshift is of order $0.1 \div 0.2$, where the basic parameter which controls redshift dependence is the deceleration parameter q_0 (see Appendix C where we discuss the main features of the different FRW models). In most of what follows we make use only of (8.6) as we discuss primarily low redshift samples ($z \leq 0.1$ – see Fig. 8.1) for which the linearity of the Hubble law is

---
[1] $1 \, \text{Mpc} \simeq 3 \times 10^{22}$ m.

well established experimentally [226]. When we consider deeper samples (in particular in Chap. 10) we discuss the effect of cosmological corrections to this relation and how such corrections change the results of the analysis of correlation properties at very large scales.

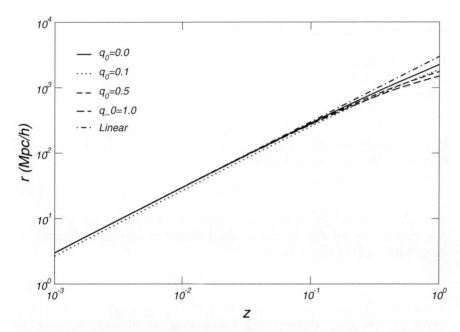

Fig. 8.1. The distance-redshift relation in different FRW models (i.e. for different values of the deceleration parameter q_0) together with the linear law. At small redshift $z \leq 0.1$ all models are very well approximated by the linear behavior

Another important correction to the Hubble law (8.6) is the effect of possible proper motions on the determination of the real distance of a given galaxy. If an object has a velocity $\boldsymbol{v}_p$ with respect to the pure Hubble flow, induced by the presence of matter in its "local" neighborhood, then this peculiar motion gives an additional contribution which adds to the (global) Hubble redshift. In the limit $v_p \ll c$, we may write [113]

$$v_p = c \frac{z_{obs} - z_{cos}}{1 + z_{cos}} \qquad (8.7)$$

where z_{obs} is the galaxy's measured redshift and $z_{cos} = z$ is its cosmological redshift given by unperturbed Hubble flow, and v_p is here the component of the motion along the line of sight. The fact that galaxies have indeed proper motions is another well-established observation [218]. At small scales these thus give rise to systematic distortions of the redshift distribution with respect to the space distribution. In most of what we describe here, this effect will be neglected and thus it needs to be born in mind that the correlation properties we derive are "in redshift space". On the relevant distance scales ($r > 10$ Mpc) the effect of peculiar motions is in general small given the fact that the measured velocities do not exceed $v_p \simeq 500 \div 1000$ km/s.

8.4 Volume Limited Samples

A galaxy of intrinsic luminosity L and at distance r from the observer will be seen to have an *apparent flux* defined as (in the three-dimensional Euclidean space)

$$f = \frac{L}{4\pi r^2}. \tag{8.8}$$

By historical convention the incoming apparent flux f from an object is given in terms of the *apparent* magnitude m defined as

$$m = -2.5 \log_{10} f + C \tag{8.9}$$

where C is a constant fixed by convention (see e.g. [185] and Chap. 10). Analogously, the *absolute* magnitude M is defined in relation to the intrinsic luminosity L of the object by

$$M = -2.5 \log_{10} L + C', \tag{8.10}$$

where the constant C' is related to C of (8.9) by a simple variable transformation (see [185]). From (8.8) it follows that the difference between the apparent and the absolute magnitudes of an object at distance r is (at relatively small distances, neglecting relativistic effects, and properly normalized [185])

$$m - M = 5 \log_{10} r + 25 \tag{8.11}$$

where r, the luminosity distance (see Appendix C), is expressed in Megaparsecs.

A redshift survey consists usually in measuring the redshift of all galaxies with a flux f greater than a certain limiting value f_{lim}, or equivalently (8.9) with apparent magnitude m brighter than apparent magnitude limit m_{lim} (i.e. $m \leq m_{lim}$), given by (8.9) with $f = f_{lim}$ in a certain region of the sky defined by a solid angle Ω. Thus in this kind of survey there is an important selection effect: the sensitivity in absolute luminosity (or equivalently in absolute magnitude) varies as a function of distance. At each distance r there is a specific lower limit to the intrinsic luminosity $L(r)$ that can be seen at that distance. This lower limit in the intrinsic luminosity (see Fig. 8.3) is usually expressed in terms of an upper limit $M(r)$ for the absolute magnitude of the faintest galaxy which can be seen at distance r by using (8.10) and (8.11):

$$M(r) = m_{lim} - 5 \log_{10} r - 25. \tag{8.12}$$

At a distance r only galaxies with absolute magnitude such that $M < M(r)$ are observed. Hence, in particular, at large distances intrinsically faint objects are not observed whereas they are detected at smaller distances.

When one sets out to analyze the spatial statistical properties of galaxy distributions, it is evidently necessary to take this effect into account. There is a simple well known procedure to do so which gives a sample that is not biased by this luminosity selection effect. This consists in the construction of the so-called VL samples (see Figs. 8.2–8.3). A VL sample is constructed by fixing a

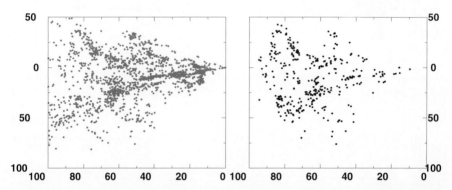

Fig. 8.2. *Left panel*: Cone diagram for a magnitude limited sample (SSRS2; distances are in Mpc/h). *Right panel*: Volume limited sample of the same survey ($M = -19.5$). The depletion at large distances in the magnitude limited sample is the selection effect we are discussing

maximal distance R_{VL} and by taking all the galaxies at a distance $r < R_{VL}$ and with absolute magnitude $M > M_{VL} \equiv M(R_{VL})$ where $M(R_{VL})$ is given by (8.12) with $r = R_{VL}$. In this way we obtain a subsample in which there is no incompleteness due to observational luminosity selection effects [47, 55], as it contains all galaxies satisfying: $m \leq m_{lim}$, $M \geq M_{VL}$ and $r \leq R_{VL}$. Note that the depth of a VL sample R_{VL} is not, in general, the effective distance R_s up to which it is possible to perform a reliable statistical analysis. This is a point we will discuss in the next chapter.

In some surveys (e.g. LCRS, SDSS, etc.) there are two cuts in apparent magnitude: the survey includes all galaxies (or a fraction of them as in LCRS) in a certain sky area, whose apparent magnitude m is in a certain interval ($m_{lim}^1 \leq m \leq m_{lim}^2$). Such a selection means that one cannot measure galaxies fainter than a certain limit, as usual, but also galaxies brighter than the second limit: A VL sample is then identified by two cuts in distances: $R_{VL}^1 \leq r \leq R_{VL}^2$. The two corresponding cuts in absolute magnitude are

$$M_{VL}^1 = m_{lim}^1 - 5\log_{10} R_{VL}^1 - 25 \qquad (8.13)$$
$$M_{VL}^2 = m_{lim}^2 - 5\log_{10} R_{VL}^2 - 25 \qquad (8.14)$$

226   8 Galaxy Surveys: An Introduction to Their Analysis

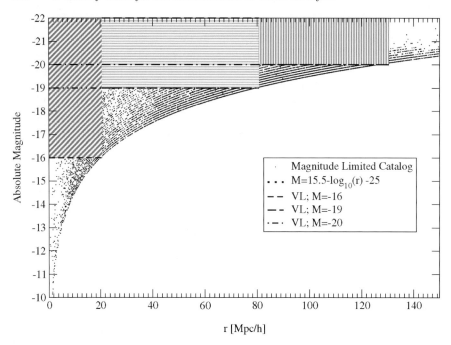

Fig. 8.3. Absolute magnitude versus distance diagram for a magnitude limited sample (in this example the SSRS2 redshift catalog). Observed galaxies lie in the region defined by the curve $M = m_{lim} - 5\log_{10} r - 25$ where $m_{lim} = 15.5$ in this case. Some volume-limited samples (with $M_{VL} = -16, -19, -21$) are also reported: they occupy a rectangular surface in this diagram

and clearly $M_{VL}^2 > M_{VL}^1$. Such a situation reduces both the number of points and the effective volume of the VL samples.

While the effect we have just discussed is the most important selection effect in redshift surveys, it is by no means the only one. The relation (8.12) is, as we noted, true only at small redshift and for deeper samples there are corrections corresponding to the modification of the relation between physical distances and redshift in different cosmological models (see Appendix C). A related further effect is the following: redshift surveys measure flux in a fixed band of wavelengths, which corresponds (because of the redshift) to a different (red-shifted) range of wavelengths in the rest-frame of the observed galaxy. When one reconstructs the absolute flux to construct a VL sample through the procedure previously discussed, one must take this into account, by relating the flux in the galaxy rest-frame to that in the observer rest-frame. To do so requires knowledge of the emission spectrum of the galaxy, and leads to a corrected ("k-corrected") expression

$$M_{VL} = m_{lim} - 5\log_{10} R_{VL} - 25 - K(z) \qquad (8.15)$$

where the term $K(z)$ is often taken also to include the relativistic corrections to the distance-redshift relation. While in the largest modern survey (SDSS [248]) spectral information is available allowing a reliable reconstruction of this correction, in most other redshift surveys the correction is applied quite crudely in an average manner which assumes some simple hypotheses about the spectral properties of the galaxy population sampled from. Like the relativistic corrections to distance, these corrections are negligible for a large part of our analysis and we will return to discuss them when we consider deep samples, and un-averaged statistics which are particularly sensitive to them.

A further related assumption made in this reconstruction of an unbiased VL sample is that we are sampling from the same distribution of galaxies as a function of depth, i.e. that the effect of galaxy evolution (including e.g. mergers of galaxies) is negligible. While again, at low redshift, this is a reasonable hypothesis well supported by observation, it is not so for deeper samples. We will return also to these corrections in Chap. 10 and Appendix D.

8.5 The Discovery of Large Scale Structure in Galaxy Catalogs

Angular catalogs are simple to make compared to redshift catalogs. They are essentially just photographs of the galaxies on the sky, in which the flux from each galaxy point can be inferred from the exposure of the plate. Redshift surveys require the measurement of galaxy spectra, and the determination of an associated redshift, which is a much more involved procedure. Thus while angular surveys with millions of points and covering substantial portions of the sky were available already in the sixties, redshift surveys containing a substantial number of points (of order of thousands) only emerged in the late seventies. Since then there has been continual and dramatic development in both types of surveys, with the number of objects rising by several orders of magnitude. Large redshift surveys from the nineties contain of the order of hundreds of thousands of galaxies (but over small regions of the sky), while the state-of-the-art SDSS survey will provide, in the coming years, about one million galaxy redshifts over a quarter of the sky.

It has been the advent of redshift surveys of galaxies since the late seventies that defined and opened up the problem of "large scale structures". While the angular catalogs which had been previously available showed a very isotropic distribution, with structures/clustering evident only on a relatively small scale (a few Mpc^2), the three dimensional surveys revealed structures

---

[2] For reference: the size of a typical galaxy is less than 0.1 Mpc/h, a distance which is of the order of the typical mean separation between neighboring galaxies. By the "local universe" – where relativistic effects are negligible – is usually meant a region of radius of order of 100 Mpc/h about the earth. The Hubble radius

at much larger scales, something which had not been anticipated from the angular data. Indeed up to scales of some tens and even hundreds of Mpc structures have been observed in the three-dimensional distribution of galaxies – which appear to form a network of filaments with large voids in between. In Fig. 8.4 is shown the distribution of galaxies in a sample from the SSRS2 survey [52], one of the important redshift surveys of the nineties. Alongside it is shown what the same sample would be expected to look like in the absence of spatial correlations. In Fig. 8.5 is shown a sample from the CfA2 redshift survey [158], also from the nineties. This was a larger version of the CfA1 survey [119] which was one of the catalogs which revealed the first evidence for large structures (see also Fig. 1.3). In Fig. 8.6 is shown a sample from the data publically available at the time of writing from the SDSS survey (see also Fig. 1.4).

To date there is no clear visual evidence, as one can see from these figures, of an upper cut-off to the size of galaxy structures. We show also in Fig. 8.7 an image of the APM angular catalog. One sees in this case little apparent sign of large scale structures in the three dimensional maps, which one infers must be "washed out" by the projection onto the celestial sphere.

8.6 Standard Characterization of Galaxy Correlations and the Assumption of Homogeneity

The first studies of galaxy distributions produced the primary result that the reduced two-point correlation function is well approximated, in the range of scales from about 0.1 Mpc/h to 10 Mpc/h, as a simple power-law [232]

$$\xi(r) \sim \left(\frac{r}{r_0}\right)^{-\gamma} \tag{8.16}$$

with $\gamma \approx 1.8$ and $r_0 \approx 4.7$ Mpc/h. This result, obtained in the late sixties, has been confirmed by many other authors in many different redshift surveys (e.g. [55]), in particular in the much larger samples we have just discussed in which larger scale structures became evident.

In terms of the discussion presented in Chap. 4 (see also Chap. 9) the relation (8.16) implies that (i) at small scales, from 0.1 Mpc/h to about 5 Mpc/h, galaxy structures are fractal-like and that (ii) the homogeneity scale λ_0, which marks the transition from the highly fluctuating regime to a regime of small amplitude fluctuations and weak correlations, is $\lambda_0 \approx 5$ Mpc/h. This trend to homogeneity at such a scale is in apparent agreement with the structure-less angular data (see Fig. 8.7), but it is puzzling with respect to the much larger

---

i.e. the scale corresponding to the size of the visible universe in FRW models, is of the order of a few times 10^3 Mpc/h. The new generation of galaxy surveys will sample well the distance scale between one hundred and one thousand Mpc/h.

8.6 Standard Characterization of Galaxy 229

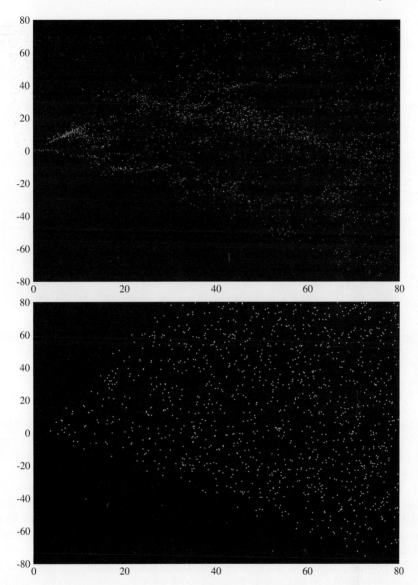

Fig. 8.4. *Upper panel:* Projection in galactic coordinates of the southern sky redshift survey (SSRS2): The earth is in the point (0,0). The survey covers a solid angle in the sky of about 1.13 sr. and it contains 3600 galaxies. The large scale structure visible in the middle of the survey is known as the "southern wall" [52]. (Distances are in Mpc/h). *Bottom panel:* An artificial realization of a random point distribution with the same selection effects as those in the SSRS2 catalog. To mimic the effect of a magnitude-limited sample, each point has been assigned an absolute magnitude using a Schechter probability distribution. The point is then included in the "survey" if its apparent magnitude for the "observer" point is brighter than the survey limit 15.5

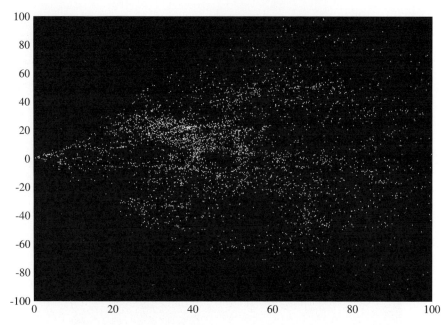

Fig. 8.5. As the upper panel of the previous figure but for the CfA2 South redshift survey: in this case there are 4390 galaxies with magnitude $m_{B(0)} \leq 15.5$ covering $20^h \leq \alpha \leq 4^h$ in right ascension and $-2.5° \leq \delta \leq 90°$. The large scale structure in the middle is called the "Perseus-Pisces chain" [120]

structures observed in the three-dimensional data where, as shown in Figs. 8.4, 8.5 and 8.6, one may see voids of size one order of magnitude larger than 5 Mpc/h: From this perspective it seems that the presence or absence of structures in the data is irrelevant for the determination of r_0.

Thus, according to this analysis, the length r_0 identified through the analysis with the amplitude of the reduced correlation function is a real physical length scale characterizing the galaxy distribution. This interpretation was, however, soon complicated by the fact that while the exponent γ was observed to be relatively stable, the scale r_0 was not measured always to be the same. By the nineties it became evident that this variation was real, and it was observed that it was primarily associated with a variation of the absolute magnitude cut in the sample. A systematic tendency was noted for brighter galaxy samples to show a larger value of r_0 [28, 56, 168, 173, 250]: in the SSRS2 catalog [28], for example, values of r_0 in the range $4 \div 15$ Mpc/h have been measured. For galaxy clusters – which are brighter still – values of r_0 in the range $20 \div 25$ Mpc/h were found [9]. This led to the modified interpretation of r_0 as a real physical scale characterizing galaxy distributions, but with a value that depends on galaxy luminosity. This is the so-called "luminosity bias" effect. Alongside these observations a theory was developed (the theory

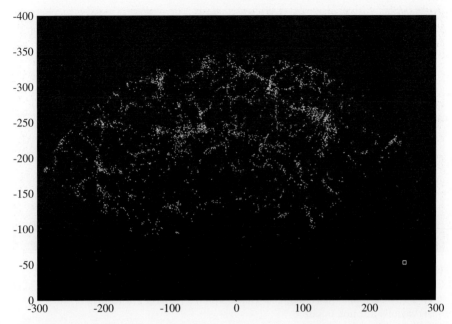

Fig. 8.6. This is a two dimensional slice of thickness 5 Mpc/h of a volume limited sample extracted from the first data release of the SDSS [250]: The earth is in the point (0,0) and the empty region up to $\sim$100 Mpc/h is due to a selection effect and hence it is artificial. In this case there is no bias neither due to luminosity selection effects, nor to orthogonal projection distortions: Structures of hundreds Mpc are still visible (distances are in Mpc/h). The *small square* at the bottom right has a side of 5 Mpc/h

of "bias") which attempted to explain such an effect "biasing" the correlation properties of different objects (see Chap. 13).

We will return in Chaps. 9–12 to the general question of the dependence of the clustering in the galaxy distribution on galaxy luminosity. We will also discuss in Chap. 13 properties of the simplest model of "biasing" in the literature on galaxy correlations. Here we wish to put aside such an interpretation of these data, and return to a more fundamental point: the intrinsic assumption of homogeneity built in when one uses the reduced two-point correlation function to characterize the data of a finite sample. The variation of r_0 can, as we will now explain, be easily understood also as a manifestation of inhomogeneity at the scale of the corresponding samples. This does not discount the possible dependence of such a scale (if it is well defined and independent of the sample size) on galaxy luminosity, but it makes clear that the question of homogeneity must be disentangled from question of the dependence of correlation properties on luminosity.

The reduced two-point correlation function is (see Chap. 9 and Appendixes F-G)

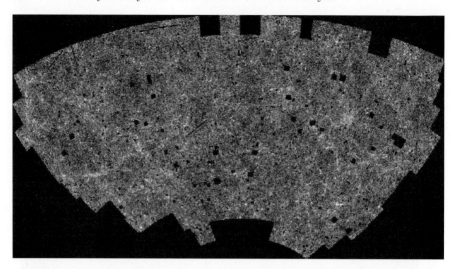

Fig. 8.7. The APM angular galaxy catalog [149]. Note that from this catalog the large scale three-dimensional structures are washed out by projection effects. This survey covers 4300 square degrees. Black squares correspond to bright stars where observations are not possible

$$\xi(r) = \frac{\Gamma(r)}{\langle n \rangle} - 1 \;, \tag{8.17}$$

where $\Gamma(r) = \langle n(\boldsymbol{r}) \rangle_p$ is the average conditional density. When the ensemble mean density $\langle n \rangle$ is estimated in a finite sample it is estimated through the actual density in the sample, i.e. N/V, where N is the number of points in the sample of volume V.

It is simple to show that an increase of the measured r_0 in deeper samples can be due simply to the fact that homogeneity is not yet reached within the limits of the samples. Consider a cubic sample of identical galaxies of size L and suppose that the system reaches homogeneity at scales larger than L. Let us suppose we have two subsamples of respective size L_1 and L_2 with $L_1 < L_2 < L$. Since the system is assumed to be clustered up to scales larger than L the estimate of $\langle n \rangle$ in the two subsamples, respectively n_1 and n_2, will satisfy $n_1 > n_2$. On the other hand the estimate of $\Gamma(r)$ does not depend strongly of the subsample size because it requires only local measurements around each galaxies, and moreover, since the system is clustered and tending far away to homogeneity, it is a decreasing function of r. The two values of r_0 inferred in the two subsamples, defined by $\xi(r_0) = 1$, will be given by

$$\Gamma(r_0^1) = n_1 \qquad \Gamma(r_0^2) = n_2 \;. \tag{8.18}$$

Therefore, as $\Gamma(r)$ is a decreasing function of r, we obtain $r_0^1 < r_0^2$.

Thus it is clear – beyond the simple visual evidence that fluctuations remain large up to scales comparable to the size of galaxy samples – that it is essential in the approach to the characterization of galaxy distributions to start from a framework which does not assume a priori homogeneity of the distribution within the sample limits. Only in this way we will be able to properly characterize the distributions and disentangle the question of luminosity dependence from that on sample depth. Further evidently it is crucial to have a correct characterization of galaxy clustering before setting out to build a theory to explain it.

8.7 Summary and Discussion

We have briefly described here the way in which galaxy surveys probe the underlying galaxy distribution. Redshift surveys have revealed structures comparable to the size of these surveys, with a visual impression of the presence of significant fluctuations up to their limits. This fact alone makes it clearly interesting to place in question the assumption usually made implicitly in analyzing such samples that the distribution approximates very well homogeneity at scales smaller than those of the samples. Further there are clear indications coming from the standard analysis that this implicit assumption can be questioned: as we have explained the observed variation of the scale r_0 has a very simple explanation in terms of a finite size effect in a distribution that has not reached homogeneity at the relevant scales. In the following chapters we describe the analysis of galaxy surveys with statistical instruments which do not depend on the assumption of small scale homogeneity, and which allow one to test for such homogeneity subsequently. We will return after this to the question of luminosity dependence in galaxy clustering.

9 Characterizing the Observed Distribution of Visible Matter I: The Conditional Average Density in Galaxy Catalogs

9.1 Introduction

In this chapter we discuss in detail the analysis of the basic properties of the spatial distribution of galaxies, focusing in particular on the use of, and results obtained with, the conditional average density.

We first address the purely methodological question of how to analyze finite samples without the a priori assumption of homogeneity, and in doing so how to test the hypothesis of homogeneity. We describe how the conditional average density can be used to characterize the correlation properties of distributions, irrespective of whether the condition of spatial uniformity (i.e. homogeneity) is fulfilled or not. We consider various simple cases which can be envisaged – a simple fractal behavior up to the limits of the sample, a simple fractal with a cross-over to homogeneity within the scales probed by the sample – and how these are distinguished by this statistic. In each case an appropriate procedure for the determination of the two-point correlation properties is outlined.

In the second part of this chapter we turn to the actual estimation of the conditional average density. We discuss finite size effects, and how they should be taken into account: we argue that a specific estimator (the so-called *full-shell* (FS) estimator) is an appropriate estimator for an analysis in which the assumption of homogeneity is not built in. Other widely used estimators build in implicitly assumptions about the distribution at the scale of the sample which can give rise to uncontrolled observational errors (see also Appendix G).

We then discuss some results of the application of this estimator of the conditional density to real galaxy catalogs. Up to the scale which can be currently very robustly probed in this way – of the order of $20\,\mathrm{Mpc}/h$ – a behavior consistent with a simple fractal is found, with no definite indication for the existence of a cross-over to homogeneity. The SDSS survey which is currently underway will provide much larger samples appropriate for this analysis, which will allow the homogeneity scale to be definitively detected in this way (or extend the lower bound we give to much larger scales).

9.2 The Conditional Average Density in Finite Samples

As already discussed in the first part of the book, the conditional average density $\Gamma(r) \equiv \langle n(r) \rangle_p$ (defined by (2.51)) is the basic two-point correlation function which is well defined in both the case of distributions with a well defined positive mean density, and those (like fractals) in which there is no such sample independent mean density. It is thus the primary instrument for the analysis we wish to perform of galaxy samples, in which we do not make the prior assumption that the underlying galaxy distribution is homogeneous.

We consider its behavior for particle distributions belonging to the following quite broad class. We suppose them to be characterized principally by two length scales: the *homogeneity scale* λ_0 and the *correlation length* r_c. The former is the scale characterizing the transition to homogeneity, or equivalently, separating a regime of large fluctuations from a regime of small fluctuations. We suppose further that, in the large fluctuation regime, the correlations have a power-law behavior, i.e. the particle distribution has at scales $r \ll \lambda_0$ the features of a simple fractal (see Fig. 9.1). The second scale r_c is the length, as understood generally in statistical physics [116] and defined in Sect. 2.3.2, measuring the spatial persistence of correlation between density fluctuations independently of their amplitude. We do not assume a priori that either of these scales are finite, but from their definition it follows that $r_c > \lambda_0$. Thus we can have:

(i) $\lambda_0 = \infty$, $r_c = \infty$: pure fractal distributions.

(ii) $\lambda_0 < \infty$, $r_c = \infty$: distributions with a well defined mean density, relative to which there are fluctuations which are "critical"-type i.e. long-range correlated (see classification in Sect. 6.6.1). In this case it is the fluctuation field, rather than the whole particle distribution, which has fractal properties.

(iii) $\lambda_0 < \infty$, $r_c < \infty$: distributions with a well defined mean density, relative to which there are fluctuations which are uncorrelated ("substantially Poisson" in the classification in Sect. 3.4) beyond the scale r_c. In this category we can also formally include the case of "super-homogeneous" distributions (see Sect. 3.5), if we identify r_c as the characteristic scale at which fluctuations attain the "super-homogeneous" behavior (i.e. surface fluctuations).

Included in this class of distributions is therefore both the sub-class of fluctuations described by standard cosmological models (as discussed in Chap. 6), and intrinsically irregular distributions such as fractals. The results we now derive can easily be generalized to distributions with a more complex sequence of scaling behaviors, with the introduction of appropriate additional length scales e.g. to include distributions in which the regime of strong clustering is not simply fractal.

Given these assumptions on the class of distributions, we can write more explicitly the form of the conditional density $\Gamma(r)$:

$$\Gamma(r) = A \left[\left(\frac{r}{\lambda_0}\right)^{D-d} f\left(\frac{r}{r_c}\right) + 1 \right], \tag{9.1}$$

9.2 The Conditional Average Density in Finite Samples

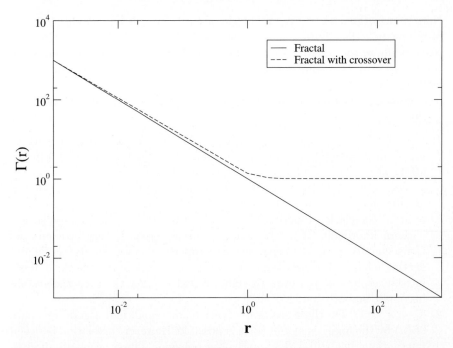

Fig. 9.1. Behavior of the average conditional density for the two principal cases considered: (1) *Dashed line*: Fractal with a crossover toward homogeneity $\lambda_0 < R_s$ (where R_s is the sample size). (2) *Solid line*: Pure fractal structure in the sample, i.e. $\lambda_0 > R_s$

where d is the spatial dimension and $0 < D < d$ is the fractal dimension characterizing the region of strong fluctuations. The function $f(x)$ is called the *cut-off* function and satisfies the following limit conditions: $f(0) = 1$, and it decreases rapidly to zero for $x \gg 1$, e.g. $f(x) = e^{-x}$. In the case of a super-homogeneous distribution, we have an additional condition on $f(x)$ in order to guarantee that the integral of $\Gamma(r) - \langle n \rangle$ (where $\langle n \rangle > 0$ is the unconditional average density) over all the space is zero (see Sect. 3.5). In this case, therefore, r_c can be seen also as the distance beyond which this integral converges sufficiently rapidly. The meaning of the constant A changes in the two cases in which (i) $\lambda_0 < +\infty$ or (ii) $\lambda_0 \to +\infty$. In the first case $A = \langle n \rangle$, i.e. it is given by the "unconditional" average density (see Chap. 2). Instead in the second "fractal" case we have (see Chap. 4)

$$A = \frac{BD}{4\pi \lambda_0^{d-D}}$$

so that

$$\Gamma(r) = \frac{BD}{4\pi} r^{D-d} . \tag{9.2}$$

Note that in the case (i) the amplitude of $\Gamma(r)$ depends on "global" features of the particle distribution such as the average density and the homogeneity scale, while in the fractal case (ii) the amplitude depends on the small scale properties of the distribution, as (see Chap. 4) B is roughly a measure of the average distance between nearest neighbors.

Therefore we can divide the total range of spatial scales roughly into three regimes:

1. $r \ll \lambda_0$: In this range the particle distribution displays fractal features (or, in more general, is strongly irregular with large fluctuations). Fluctuations are so large and strongly correlated that it is impossible to obtain a good estimate of the intrinsic unconditional average density $\langle n \rangle$ by analyzing only this range of scales. This implies that the estimate of the reduced correlation function $\xi(r)$ obtained by studying only these scales, without knowledge of the behavior at larger scales, is very problematic. Consequently, in order to characterize properly the intrinsic statistical and spatial behavior it is necessary to use the framework of the fractal analysis introduced in Chap. 4 even though, on scales larger than λ_0, the system becomes "smooth".
2. $\lambda_0 \ll r \ll r_c$: On these scales fluctuations are small and a good estimate of $\langle n \rangle$ by the analysis at this scale is possible. However correlation between fluctuations has still a slow power-law behavior. Observations limited to this range of scales cannot give a good estimate of the cut-off of correlations. From the analysis of such observations an artificial cut-off due to the *integral constraint* (see Sect. 6.5) can, however, be introduced at the larger accessible scales. More details on this point can be found in the analysis below of finite size effects.
3. $r \gg r_c$: If $P(0) > 0$, the system has in this range of scales essentially the same statistical properties as a Poisson distribution (substantially Poisson case), with fluctuations on these scales which can be considered uncorrelated and completely random. In the super-homogeneous case (i.e. $P(0) = 0$) this is the range of scales over which the long-range ordered (lattice-like) nature of fluctuations starts to be evident.

Before passing to the discussion of the statistical analysis in finite samples, it is important to point out that even though we dispose of data about the particle distribution over a wide set of scales ranging from well below λ_0 to well beyond r_c, the best statistical tools to characterize the distribution can change with scale (see Fig. 9.2). For instance if we are interested in a complete statistical and morphological characterization of the behavior on distances smaller than λ_0 the fractal framework (e.g. mass-length, box counting, conditional variance, void-lacunarity analysis, etc.) is the most appropriate.

We now move to the analysis of the behavior of estimates of $\Gamma(r)$ in a finite sample of size R_s. In this discussion we consider only the *average* behavior of the estimator in many such samples (of size R_s with N points): in particular we neglect the effects of variance and bias (possible offset between

9.2 The Conditional Average Density in Finite Samples

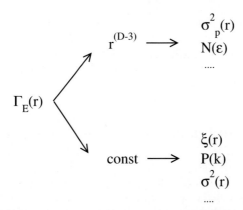

Fig. 9.2. Analysis with the conditional density discriminates between systems with intrinsically large fluctuations at all scales, and those in which there is a scale λ_0 characteristic of homogeneity (defining a regime of small fluctuations). The subsequent step in the study of fluctuations is to use the appropriate statistical tools for each case: for strong clustering this means the conditional variance, box counting, and other suitable statistical quantities introduced in Chaps. 4–5; for weak clustering the reduced two-point correlation, unconditional variance, power spectrum, and other quantities defined in Chaps. 2–3

and estimator and its ensemble value) which we will discuss in Sect. 9.5 below, where we discuss the practical question of estimation in a single sample (see also Appendixes F and G).

In general if R_s is the sample size and N the typical number of particles in it seen by the observer (e.g. from the earth in case of galaxy distribution) we can write the following general form for the "average" estimator of the conditional density:

$$\Gamma_E(r) = A\left(N, R_s, \frac{R_s}{\lambda_0}\right)\left[\left(\frac{r}{\lambda_0}\right)^{D-d} f\left(\frac{r}{r_c}\right) + 1\right], \quad (9.3)$$

where now the amplitude A depends on the characteristics of the sample. In particular, following our discussion above, we expect that dimensionally it has the following behavior

$$A\left(N, R_s, \frac{R_s}{\lambda_0}\right) = \frac{N}{\Omega R_s^d} A'\left(\frac{R_s}{\lambda_0}\right), \quad (9.4)$$

where Ω is the solid angle covered by the sample, which we take to be a portion of a sphere. The function $A'(y)$ will have the following limit behaviors:

$$A'(y) \to 1 \text{ for } y \to \infty. \quad (9.5)$$

and

$$A'(y) \to \frac{\Omega}{\Omega'} y^{d-D} \text{ for } y \to 0, \tag{9.6}$$

where Ω' is a geometrical factor connected to Ω, (which in a spherical sample of radius R_s gives $\Omega' = (3/D)\Omega$) since the average number of points in such a sample is given by $N(R_s) = BR_s^D$).

9.3 Sample Size Smaller than the Homogeneity Scale

We consider first the case that $R_s < \lambda_0$ (see Fig. 9.1). This includes both the case of a pure fractal distribution and the case of a distribution with a well defined homogeneity scale, but when the sample considered is smaller than this scale. Hereafter we will study the three-dimensional case i.e. $d = 3$, and suppose that the sample is a sphere of radius R_s. We have seen that in this case we expect the simple power-law behavior of (9.2), with a sample size *independent* pre-factor $B = \langle N(R_s) \rangle_p / R_s^D$. As discussed in Chap. 4, this factor B is related in a simple way to the average distance between nearest neighbor points in the system. This is again the *average* behavior in finite samples of this quantity, and neglects the systematic and stochastic errors there will be in estimations from a single sample, which are discussed in Sect. 9.5 below.

9.3.1 The Reduced Correlation Function for a Particle Distribution with Fractal Behavior in the Sample

We consider now what one obtains in this same case ($R_s \ll \lambda_0$) by estimating the reduced two-point correlation function $\xi(r)$. This is an extension of the discussion in Chap. 8 of the variation of the scale r_0 (defined by $\xi(r_0) = 1$) in real galaxy catalogs.

A typical estimator $\xi_E(r)$ of the reduced two-point correlation function $\xi(r)$ is given by

$$\xi_E(r) = \frac{\Gamma_E(r)}{\bar{n}_{R_s}} - 1. \tag{9.7}$$

where $\bar{n}_{R_s}$ is the estimate of the mean density in the sample. To get the average behavior of the estimator we take the mean value of $\bar{n}_{R_s}$ in a spherical sample of size R_s which, given that we are in the limit $R_s < \lambda_0$, is simply given by

$$\bar{n}_{R_s} = \frac{\langle N(R_s) \rangle_p}{V(R_s)} \equiv \Gamma^*(R_s) = \frac{3B}{4\pi} R_s^{D-3} \tag{9.8}$$

where we have used (9.2) in writing the last equality. Therefore, using also the expression for $\Gamma_E(r)$ we have

$$\xi_E(r) = \frac{D}{3} \left(\frac{r}{R_s} \right)^{D-3} - 1. \tag{9.9}$$

9.3 Sample Size Smaller than the Homogeneity Scale

From (9.9) it follows that r_0 (defined as $\xi_E(r_0) = 1$) is a linear function of the sample size R_s

$$r_0 = \left(\frac{D}{6}\right)^{\frac{1}{3-D}} R_s \,. \tag{9.10}$$

Thus, in the case that r_0 is measured in samples whose size is smaller than the homogeneity scale λ_0, it is a quantity which is sample dependent, related in a simple way to this size. We stress that this is an *average* behavior which should hold as a mean over many samples, not a deterministic relation between r_0 measured in individual samples of differing sizes. Note that the amplitude of $\Gamma_E(r)$ is related to the lower cut-off (roughly given by average distance between nearest neighbors $\langle \Lambda \rangle$), while the amplitude of $\xi_E(r)$ is related to the upper cut-off (sample size R_s) of the distribution (see Fig. 9.3).

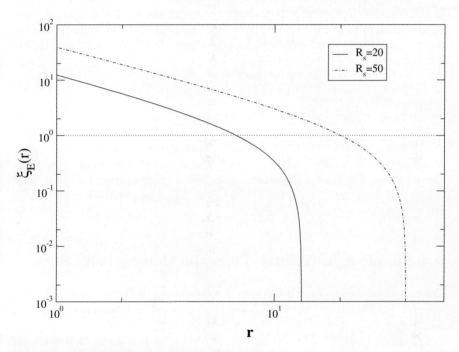

Fig. 9.3. The estimator of the reduced two-point correlation function for a fractal with dimension $D = 2$ in $d = 3$, in two different samples: the first one has size $R_s = 20$ and the second $R_s = 50$. The amplitude of $\xi_E(r)$ depends on the sample size and clearly also the distance-scale r_0 where $\xi_E(r_0) = 1$

Another point which is important to note is the following: if one estimates in this case the fractal dimension by fitting $\xi_E(r)$ with a power-law, there can be a systematic effect which alters the value of D. In fact, while $\Gamma_E(r)$

is a simple power-law over the entire range probed ($r < R_s$), $\xi_E(r)$ is a power-law only for $r \ll r_0$ or $\xi_E \gg 1$, as can be seen from (9.9). At larger separations there is a clear deviation from the power-law behavior. This break in the behavior is due to the finite size of the sample, and evidently does not correspond to any real change in the correlation properties. Note indeed that for any spherical volume V the estimator (9.9) of the reduced two-point correlation function has to satisfy the integral-constraint

$$\int_V \xi_E(r) r^2 dr = 0 \qquad (9.11)$$

which is just a boundary condition in this case. In general any estimator of $\xi(r)$ will be subject to similar constraints, reflecting the fact that the density estimate comes from the sample itself. It is easy to see that if one estimates the exponent at distances $r \lesssim r_0$, one systematically obtains a higher value of the correlation exponent (i.e. smaller value of the fractal dimension) due to the break in $\xi_E(r)$ in a log-log plot. More precisely we can compute the log derivative of (9.9) with respect to $\log(r)$, writing $D - 3$ as γ and its estimate with γ':

$$\gamma' = \frac{d(\log(\xi_E(r)))}{d\log(r)} = \frac{2r_0^\gamma r^{-\gamma}}{2r_0^\gamma r^{-\gamma} - 1} \gamma \qquad (9.12)$$

where r_0 is defined by (9.10). The tangent to $\xi_E(r)$ at $r = r_0$ has a slope $\gamma' = 2\gamma$.

This is a point which, as we discuss in Sect. 9.6 below, is important in the analysis of real galaxy catalogs: It explains a notable discrepancy between the values for the fractal dimension obtained in the literature on galaxy and cluster catalogs: $\gamma \sim 2$ from the $\xi(r)$ analysis [55], compared with $\gamma \sim 1$ with the $\Gamma(r)$ analysis [223].

9.4 Sample Size Greater Than the Homogeneity Scale

We now consider the case that there is a cross-over to homogeneity i.e. $\lambda_0 < \infty$, and that the sample size is sufficiently large to probe this scale i.e. $R_s > \lambda_0$ (see Fig. 9.1).

Using (9.1)–(9.5), and defining $\bar{n}_{R_s} = N/V$ with $V = 4\pi R_s^3/3$, it is simple to see that the behavior of $\Gamma_E(r)$ is a power-law up to the scale λ_0, and then flattens i.e. we have

$$\begin{cases} \Gamma_E(r) = A' \left(\frac{R_s}{\lambda_0}\right) \bar{n}_{R_s} \left(\frac{r}{\lambda_0}\right)^{D-3} & \text{for } \langle \Lambda \rangle \leq r \ll \lambda_0 \\ \Gamma_E(r) \simeq A' \left(\frac{R_s}{\lambda_0}\right) \bar{n}_{R_s} \simeq \langle n \rangle & \text{for } \lambda_0 \ll r \leq R_s \end{cases} \qquad (9.13)$$

Note that

9.4 Sample Size Greater Than the Homogeneity Scale

$$A'\left(\frac{R_s}{\lambda_0}\right)\overline{n}_{R_s}$$

is the estimate from the sample of the asymptotic mean density $\langle n \rangle$. It is, in general, different from $\overline{n}_{R_s}$ as $A'\left(\frac{R_s}{\lambda_0}\right) = 1$ only asymptotically. The difference comes from the fact that the density in the sample is implicitly conditioned on the fact that there is a point at the origin where is the observer (e.g. our galaxy for galaxy surveys). Thus this difference is one which depends on the correlation properties which one does not know. It is a difference, however, which decreases as the ratio R_s/λ_0 increases.

Taking the estimate of $\langle n \rangle$ to be given as in (9.8), an estimator of the reduced two-point correlation function is

$$\xi_E(r) = \frac{\Gamma_E(r)}{A'\left(\frac{R_s}{\lambda_0}\right)\overline{n}_{R_s}} - 1 = \left(\frac{r}{\lambda_0}\right)^{D-3} f\left(\frac{r}{r_c}\right). \quad (9.14)$$

Note that we now have that the amplitude of $\xi_E(r)$ is determined by the homogeneity scale λ_0, which has been previously extracted from the functional behavior of $\Gamma_E(r)$. Therefore it is a real physical scale, rather than being sample dependent as in the previous case ($R_s < \lambda_0$). The relation between the scales r_0 and λ_0 is determined by the function $f(x)$, and will generally be a coefficient of order one.

We now consider case by case the difference generic behaviors of $f(x)$.

9.4.1 Critical Case

We consider first the case of a critical system, i.e. $\lambda_0 \ll R_s < r_c$, which includes both the case of a truly critical system ($r_c = \infty$) or the case of a system with a finite correlation length greater than the sample size.

In this case $\xi_E(r)$ shows a power-law behavior on scales $r \ll R_s$

$$\xi_E(r) \approx \left(\frac{r}{\lambda_0}\right)^{-\gamma}: \quad (9.15)$$

The correlation between (positive and negative) fluctuations about the average density has no intrinsic characteristic scale beyond which is effectively cut-off. The only intrinsic scale of the system is then λ_0, the length-scale around which $\Gamma(r)$ flattens and the fluctuations with respect to the average become small.

Note that any estimator of $\xi_E(r)$ introduces an artificial cut-off at $r \simeq R_s$. For example if we use (9.7) we have the constraint coming from the estimation of the sample average: if we use $\overline{n}_{R_s} = \Gamma(R_s)$ then we get $|\xi_E(R_s)| = 0$. If instead we use $\overline{n}_{R_s} = \Gamma^*(R_s)$ we have that the integral constraint (9.11) must be satisfied: This is simply, as we have discussed, an effect of the boundary

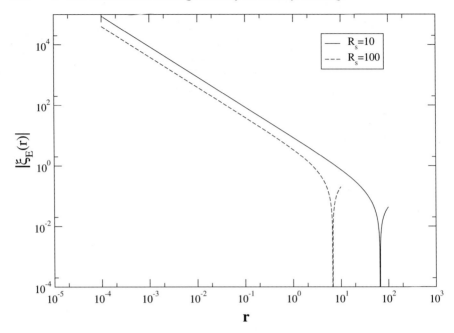

Fig. 9.4. Behavior of the absolute value of the estimator $|\xi_E(r)|$ of the two-point reduced correlation function for a critical system in two different samples with size, respectively, $R_s = 10, 100$. Note that the negative part (i.e. for distances larger than the sharp cut-off) is due to a finite-size effect (integral constraint) which becomes important at scales of order R_s

condition: the fact that $\xi_E(r)$ must be negative beyond a certain scale *does not* imply that the system becomes anti-correlated.

As a simple example, consider the case of a power-law correlation function of the type

$$\xi(r) = \frac{A}{1 + (r/\Lambda)^\gamma} \tag{9.16}$$

for which we show in Fig. 9.4 the behavior of $\xi_E(r)$ obtained using (9.7) and $\bar{n}_{R_s} = \Gamma^*(R_s)$ in two spherical samples respectively with size $R_s = 100, 200$ (the cut-off scale is $\Lambda = 0.01$, the power-law exponent $\gamma = 1$, and the amplitude $A = 100$). The artificial distortion of the estimator occurs always at a scale comparable with the sample size.

9.4.2 Substantially Poisson Case

Next we consider the case of a substantially Poisson distribution, i.e. $\lambda_0, r_c < R_s$. For the estimated reduced correlation function we have again (9.14) with $f(r/r_c)$ now a rapidly decaying function for $r > r_c$. The aim of the analysis with $\xi_E(r)$ would then be to determine the scale r_c at which the function

$f(r)$ is cut-off. In such analysis it is crucial to keep in mind the boundary effects in the estimation of $\xi_E(r)$ mentioned above, coming from the integral constraint. Such constraints force the estimated correlation function to decay rapidly, at the scale of the sample. It is thus crucial in identifying the cut-off length r_c to determine that it is robust to changes in the sample size.

9.4.3 Super-Homogeneous Case

Finally for the case of super-homogeneous distributions we can also determine a scale r_c characterizing the behavior of the reduced correlation function, above which it decays faster that $1/r^d$, e.g. as $1/r^4$ in standard cosmological models (see Chap. 6). This alone is, however, not enough to determine that the distribution is super-homogeneous: for this one needs to study the behavior of integrated quantities, as discussed in Chap. 3. In particular the most characteristic real space feature of these distributions is manifest in the sub-Poisson behavior of the unconditional normalized mass variance. This quantity $\sigma^2(R)$ (see Chaps. 2–3) shows the behavior $R^{-\alpha}$ with $d < \alpha \leq d+1$ and where d is space dimension (see Fig. 9.5). This corresponds in k space to that fact that the power spectrum presents a tail for $k \to 0$ which goes as k^n with $0 < n < 1$ if $d < \alpha < d+1$, and $n \geq 1$ if $\alpha = d+1$.

9.4.4 Some Remarks

Following this lengthy analysis of the behavior of the average conditional density in a finite sample we can draw some important conclusions. In a finite sample with sufficient statistics the average conditional density is the primary tool suitable to investigate the two-point correlation properties of the particle distribution independently of whether the homogeneity scale of the system is greater or smaller than the sample size (see Figs. 9.1–9.2). This is due to the fact that, as shown better in the next section, the estimator of $\Gamma(r)$ at least in full shells, is an average of only local quantities (the conditional densities from single observers) with no a priori assumptions on their behavior, and consequently it is only weakly affected by the finiteness of the sample. Instead the estimate of the reduced correlation function $\xi(r)$, being strongly dependent at any r on the estimate of the "global" average density (which can be strongly dependent on the sample size as it is well defined only in the limit of an infinite sample) can be affected and distorted from the real behavior by the finite size of the sample. More precisely, if the estimator of $\Gamma(r)$ does not display a clear flattening toward the asymptotic average density before the maximal available distance, the estimator of $\xi(r)$ will show spurious cut-off properties. Therefore a correct procedure for the statistical analysis of a given finite sample is to study, first of all, the estimator of $\Gamma(r)$, which gives good information about the correlation properties in the region of strong clustering (i.e. well before the flattening region). Only

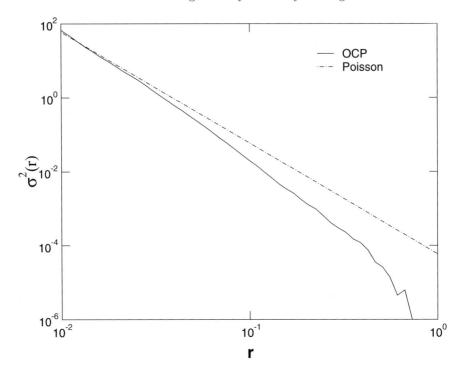

Fig. 9.5. Behavior of unconditional normalized mass variance for a realization of the one-component plasma model (see Chap. 3) which shows its super-homogeneity in decreasing faster than for a Poisson distribution

as a second step, and only if the estimator of $\Gamma(r)$ flattens well before the maximal size available for the statistical analysis, should we estimate $\xi(r)$ in order to investigate the correlation properties also in the flattening region of the average conditional density. Finally, we recall that in all cases the estimate of $\xi(r)$, as explained in Sect. 6.5, is always affected, at the largest scales available, by the distortion due to the so-called "integral constraint".

9.5 Estimating the Average Conditional Density in a Finite Sample

We now consider the question of how practically to estimate the average conditional density in a finite sample without introducing artificial distortions to the intrinsic behavior. The central question here is what the properties are of different estimators i.e. how well they approximate the true underlying behavior of this quantity as a function of scale. For the case (which is ours) where one wishes to analyze the sample without making the assumption of

9.5 Estimating the Average Conditional Density in a Finite Sample

homogeneity, we explain that the most controlled procedure (the standard one in statistical physics [102]) is to use the so-called full-shell (FS) estimator.

9.5.1 Estimators of the Average Conditional Density

An estimator of a statistical quantity should have the property that it converges to the ensemble average value of the quantity as the volume of the sample becomes infinite. A stronger requirement of an estimator is that it should be *unbiased*, which means that also the ensemble average of the estimator in the finite volume is equal to the true ensemble average of the estimated quantity. Typically estimators are not unbiased, but show systematic offsets (i.e. *bias*) with respect to the true (ensemble) value of the quantity [126]. Such effects will depend in some way on the relation between the scale at which the estimator is calculated and the size of the sample, becoming important as the scale of the sample is approached.

We introduce now two classes of estimators of the conditional density $\Gamma(r)$ in a finite sample respectively with and without bias. This discussion gives the basis for our conservative choice of using FS estimates for this quantity.

1. Given a finite sample of N points in a volume V one class of estimators of the average conditional density is defined by

$$\Gamma_E^{(1)}(r) = \sum_{i=1}^{N} n_i(r) w_i \qquad (9.17)$$

where the sum is over all the N points in the sample, and $n_i(r)$ is the numerical density in the spherical shell between radii $r - \frac{\Delta}{2}$ and $r + \frac{\Delta}{2}$ around the ith point. i.e.

$$n_i(r) = \frac{\Delta N_i(r)}{4\pi r^2 \Delta},$$

where $\Delta N_i(r)$ is the number of point in the spherical shell, and where we have used the approximation $\frac{4\pi}{3}[(r+\Delta/2)^3 - (r-\Delta/2)^3] \simeq 4\pi r^2 \Delta$. The coefficient w_i in (9.17) is a weight (which can also depend on r) which is normalized so that

$$\sum_{i=1}^{N} w_i = 1 \ \forall r > 0 \ .$$

It is clear that, for any reasonable weighting scheme, this estimator is a good one for the conditional density, in the weak sense that it approaches the true value of this quantity as the sample becomes infinite (because of the assumption of ergodicity), and as the shell thickness Δ goes to zero. It is easy to see, however, that the estimator is not in general unbiased. This is so because the density in spherical shells is calculated in the same way, irrespective of whether they lie completely inside the sample volume or not. If any of the weights w_i is different from zero for particles i for which the spherical shell is not completely contained in the sample, the estimator is clearly negatively biased as it tends systematically to *underestimate* the conditional density. For instance this is the case in which we give an equal weighting to all the sample points, i.e. $w_i = 1/N$ for $i = 1, \ldots, N$. Clearly this bias becomes more and more important as the distance r approaches the sample size (as more and more shells fall outside the sample volume). Estimating the effect of this negative bias is not possible without making a priori assumptions about the correlation properties of the system. A systematic bias whose amplitude is in general unknown in thus introduced.

2. A second class of estimators can be written in the same form:

$$\Gamma_E^{(2)}(r) = \sum_{i=1}^{N} n_i(r) w_i , \qquad (9.18)$$

but where, now, $n_i(r)$ is the numerical density only in the portion of the spherical shell around the point i contained in the sample limits, i.e.

$$n_i(r) = \frac{\Delta N_i(r)}{\Delta V_i(r)} , \qquad (9.19)$$

where $\Delta N_i(r)$ is the same as in the previous class and $\Delta V_i(r)$ is the volume of the portion of the spherical shell contained in the sample. The weights w_i have the same general properties as in the previous case.

If we assume the hypothesis of statistical stationarity and isotropy of the particle distribution at all scales (which is in general an a priori hypothesis), we can say that, for any choice of w_i, the estimator (9.18) is not only a good one in the infinite sample volume limit, but is also unbiased[1].

However without knowing a priori the correlation properties of the system at all scales (the determination of which is in fact the target of this kind of study!), it is impossible to control the statistical errors if w_i are taken non-zero for those spherical shells only partially contained in the sample. This problem is of course, along with the bias, also present for the previous class of estimators.

We consider this problem a little more explicitly. Let us suppose that we know a priori that the particle distribution is Poisson-like on the spatial scale r. We then expect that the typical statistical fluctuations of $n_i(r)$, given by (9.19), with respect to its average value $\Gamma(r)$, is proportional to $[\Delta V_i(r)]^{-1/2} = [\Omega_i r^{d-1} \Delta]^{-1/2}$, where Ω_i is the solid angle of the portion of the spherical shell contained in the sample.

On the other hand let us suppose now that we know a priori that the particle distribution on the same scale r is fractal-like with dimension $D <$

---

[1] This is true strictly only when we neglect the implicit condition that there is a point (our galaxy) at the apex of the sample i.e. we assume that three point correlation properties are sufficiently weak at the scale of the sample.

9.5 Estimating the Average Conditional Density in a Finite Sample

d. In this case, as a fractal is strongly fluctuating at all scales (see Chap. 4), we expect to have a typical statistical fluctuation of $n_i(r)$ with respect to $\Gamma(r)$ proportional to r^{D-d} (i.e. with the same scale dependence of $\Gamma(r)$ itself), and depending on the solid angle Ω_i of the part of the spherical shell contained in the sample in such a way that the fluctuation increases if this angle is decreased. Therefore we find a very different behavior to that in the Poisson case. Moreover, in order to evaluate the typical error between $\Gamma_E^{(2)}(r)$ and $\Gamma(r)$, it is necessary to know the correlation matrix $C_{ij}(r)$ between $n_i(r)$ and $n_j(r)$ depending on both distances $r_{ij} = |\mathbf{r}_i - \mathbf{r}_j|$ and r. This quantity can change essentially arbitrarily with the correlation properties of the system. In the fractal case we expect $C_{ij}(r)$ for fixed r_{ij} and r to be in general larger than in a Poisson-like distribution.

Thus we see that, in the case that the correlation properties of the system are a priori unknown, the statistical errors of estimators of the conditional density (or indeed any other quantity) become intrinsically uncontrollable when r approaches the sample size.

There is then a simple choice to make in choosing a weighting scheme: either one excludes or includes shells which partially overlap with the sample volume [47, 223]. The former choice leads naturally to the so-called *full-shells* estimator, which is (9.18) (or equivalently (9.17)), with $w_i(r) = \frac{1}{N_c(r)}$ if the spherical shell about the point i is fully included in the volume, and $N_c(r)$ is the number of such points at radius r, and $w_i = 0$ otherwise. The variance in this estimator can be estimated from the sample itself, without using any further assumption about the underlying distribution, by considering the measured fluctuations in the full shell density about different points. No particular hypothesis on the spatial correlation needs to made about the underlying distribution. In particular no hypothesis is made about whether the distribution becomes homogeneous (i.e. uniformly distributed) at any scale.

The use of this estimator is very restrictive: it allows us only to extract information up to, at most, the scale corresponding to the largest sphere which can be inscribed in the survey volume. Given the non-spherical geometry of almost all surveys this corresponds to a much smaller scale than the simple depth of the survey. Estimators which incorporate information from shells partially overlapping the survey volume allow one to get estimates of the average conditional density at much deeper scales. Further such estimators are practically much easier to implement, as they can be calculated by pair counting algorithms (see Appendix G). To estimate their variance one must necessarily make implicit assumptions about the correlation properties of the distribution on scales where only partially contained shells give a significant contribution to the estimator. It is evidently not incorrect in itself to use such estimators – practically they are usually easier to implement – but great caution should be used in interpreting the results obtained with them beyond the scale up to which the FS estimator can be calculated i.e. beyond the radius of the largest sphere which can be inscribed in the sample.

9.5.2 Effective Depth of Samples

In the discussion above we have introduced a scale R_s characterizing the size of the finite sample, without specifying what this scale is. Galaxy samples are never spherical, so what this scale is must be specified. Effectively the meaning of R_s is simply the scale up to which the two-point correlation function can be reliably inferred from the sample. Evidently what scale this corresponds to depends on what one means by "reliably". In the discussion just above we have given an answer to this question: if one does not want to introduce strong a priori assumptions about the nature of the underlying distribution, the scale R_s is determined by the radius of the largest sphere about a galaxy which can be completely inscribed within the sample boundaries (see Fig. 9.6) [47, 223]. It is not necessarily identical with this radius, since of course there is a rapid diminution in the number of independent spheres as this scale is approached, which means an increase in variance of the estimator. We will give a stricter criterion for determining practically in a real sample the upper cut-off for the reliable estimation of the conditional density in the next section.

For the moment we will identify R_s with the radius of the largest inscribed sphere, which we will refer to as the "effective depth" of the sample for the analysis of its correlation properties which we will perform. Different VL samples, extracted from the same catalog, have different values of R_s. The deeper is the VL sample, the greater is its R_s. The effective depth for the whole catalog is then the largest R_s in any VL sample. It is not difficult to calculate R_s for the case of a sample which is a single angular slice. For a catalog with the limits in right ascension $\alpha_1 \leq \alpha \leq \alpha_2$ and declination $\delta_1 \leq \delta \leq \delta_2$ we have that

$$R_s = \frac{R_d \sin(\delta\theta/2)}{1 + \sin(\delta\theta/2)} \qquad (9.20)$$

where $\delta\theta = \min(\alpha_2 - \alpha_1, \delta_2 - \delta_1)$, and R_d is the real radial depth of the sample.

9.6 The Average Conditional Density (FS) in Real Galaxy Catalogs

We now turn, finally, to real data. Our aim here is not to treat this question exhaustively – we refer the reader to [47, 193, 223] and [40, 46, 49, 57, 157, 210, 247] for an extended discussion – but rather to give an example of the implementation of the methods we have discussed. The example we choose

9.6 The Average Conditional Density (FS) in Real Galaxy Catalogs

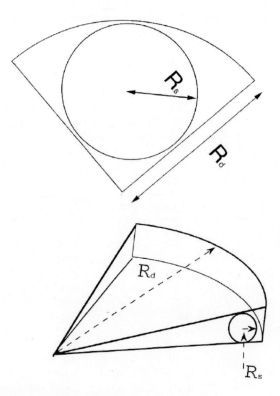

Fig. 9.6. *Upper part:* A typical cone diagram for a wide angle galaxy catalog (e.g. CfA, SSRS, Perseus-Pisces). The depth of the VL sample considered is $R_{VL} = R_d$. The *effective depth* is R_s and it corresponds to the radius of the maximum sphere fully contained in the sample volume ($R_s \lesssim R_d$). *Bottom part:* A typical cone diagram for a narrow angle galaxy catalog (e.g. LCRS, ESP). In this case $R_s \ll R_d$ (from [223])

– the CfA2 galaxy catalog – is also a catalog which allows us to obtain robust bounds on the scale of homogeneity which are representative of what is feasible with current data, using these methods. Other more recent and larger catalogs, in particular the 2dF catalog and early SDSS data, do not allow us actually to derive stricter limits, despite the much larger number of measured redshifts. This is because they contain contiguous pieces of very small solid angle. Thus the effective depth we have defined above remains small. In the case of 2dF there is the further problem that the selection criteria and completeness of the survey are very inhomogeneous over the angles covered, making it intrinsically difficult to measure correlation properties without making the a priori hypothesis of homogeneity. This limitation on this analysis will be relaxed radically in the next years, with the advent of large samples from the SDSS survey.

In Table 9.1 we report some characteristics of the main galaxy redshift surveys completed up to now. There are several important redshift surveys

Table 9.1. Volume limited catalogs characterized by the following parameters: Ω (in steradians) is the solid angle R_s (in Mpc/h) is the radius of the largest sphere that can be contained in the catalog volume (which gives the limit of statistical validity of the sample), N is the number of objects, and $\langle R \rangle$ (Mpc/h) is the average depth (defined for example to be the typical distance of an M^* galaxy) of the catalog. (Adapted from [47, 140, 221, 223])

Sample	Ω (sr)	R_s	N	$\langle R \rangle$
CfA1	1.6	20	1845	30
CfA2	1.23	30	4390	60
Perseus-Pisces	0.9	25	3301	60
SSRS1	1.75	25	1773	30
SSRS2	1.13	30	3600	60
Stromlo-APM	1.3	40	1797	150
LEDA	4π	50	~8000	60
IRAS1.2Jy	4π	40	5313	60
ESP	0.006	10	3175	300
LCRS	0.12	20	~$3 \cdot 10^4$	300
SDSS-DR1	0.14	20	~10^5	300
2dF	0.05	20	~$2.5 \cdot 10^5$	300

(LCRS, ESP, 2dF and the early data release of SDSS) whose average depth (i.e. the distance at which the selection function of the survey has its maximum – see Sect. 10.4 Chap. 10) is about 300 Mpc/h. Their effective depth as we have defined it, however, does not substantially differ with respect to that of the surveys with a much lesser depth (e.g. CfA2 and SSRS2), but greater coverage in solid angle. The great improvement we can anticipate with SDSS is due to the fact that it will cover a solid angle $\Omega \sim \pi$, at a large average depth. Further it will do this with an accurate photometry in 5 bands.

The CfA2 South galaxy sample [120] contains 4390 galaxies with magnitude $m_{B(0)} \leq 15.5$ covering $20^h \leq \alpha \leq 4^h$ in right ascension and $-2.5° \leq \delta \leq 90°$ in declination. This part of the sky includes regions where the galactic extinction has been measured to be important, which needs to be corrected for. For the purposes of comparison of our results with the analysis carried out by [173] we perform these corrections in exactly the same way as these authors, by excluding the same regions: $20^h \leq \alpha \leq 21^h$, $3^h \leq \alpha \leq 4^h$, $21^h \leq \alpha \leq 2^h$, $b > -25°$, and $2^h \leq \alpha \leq 3^h$ and $b > -45°$ where b is the galactic latitude. The catalog contains a large super-cluster called *the Perseus-Pisces chain* and it reveals large voids in the foreground

9.6 The Average Conditional Density (FS) in Real Galaxy Catalogs

Table 9.2. Volume limited samples of the CfA2-South survey. R_{VL} (in Mpc/h) is the depth of the VL sample, M_{VL} is the absolute magnitude limit of the VL sample, N the number of galaxies in the sample $\bar{n} = N/V$ where $V = (\Omega/3)R_{VL}^3$ and Ω is the solid angle of the catalog, R_s (Mpc/h) is the effective depth and D is the estimated fractal dimension in the range $[\langle \Lambda \rangle, R_u]$. (From [137])

Sample	R_{VL}	M_{VL}	N	R_s	$\bar{n}$	$\langle \Lambda \rangle$	R_u	D
VL185	60.2	−18.5	724	13.0	$1.3 \cdot 10^{-2}$	0.5	7	2.0
VL19	74.9	−19.0	622	18.0	$6.0 \cdot 10^{-3}$	0.7	8	1.8
VL195	92.9	−19.5	520	18.5	$2.5 \cdot 10^{-3}$	1.2	12	1.9
VL20	115	−20.0	292	24.0	$7.5 \cdot 10^{-4}$	3.0	15	1.9
VL205	141.7	−20.5	132	29.6	$2.0 \cdot 10^{-4}$	6.0	20	1.8

and background of this super-cluster. The complex galaxy structures found in this catalog have been described by [120]. Note that all the wide angle redshift surveys (i.e. CfA1, SSRS1, SSRS2, Perseus-Pisces) have shown comparable fluctuations.

In Table 9.2 we report the features of the considered VL samples, characterized by successive limits in absolute magnitude with a step of 0.5 magn. The distances have been computed from the linear Hubble law

$$r = cz/H_0 \text{ where } H_0 = 100h \text{ km/sec/Mpc and } 0.5 < h < 1,$$

and the absolute magnitude from the standard relation $M = m - 5\log_{10} r(1+z) - 25 - Kz$, where for the k-correction we have taken $K = 3$ (as in [173]). The results of our analysis are extremely weakly dependent on this latter correction, as they are on modifications to the $r - z$ relation corresponding to different cosmological models, simply because the maximum redshift is so small ($z < 0.05$).

Before discussing the results, let us recall that there are two important physical scales defining the range in which we can reliably infer the underlying correlation properties: (i) The upper cut-off R_s (or effective depth) which we have discussed at length above; and (ii) A lower cut-off $\langle \Lambda \rangle$, which is the average distance between nearest galaxies, and which is related to the number of points contained in the sample. It is simply the scale below which the behavior of the conditional density is dominated by the sparseness of the points (i.e. the shot noise). In the average conditional density $\Gamma(r)$ this regime is characterized by highly fluctuating behavior, while in the integrated conditional average density $\Gamma^*(r)$ (i.e. conditional density in complete spheres) it is manifest as $1/r^3$ decay away from any finite value.

In Fig. 9.7 $\Gamma^*(r)$ is plotted for each of the five samples reported in Table 9.2. The error bars displayed correspond to the variance on 20 "bootstrap" re-samplings of each sample[2]. The behavior of these errors can also

---

[2] For a sample with N points, a random integer is generated between 1 and N for each point. Those points with the same number are discarded to produce

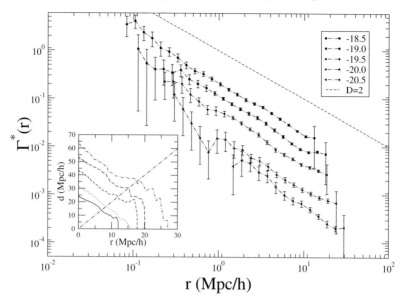

Fig. 9.7. The conditional average density $\Gamma^*(r)$ in complete spheres in the different VL samples of the CfA2-South galaxy sample. The error bars displayed correspond to the variance measured for 20 bootstrap re-sampling of each sample. The values of the power-law fit are reported in Table 9.2. In the inserted panel we show the average distance between the centers of the spheres (see text): the *dashed-dotted line* corresponds to the curve $\langle d \rangle = 2r$ where $\langle d \rangle$ is the average separation between the centers of the spheres and r is the depth. (Adapted from [137])

be understood in terms of the two limits we discussed above: Since there are typically no points at sufficiently small distances in the neighborhood of any given one, we expect $\Gamma(r)$ to fluctuate back and forth to zero, and $\Gamma^*(r)$ to decay away from any finite value as the inverse of the volume, i.e., $\Gamma(r) \sim 1/r^3$. In some cases fluctuations are large also at $r \sim R_s$ because here one is averaging about just a few well separated galaxies, one of which is removed in the re-sampling. It is important to note that although the behavior of these errors is related to the two cut-offs, they *cannot* be taken to be measurement errors for these effects, which are systematic. At small scales the real behavior of the average conditional density is determined by the real lower cut-off and is intrinsically highly fluctuating no matter how well sampled; at large scales no re-sampling of the points around a single galaxy can tell us what the intrinsic variance is in the quantity measured from different independent points.

---

the sub-sample. An alternative way of estimating errors of $\Gamma(r)$ is to measure directly the dispersion in the different $n_i(r)$ (9.17) over which the average is performed.

9.6 The Average Conditional Density (FS) in Real Galaxy Catalogs

An examination of the Fig. 9.7 shows that beyond a scale, which grows with the depth of the VL sample, there is in each case a rather well defined power-law, until a scale near to the upper cut-off R_s at which, in some samples, it shows a deviation toward a flatter behavior. The first scale is just the lower cut-off $\langle \Lambda \rangle$ due to sparseness discussed above. It can be checked quantitatively that its increase in the deeper samples scales with the growing mean distance between points. To perform a fit to these curves we also need to take account of the systematic effect as one approaches $r \sim R_s$, not included in the bootstrap errors, due to the non-averaging. In principle we cannot know how large the error at this scale is since we do not know the real variance in the density at this scale. The criterion we may use to place an upper cut-off up to which we assume this systematic effect is not important is a simple one: We require that in the sample there are a sufficient number of non-overlapping spheres we can average over. The quantitative meaning of this can be read off from the figure inserted in Fig. 9.7 as the point where the average distance becomes equal to twice the depth. Obviously this scale R_u grows with sample size and we see it reaches a maximum of about 20 Mpc/h in our deepest VL sample.

The striking feature of all the samples is that they exhibit a fairly well defined behavior, consisting of two regimes. Beyond a certain scale, which varies from sample to sample, the behavior in each of them is fairly well defined and, we find, well-fitted by a power-law with effective fractal dimension $D \approx 2$ for all the samples. The behavior at smaller r can be well understood in terms of the lower cut-off discussed above. This can be clearly seen by looking also at $\Gamma^*(r)$ in each of the samples. The break in each plot occurs in the same range as that in which $\Gamma(r)$ changes from a highly fluctuating behavior to a well-defined one. In each of the samples we perform a best fit to the dimension D in the range of scales $[\langle \Lambda \rangle, R_u]$ (see Table 9.2). Our result is that the dimension is $D = 1.9 \pm 0.1$ in this range of scales probed by the samples i.e. from 0.5 Mpc/h to 20 Mpc/h.

The normalization of the conditional average density in different VL samples depends on the luminosity selection function of the sample considered. We will discuss this procedure further in the next section. Essentially, making the assumption of independence of the space and luminosity distributions given (8.2) in Chap. 8, i.e.

$$\nu(M,r) = \Gamma(r) \cdot \phi(M) = \frac{DB}{4\pi} r^{D-3} \cdot \phi(M) \tag{9.21}$$

where the luminosity function $\phi(M)$ (now in terms of magnitudes) has been normalized to unity, i.e.

$$\int_{-\infty}^{M_{min}} \phi(M) dM = 1$$

and M_{lim} is the faintest absolute magnitude contained in the sample. We can associate to a VL sample limited at M_{VL} a luminosity factor:

$$\Phi(M_{VL}) = \int_{-\infty}^{M_{VL}} \phi(M)dM \ . \qquad (9.22)$$

The normalization of the space density is then (see Fig. 9.8).

$$\Gamma_{norm}(r) = \frac{\Gamma(r)}{\Phi(M_{VL})} = \frac{DB}{4\pi} r^{D-3} \cdot \frac{1}{\Phi(M_{VL})} \ . \qquad (9.23)$$

From (9.23), we can estimate the parameter B of the distribution. We find $B \approx (12 \pm 2 \mathrm{Mpc/h})^{-D}$, which agrees very well with the value found in various other catalogs [223].

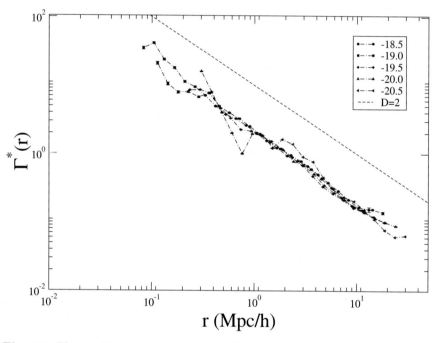

Fig. 9.8. The conditional average density $\Gamma^*(r)$ in complete spheres shown in the previous figure but normalized as discussed in the text

Beyond a distance between 20 and 30 Mpc/h we cannot reliably infer average properties of the galaxy distribution from the CfA2 South catalog alone. We will discuss in the next chapter what can be inferred at larger scales from the analysis of un-averaged quantities like the number counts $N(<r)$ from the origin.

Let us return now to the standard $\xi(r)$ analysis: Our findings of power-law correlations without evidence for a cut-off imply that the normalization to a mean density to derive a "correlation length" is *conceptually* flawed. Calculationally, however, there is nothing wrong with deriving such a scale,

9.6 Average Conditional Density in Volume-Limited Samples

and the results should be perfectly consistent (numerically) with those given here. In Table 9.2 we list the estimations of the volume density $\overline{n}$ in each of the VL samples (see Fig. 9.9). Using this as our normalizing density and

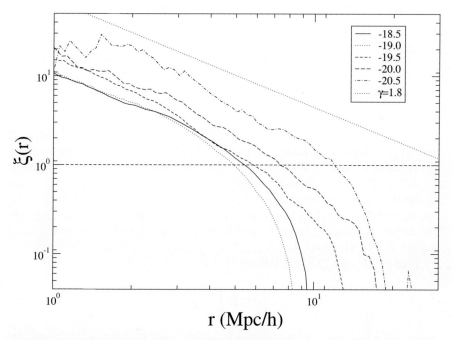

Fig. 9.9. Standard $\xi(r)$ analysis for the same sample shown in Fig. 9.7 obtained by using the full-shell estimator, and using the normalizing average density reported in Table 9.2. This gives values for r_0 in the VL samples shown of approximately from 4 to 10 Mpc/h. Note that the *dotted line* has slope of $-\gamma = -1.8$, while by considering the conditional density one gets $D = 2$. This explanation for this discrepancy is given in Sect. 9.3.1 above

the values of B and D from the measured $\Gamma^*(r)$, we obtain r_0 in the range $5 \div 10$ Mpc/h. To invoke the additional hypothesis of luminosity selection bias (see Chap. 13) to explain all the variation of the observed r_0 is very problematic unless one has first clarified the role of the intrinsic variance in the densities to which one is normalizing to obtain r_0.

9.6.1 Normalization of the Average Conditional in Different VL Samples

We return in this section to give a little more detail on the procedure used for normalizing the different VL samples. The question we address here concerns the estimation of the average conditional density in different VL samples characterized by different M_{VL} and then its normalization to the luminosity function[3]. If we are able to normalize the amplitude of the average conditional density in different VL samples by considering the appropriate luminosity factor (which depends on the integral of the luminosity function and not on the pre-factor of the space conditional density), this implies that the assumption described by (8.2) is a good working hypothesis.

Using (8.2) we may then write, for a constant fractal dimension D, the conditional average number of galaxies as a function of distance as

$$\langle N(<R;>L_{VL})\rangle_p = \int_0^R \int_{L_{VL}}^\infty \langle \nu(r,L)\rangle_p dL d^3 r = B_{VL} R^D \qquad (9.24)$$

where $N(<R;>L_{VL})$ is the number of galaxies in a sphere of radius R and with intrinsic luminosity larger than L_{VL}, and B_{VL} is the amplitude of the number counts in a VL sample with absolute faint luminosity limit at $L = L_{VL}$. From the behavior of (9.24) one may easily determine the value of the fractal dimension. Clearly one may compute $\Gamma^*(r)$ (see Chap. 4) by simply dividing (9.24) for the sample volume. Alternatively one may compute the differential counts in shells which are simply related to the conditional density $\Gamma(r)$ (see Chap. 4).

From (8.2) and (9.24) it follows that

$$B = B_{VL} \frac{1}{L_*^{\delta+1}} \left(\int_{y_{VL}}^\infty y^\delta e^{-y} dy \right)^{-1}. \qquad (9.25)$$

where we have put $y = L/L_*$ and hence $y_{VL} = L_{VL}/L_*$. In a given redshift survey we have seen that one may construct several VL samples which are defined by different cuts in absolute magnitude M_{VL} (i.e. in L_{VL}) and hence different cuts in distance R_{VL}. In Table 9.3 we report some values of B_{VL} determined in the CfA2 galaxy sample (from [140]). The fact that B_{VL} decreases when the modulus of the absolute magnitude is increased, has a straightforward meaning: given a certain volume of space, the probability to find only very bright galaxies (e.g. $M < -20$) is lower than the probability to find both faint and relatively bright galaxies (e.g. $M < -19$).

In such an analysis, one has to consider that, the larger is the distance cut, the brighter is the average absolute magnitude of the galaxies contained in the sample. This is a systematic effect which is intrinsic to the way VL samples have been built. In such a situation it may happen that the measured statistical property changes in different VL (for example the amplitude of the correlation function). Then one would like to establish whether such differences are due to the difference in the luminosity of the galaxies considered *or* due to some sample-size dependent effects. In order to distinguish

---

[3] As shown by (8.10) a upper limit in absolute magnitude corresponds to an lower limit in absolute luminosity L_{VL}.

9.6 Average Conditional Density in Volume-Limited Samples

Table 9.3. Values of B_{VL} estimated in volume limited samples. The absolute magnitude cut is at M_{VL}. In the third column is shown the determined value for the parameter Φ_{VL} (see Sect. 9.6.2) where we have used $M_* = -19.1$ and $\delta = -1.00$ as parameters of the luminosity function. (From [140])

M_{VL}	B_{VL}	Φ_{VL}
-18.0	0.7 ± 0.1	0.9 ± 0.1
-19.0	0.4 ± 0.05	1.6 ± 0.2
-19.5	0.17 ± 0.02	1.5 ± 0.2
-20.0	0.06 ± 0.01	1.8 ± 0.3

between these two effects one may proceed as follows. Given a single VL sample, one may consider additional cuts in absolute magnitude and/or distance if one wants to check, for example, some possible distance versus luminosity effects. This is actually the basic test for what concerns the so-called "luminosity bias" (see below): to cut a VL sample into two distance slices (e.g. $0 < r < R_{VL}/2$ and $R_{VL}/2 < r < R_{VL}$) and then to consider some statistical measure in the two sub-samples obtained in this way. If there is a luminosity dependent effect one should see no change of the statistical properties, while if some statistics depend on the depth of the sample, the effect should be clearly visible.

9.6.2 Estimation of the Conditional Average Luminosity Density

An important quantity which can be determined in a VL sample is the conditional average luminosity density in a sphere of radius R and volume $V(R)$ defined as

$$\langle j(<R)\rangle_p = \frac{1}{V(R)} \int_0^R \int_0^\infty L\langle \nu(r,L)\rangle_p dL d^3r \qquad (9.26)$$

which is R dependent as long as the space average conditional density does not flatten clearly. Considering the simple hypothesis (8.2) we obtain that (in the case $D = $ const.)

$$\langle j(<R)\rangle \equiv j(10)\left(\frac{R}{10h^{-1}}\right)^{D-3}, \qquad (9.27)$$

in $L_\odot \times$ Mpc^{-3}, where $L_\odot$ is the solar luminosity and

$$j(10) = \frac{3}{4\pi} L_*(10h^{-1})^{D-3} \Phi_{VL} \int_{y_{VL}}^\infty y^{\delta+1} e^{-y} dy, \qquad (9.28)$$

and we have defined

$$\Phi_{VL} = \frac{B_{VL}}{\int_{y_{VL}}^\infty y^\delta e^{-y} dy}. \qquad (9.29)$$

Note that, according to (9.27), if, in the considered range of scales, the fractal dimension is smaller than 3 the conditional average luminosity density is a sample-size dependent quantity. Otherwise for $D = 3$, with weak fluctuations and correlations, one obtains the estimator of the unconditional average luminosity density.

In Table 9.3 we report an example of the determination of B_{VL} and Φ_{VL} in various VL samples of the CfA2-South redshift survey. While B_{VL} varies over more than one order of magnitude, Φ_{VL} is almost constant confirming the fact that B, defined by (9.25), is almost independent of the cut in magnitude considered. The results very weakly depend on the cosmological parameters assumed in the reconstruction of distances and absolute magnitudes from measured redshifts and apparent magnitude (see below). The values quoted correspond to the Mattig relation with $q_0 = 0.5$ (see Appendix C), but the results do not sensibly change for any other reasonable choice of q_0 as the redshifts involved are very small ($z \leq 0.05$). Note that a significant dependence of $j(10)$ on the cut in magnitude of the VL can come only from the factor $\int_{y_{VL}}^{\infty} y^{\delta+1} e^{-y} dy$ in (9.28). In fact if $\delta < -2$ this integral is divergent for $y_{VL} \to 0$. Instead if $\delta > -2$ the same integral depends only marginally on y_{VL} [140]. For the relevant values considered here ($\delta \simeq -1$), therefore, this dependence is marginal.

9.6.3 Measuring the Average Mass Density Ω from Redshift Surveys

In cosmology a quantity of importance is the mean density of visible matter. Evidently this is a quantity which is well defined only in the case that there is a well determined transition to homogeneity. What we do here is make this explicit, giving an estimate of this parameter as a function of the homogeneity scale λ_0 (defined appropriately below). Further details can be found in [140]. Essentially we use the previous results on the average conditional luminosity density, making the same assumptions, to derive an estimate of the average mass density. As in the previous section, given a value of (or lower bound on) the homogeneity scale in a clustered mass distribution, it is straightforward to obtain the corresponding value (or upper bound on) the total mass density, once one has an appropriate estimate of the global mass to luminosity ratio.

Taking the values quoted in Table 9.3, from which we infer the average value $\langle \Phi_{VL} \rangle = 1.4 \pm 0.4$, we obtain the numerical value (9.28)

$$j(10) \approx (2 \pm 0.6) \times 10^8 \, hL_\odot/\text{Mpc}^3 \; . \tag{9.30}$$

The fractal dimension D is given by the slope of $\langle N(<r) \rangle_p$ as a function of r in a VL sample. Hereafter we adopt for illustrative purpose the following simplified form:

$$\Gamma(r) = \frac{DB}{4\pi} \begin{cases} r^{D-3} & \text{for } r < \lambda_0 \\ \lambda_0^{D-3} & \text{for } r \geq \lambda_0 \end{cases}$$

9.6 Average Conditional Density in Volume-Limited Samples

with $D = 2$. Note that by changing the definition of the homogeneity scale λ_0, according to the different possibilities give in Chap. 2, we will obtain slightly different numerical estimates.

For a given λ_0 we now find the mass density parameter in units of the critical density $\rho_c = 2.78 \cdot 10^{11} h^2 \, M_\odot/\text{Mpc}^3$ where $M_\odot$ is the solar mass, and as a function of a specified global mass-to-luminosity ratio (in solar and h units), to be

$$\Omega_m\left(\lambda_0, \frac{M}{\mathcal{L}}\right) = \langle j(<\lambda_0)\rangle \frac{M}{\mathcal{L}} \frac{1}{\rho_c}$$

$$= [(6 \pm 2) \times 10^{-4}]\frac{M}{\mathcal{L}} h^{-1} \left(\frac{10h^{-1}}{\lambda_0}\right), \qquad (9.31)$$

where we have used (9.27). Note that because estimates of $M/\mathcal{L}$ are linearly dependent on h, and λ_0 is measured in units of Mpc/h, (9.31) is in fact independent of the Hubble constant.

Let us now consider further our estimate of Ω_m. Taking first the estimate $M/\mathcal{L} \approx 10h$ in the B-band as derived by [81], which corresponds to a global mass to luminosity ratio typical of spiral galaxies, we obtain

$$\Omega_m(\lambda_0) \approx 6 \times 10^{-3} \left(\frac{10h^{-1}}{\lambda_0}\right). \qquad (9.32)$$

With $\lambda_0 \approx 10 \, \text{Mpc/h}$ we obtain the value $\Omega_g \approx 6 \times 10^{-3}$ of the standard treatment (see e.g. [185]). On the other hand we can determine the mass to luminosity ratio which would give a critical mass density Universe. Again, for a given λ_0 we find

$$\left(\frac{M}{\mathcal{L}}\right)_{crit} \approx 1600h \left(\frac{\lambda_0}{10h^{-1}}\right), \qquad (9.33)$$

so that again the canonically quoted value of $(M/\mathcal{L})_{crit} \approx 1600h$ corresponds to the homogeneity scale $\lambda_0 \approx 10 \, \text{Mpc/h}$ (e.g. [185]).

Galaxy clusters have been much studied in recent years, and they are believed to probe well the global mass to luminosity ratio, for which the observed value is $(M/\mathcal{L})_c \approx 300h$ in the B-band (i.e. [114]). Taking this value one obtains

$$\Omega_m(\lambda_0) \approx (0.18 \pm 0.06) \left(\frac{10h^{-1}}{\lambda_0}\right). \qquad (9.34)$$

The value resulting for the same reference value $\lambda_0 = 10 \, \text{Mpc/h}$ is thus $\Omega_m \approx 0.2$ [74, 185]. Note that by using such an high value for the $M/\mathcal{L}$ ratio, we are assuming that this is the typical value associated to all the galaxies, instead of using the number density (or luminosity density) of galaxy clusters in (9.34).

Using the lower bound $\lambda_0 \approx 20 \, \text{Mpc/h}$ obtained above from the analysis of the CfA2 South redshift survey, we obtain the upper bound $\Omega_m \leq 0.1$ on the total mass density (see Fig. 9.10).

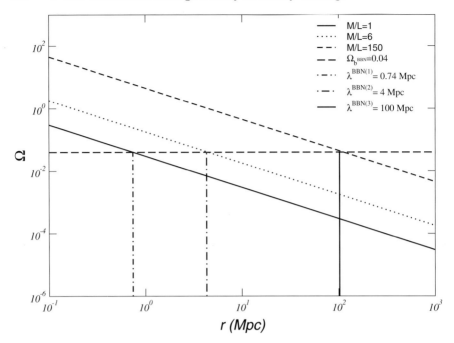

Fig. 9.10. Estimation of the average mass density in the fractal case with a crossover to homogeneity at a generic scale λ_0. The behavior of $\Omega(r)$ is reported for different value of the mass-to-luminosity ratio in the case $h = 0.5$. The *horizontal dashed line* represents the SBBN value for the value of Ω_b

There are some interesting remarks that can be made given this result. One of the important cosmological implications of the measurement of the mass density comes from the comparison of its value with the standard Big Bang nucleosynthesis (SBBN) limits on the baryon density of the universe, which give

$$\Omega_b^{SBBN} h^2 \approx 0.019 \pm 0.004$$

[169]. While this comparison results in the inference of the existence of non-baryonic dark matter given the usually supposed value for the homogeneity scale, as $\Omega_m \gg \Omega_b^{BBN}$, it is interesting to view our result as suggesting another possible solution for this difference: it provides a possible "window of consistency" for the two values. Using the estimate obtained above in (9.34), we find that for the homogeneity scale

$$\lambda_0^{SBBN} = (0.3 \pm 0.15)\mathcal{M}/\mathcal{L}\,\text{Mpc}$$

the dark matter in the Universe can be purely baryonic with its global density satisfying the constraints of SBBN. Conversely an homogeneity scale larger than this value would be inconsistent with the theory of SBBN. Adopting the value $(\mathcal{M}/\mathcal{L})_c \approx 300h$, we find

$$\lambda_0^{SBBN} = (90 \pm 45)h\,\mathrm{Mpc}$$

which, for $h = 0.65$, corresponds to

$$\lambda_0^{SBBN} = (60 \pm 30)\,\mathrm{Mpc},$$

which allows potential compatibility even for values of λ_0 as small as our estimation $\lambda_0 = 20\,\mathrm{Mpc/h} \approx 30\,\mathrm{Mpc}$ (for $h = 0.65$).

Various other methods are commonly used to estimate the mass density of the Universe. One based on galaxy clusters is obtained by observations which constrain the fraction of hot baryonic X-ray emitting gas to the total mass in clusters. By adopting the hypothesis that the rest of the mass may be non-baryonic this gives, when one uses the nucleosynthesis upper bound on Ω_b, a cosmological upper bound on the total mass $\Omega_m \leq 0.3$. Further, taking most of this mass to be non-baryonic one infers a value of Ω_m consistent with the value from the direct estimate. In the present context we note simply that, if the scale λ_0 is larger than the value standardly assumed, the total mass density may be much lower and the dark mass in clusters may quite consistently be baryonic [140].

9.7 Summary and Discussion

We have considered here the use of the average conditional density to describe the correlation properties of galaxies, without the prior assumption of homogeneity. We have explained that this analysis allows one to probe for various generic behaviors – fractality pure and simple, fractal or more general strongly irregular and clustered behavior followed by a cross-over to homogeneity. We have then discussed the practical aspects of the estimation of this quantity in real galaxy samples, explaining the motivation for using the so-called FS estimator which restricts generally the effective depth one can probe to much smaller than the simple depth of the catalog. Turning to real galaxy samples we have given results for the conditional density in the CfA2-South catalog, which shows a behavior clearly indicative of a simple fractal clustering up to a scale of at least $20\,\mathrm{Mpc/h}$, with fractal dimension close to two. This is an extremely conservative conclusion, based on only the most statistically robust results. We refer the reader to [223] for a discussion of a similar analysis in other catalogs. In the last part of the chapter we have also shown several further applications of the conditional density to real data, in particular showing how our analysis modifies standard quantities like the luminosity density and mass density of visible matter into ones which explicitly depend on the homogeneity scale (which we have failed to detect with our methods). In the next chapter we turn to un-averaged statistics, which give less statistically robust results, but allow one to access much larger scales in real galaxy catalogs.

10 Characterizing the Observed Distribution of Visible Matter II: Number Counts and Their Fluctuations

10.1 Introduction

In this chapter we continue our discussion of the analysis of the correlation properties of galaxy distributions, with methods that do not make (but can test for) the assumption that the underlying distribution of galaxies is homogeneous at the scales probed by the galaxy survey. We consider here statistics which do not contain averages over different galaxies as centers, but which are un-averaged, i.e. all with respect to a single observer (on earth). In particular we consider galaxy counts as a function of radial distance, and as a function of apparent magnitude.

Following the same structure as the previous chapter, we first discuss the theoretically predicted behavior of these quantities – both the average ensemble value and the variance about this average. As noted above, in this context the data are given with respect to a single observer, and the average and variance calculated observationally are with respect to a set of samples constructed by considering different parts of the sky. Just as in the previous chapter we consider the theoretical behaviors both in simple fractal distributions and in distributions with a well defined homogeneity scale. In this context we will focus on the generic features which can be used to probe clearly the difference between the case of a fractal and a homogeneous distribution of any type at the sample scale, but we will be less systematic than in the previous chapter in treating the latter case, limiting ourselves to the pure (uncorrelated) Poisson case and the critical case. We focus on two distinguishing tests. Firstly there is the slope of the average number counts, which is directly related to the fractal dimension (with $D = 3$ indicating a homogeneous distribution). Secondly there is the behavior of the relative fluctuations about this average: in a fractal they are (almost) independent of scale, while in any homogeneous distribution they rapidly decrease, typically exponentially.

Most of our analysis will consist in determining the expected behavior of the counts in magnitude limited catalogs. As discussed in Chap. 8 these contain much greater numbers of points than redshift catalogs in which one can compute real space counts, because the latter requires the extra step of a spectroscopic measurements to determine the redshift. We will make the considerable simplification here of limiting our analysis to Euclidean space, and of neglecting evolutionary effects and the so called k-corrections, discussed briefly in Chap. 8. This means that, just as in the previous chapter, our analysis applies well only in the local universe, up to at most a few hundred Mpc. For these magnitude counts – which can probe extremely deep samples – it is interesting and important to go beyond this approximation, and in Appendixes C-D we present some of the relevant considerations in this direction. In particular we consider the effect of cosmological and k-corrections.

In the latter part of this chapter we turn to observations in real galaxy catalogs. Given the assumptions in our analysis, it is appropriate only to consider counts at modest distances, i.e. the bright end in magnitudes. The data in real and magnitude space are very consistent: their average behavior is well fit by an effective fractal dimension $D \approx 2.5$, larger than that found in the analysis of the data we considered in the previous chapter, while their fluctuations appear to continue to be large – to the limits of the samples we consider (typically $\sim 100 \,\mathrm{Mpc/h}$). While the origin of these fluctuations may be due to effects other than large scale structures (calibration problems or photometric errors) the results can also be interpreted as indicating a continuation of the strongly irregular inhomogeneous behavior observed at smaller scales with the analysis described in the previous chapter. We discuss the apparent change in the best-fit dimension – inconsistent with a simple fractal behavior, and possibly indicative of a tendency to homogeneity – and conclude that future observations will be required to confirm whether this is a real average behavior.

10.2 Number Counts in Real Space

A number count in real space is simply the number of points found in a sample as a function of distance. We assume that we are considering VL samples in which selection effects with respect to the observer are absent. Given that the origin of the counts is on a point (our galaxy) the expected behavior is simply determined by the conditional average density integrated over the geometry of the sample. Correspondingly the variance in the number count will be determined by the variance in the conditional average density (i.e. the conditional variance). One wants therefore to compute simply the average counts $\langle N(<R;\ >L_{VL})\rangle_p$ (where $N(<R;\ >L_{VL})$ is defined in Sect. 9.6.1), and the average quadratic fluctuation $\langle \Delta N^2(<R;\ >L_{VL})\rangle_p \equiv \langle N^2(<R;\ >L_{VL})\rangle_p - \langle N(<R;\ >L_{VL})\rangle_p^2$ [90].

The expected behaviors of these quantities are given in the ensemble sense. For a Poisson distribution, using the hypothesis (8.2) for the joint spatial and luminosity distribution of galaxies, we obtain easily (see Chap. 2):

$$\langle N(<R;\ >L_{VL})\rangle_p = \frac{4\pi n_0 R^3}{3} \int_{L_{VL}}^{\infty} dL \phi(L) \tag{10.1}$$

10.2 Number Counts in Real Space

$$\sigma_p^2(R; >L_{VL}) \equiv \frac{\langle N^2(<R; >L_{VL})\rangle_p - \langle N(<R; >L_{VL})\rangle_p^2}{\langle N(<R; >L_{VL})\rangle_p^2} \sim R^{-3},$$

where n_0 is the unconditional average number density and $\phi(L)$ is the PDF of the luminosity of a randomly chosen galaxy. The typical relative fluctuation between $N(<R; >L_{VL})$ seen by a single observer and the average over all the possible observers (i.e. the first of (10.1)) increases with R as $R^{\frac{3}{2}}$. For any homogeneous point distribution (i.e. with a well defined average density and homogeneity scale λ_0) with correlations, in general, $\sigma_p^2(R)$ decreases with R, with the rate of decrease depending on the exact behavior of the two and three-point correlation function. The relevant formulas can be derived easily from the discussion of the mass variance in Chap. 2, and will be given below for the "critical" case when we discuss this case for the counts in magnitude.

For the case of a simple fractal distribution with dimension D and average conditional density

$$\langle n(\mathbf{r})\rangle_p = \frac{DB}{4\pi} r^{D-3}$$

we obtain qualitatively different behaviors (see Chap. 4):

$$\langle N(<R; >L_{VL})\rangle_p = BR^D \int_{L_{VL}}^{\infty} \phi(L)dL$$

$$\sigma_p^2(R; >L_{VL}) \equiv \frac{\langle N^2(<R; >L_{VL})\rangle_p - \langle N(<R; >L_{VL})\rangle_p^2}{\langle N(<R; >L_{VL})\rangle_p^2} \sim \text{const.},$$

(10.2)

on scales sufficiently large that shot noise can be neglected.

There are thus two different basic tests which can be performed with number counts to probe the underlying distribution. On the one hand, by measuring the average conditional counts, the fractal dimension may be estimated. On the other hand one can study the fluctuations, and we expect a qualitatively different behavior in the case of a fractal or homogeneous distribution. For a fractal we expect that fluctuations from a single observer with respect to the average are (almost) proportional to the average itself at any scale (see Chap. 4), i.e. the relative fluctuations are approximately constant as a function of scale, in contrast to the decay of this quantity in any homogeneous distribution. This difference just comes from the fact that in a fractal there are large scale-invariant fluctuations (and in particular large voids) at all scales, while in a homogeneous distributions fluctuations and correlations have a characteristic scale, and become small as one goes to scales larger than the homogeneity scale. Thus this is a very clear cut test which can in principle be used to distinguish the two cases (or variants thereof).

10.3 Number Counts as a Function of Apparent Magnitude

In this section we discuss the counts of galaxies as a function of apparent flux (or apparent magnitude) and their fluctuations. These are interesting statistical quantities to determine in view of the fact that angular galaxy catalogs contain millions of objects and can sample volumes of great depth of space. Despite the fact that there is the intrinsic convolution of space and luminosity properties which is difficult to disentangle, one can perform a series of tests to probe the intrinsic character of fluctuations in these samples. To this end we now derive the expected behavior of the average counts and their variance for different categories of distributions. For clarity we consider the Poisson case separately first, as it contains most of what is needed to understand the behavior in the more general homogeneous case with correlation. Secondly we consider both the cases of a pure fractal distribution and homogeneous distributions with critical fluctuations. In all cases we will use the approximation (8.2) of statistical independence between the position and the luminosity of a galaxy (we will go beyond this hypothesis in the next chapter). Moreover we will call $N(>f)$ the stochastic number of galaxies with an apparent flux larger than f, and with

$$n(f) = -dN(>f)/df$$

the stochastic density of galaxies with a flux f.

10.3.1 Poisson Distribution

Let us now suppose to have an ensemble of angular catalogs, where we may estimate the ensemble average. The ensemble conditional average number of galaxies $\langle n(f) \rangle_p df$ with apparent flux in the interval $[f, f+df]$ is then given with complete generality by

$$\langle n(f) \rangle_p = \int_0^\infty dL \phi(L) \int d^3r \langle n(\mathbf{r}) \rangle_p \delta\left(f - \frac{L}{4\pi r^2}\right), \quad (10.3)$$

where the spatial integral is over all space. For the conditional integrated number of galaxies with flux larger than f (averaged over an ensemble of realizations) we simply have

$$\langle N(>f) \rangle_p = \int_f^\infty df' \langle n(f') \rangle_p = \int_0^\infty dL \phi(L) \int d^3r \langle n(\mathbf{r}) \rangle_p \Theta\left(\frac{L}{4\pi r^2} - f\right), \quad (10.4)$$

where $\Theta(x)$ is the Heaviside step function.

Let us now make the assumption that the galaxies are spatially distributed as an isotropic Poisson particle distribution (see Chap. 2) of average number

10.3 Number Counts as a Function of Apparent Magnitude

density $n_0 > 0$. We know that in this case $\langle n(\boldsymbol{r})\rangle_p = n_0$ at any r, therefore from (10.4) we have that

$$\langle N(>f)\rangle_p = \int_f^\infty df' \langle n(f')\rangle_p = n_0 \mathcal{C}_1 f^{-\frac{3}{2}}, \tag{10.5}$$

where

$$\mathcal{C}_1 = \frac{1}{3\sqrt{4\pi}} \int_0^\infty dL\, \phi(L) L^{3/2}. \tag{10.6}$$

If for $\phi(L)$ we use the usual Schechter function (8.3), then $\mathcal{C}_1$ will be dependent on its luminosity cut-off L_* and the normalization amplitude A. Thus the conditional average differential counts are given by

$$\langle n(f)\rangle_p = -\frac{d}{df}\langle N(>f)\rangle_p = \frac{3n_0}{2}\mathcal{C}_1 f^{-\frac{5}{2}}. \tag{10.7}$$

The average value of the square of the integrated counts is given with complete generality by

$$\langle N^2(>f)\rangle_p = \int_0^\infty dL_1 \int_0^\infty dL_2 \phi(L_1)\phi(L_2) \tag{10.8}$$

$$\int d^3r_1 \int d^3r_2 \langle n(\boldsymbol{r}_1)n(\boldsymbol{r}_2)\rangle_p \Theta\left(\sqrt{\frac{L_1}{4\pi f}} - r_1\right) \Theta\left(\sqrt{\frac{L_2}{4\pi f}} - r_2\right).$$

Since in an isotropic Poisson distribution $\langle n(\boldsymbol{r}_1)n(\boldsymbol{r}_2)\rangle_p = n_0^2 + n_0\delta(\boldsymbol{r}_1 - \boldsymbol{r}_2)$, by using also (10.5), we obtain

$$\langle N^2(>f)\rangle_p - \langle N(>f)\rangle_p^2 = n_0 \mathcal{C}_2 f^{-\frac{3}{2}}, \tag{10.9}$$

where we have defined

$$\mathcal{C}_2 = 4\pi \int_0^\infty dL_1 \int_0^\infty dL_2 \phi(L_1)\phi(L_2) \int_0^\infty dx\, x^2 \Theta\left(\sqrt{\frac{L_1}{4\pi}} - x\right) \Theta\left(\sqrt{\frac{L_2}{4\pi}} - x\right). \tag{10.10}$$

With $\phi(L)$ as given by (8.3) $\mathcal{C}_2$ depends on L_* and A.

From (10.5) and (10.9) we obtain that

$$\sigma_p^2(f) \equiv \frac{\langle N^2(>f)\rangle_p - \langle N(>f)\rangle_p^2}{\langle N(>f)\rangle_p^2} = \frac{\mathcal{C}_2}{\mathcal{C}_1^2}\frac{1}{n_0} f^{\frac{3}{2}}. \tag{10.11}$$

Thus in the limit $f \to 0$ we find that $\sigma_p^2(f)$ decreases as $f^{\frac{3}{2}}$. This is just the flux counterpart of the well known law $\delta N/N \sim N^{-1/2} \sim r^{-3/2}$, describing the decay of the Poisson noise in real space (see Chap. 2), but now translated into the space of apparent flux.

We can easily convert these expressions into magnitude space. From (8.9) we have $f = K \cdot 10^{-0.4m}$ where K is a constant (see (10.31) below). Hence we have (see Fig. 10.1)

$$\langle N(<m)\rangle_p = n_0 C_1 K^{-\frac{3}{2}} 10^{0.6m} \ . \tag{10.12}$$

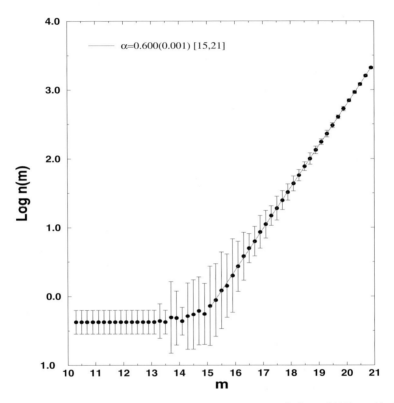

Fig. 10.1. Behavior of the differential average counts $n(m) = d\langle N(<m)\rangle_p/dm$ (in unit of $magn^{-1} deg^{-2}$ where the magnitude is measured in bins of 0.2) as a function of apparent magnitude in a Poisson distribution, where we have used $\phi(L) = \delta(L - L^*)$ as luminosity PDF. Note that the exponent of the counts is $\alpha = D/5 = 0.6$ as expected. The average is performed over 100 angular fields of a single realization

Analogously, from (10.11) we easily obtain the normalized conditional variance in magnitude space:

$$\sigma_p^2(m) = \frac{\langle N^2(<m)\rangle_p - \langle N(<m)\rangle_p^2}{\langle N(<m)\rangle_p^2} = \frac{C_2}{C_1^2} \frac{K^{\frac{3}{2}}}{n_0} 10^{-0.6m} \ , \tag{10.13}$$

10.3 Number Counts as a Function of Apparent Magnitude

This quantity is thus exponentially decreasing, a behavior which should be, in principle, easy to detect in real samples if one has a sufficiently large range of magnitudes available (see Fig. 10.2).

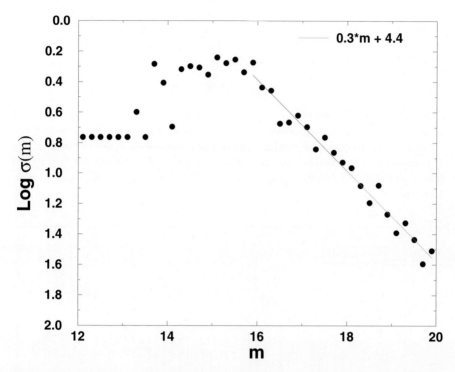

Fig. 10.2. Behavior of the square root of the normalized conditional variance (10.13) as a function of apparent magnitude in a Poisson distribution (for details see the caption of the previous figure)

10.3.2 Simple Fractal Distribution

Let us now consider the case of a fractal distribution. For this calculation we need to use the full three-point correlation function (or two-point conditional density) whose expression is unknown for the general case. Hence firstly we write the general expressions and then we use a simple approximation for the two-point conditional density as in Chap. 4 (see [34]) to determine some useful quantities to be compared with observations.

For the average counts, by considering (10.4) as for the Poisson case, and using (9.21), we readily obtain that for a fractal with dimension D

$$\langle N(>f)\rangle_p = \mathcal{Q}_1 f^{-\frac{D}{2}}, \qquad (10.14)$$

where we have defined

$$\mathcal{Q}_1 = \frac{B}{(4\pi)^{\frac{D}{2}}} \int_0^\infty dL\phi(L)L^{D/2} \, . \qquad (10.15)$$

Analogously to the Poisson case, with $\phi(L)$ given by (8.3) $\mathcal{Q}_1$ depends on A and L_*. Clearly in the limit $D = 3$ we recover (10.5) and (10.6) with $n_0 = DB/(4\pi)$ for the Poisson case. By using again $f = K\,10^{-0.4m}$, it is straightforward to derive the behavior in magnitude space

$$\langle N(<m)\rangle_p = \mathcal{Q}_1 K^{-\frac{D}{2}} 10^{\frac{D}{5}m} \, . \qquad (10.16)$$

The difference with respect to the Poisson case, and in general with respect to any correlated homogeneous particle distribution, lies in the slopes of the counts: in the fractal case the increase of the counts with apparent magnitude is slower since the system is, on average, emptier. The pre-factor in (10.16) has a well-defined constant value, as already discussed in the Poisson case (see Fig. 10.3).

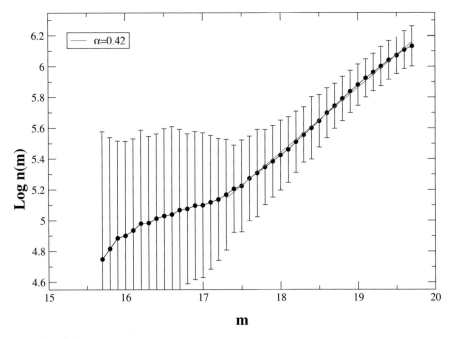

Fig. 10.3. Behavior of the differential average counts $n(m)$ as a function of apparent magnitude in an artificial fractal distribution with dimension $D = 2.1$, that is $\alpha = D/5 = 0.42$ (see Fig. 10.1 for more details)

We can now evaluate the average square value of the counts $\langle N^2(>f)\rangle_p$ using the general relation (10.8). Some hypotheses are necessary in order to evaluate the integrals in that equation, and we use the same approximations as in (4.26)–(4.28) for the two-point conditional density $\langle n(\boldsymbol{r}_1)n(\boldsymbol{r}_2)\rangle_p$ (see Chap. 4). Then from (10.4) and (10.8) in this hypothesis we obtain

$$\langle N^2(>f)\rangle_p - \langle N(>f)\rangle_p^2 = \mathcal{Q}_2 f^{-D} \tag{10.17}$$

where we have defined

$$\mathcal{Q}_2 = \left(\frac{DB}{4\pi}\right)^2 \int_0^\infty dL_1 \int_0^\infty dL_2 \phi(L_1)\phi(L_2) \tag{10.18}$$

$$\int_0^{\sqrt{\frac{L_1}{4\pi}}} dx_1 \int_0^{\sqrt{\frac{L_2}{4\pi}}} dx_2 x_1^{D-1} x_2^{D-1} \int_{4\pi} d\Omega_1 \int_{4\pi} d\Omega_2 g\left(\frac{x_1}{x_2}, \theta\right) ,$$

where $d\Omega_i$, with $i = 1, 2$, is the differential solid angle of $\boldsymbol{x}_i$ and θ is the angle between $\boldsymbol{x}_1$ and $\boldsymbol{x}_2$. Again using for $\phi(L)$ (8.3), $\mathcal{Q}_2$ will depend on A and L_*. Equation (10.17) shows the persistent character of fluctuations which are proportional to $\langle N(>f)\rangle_p$ for any value of f. By passing from f to m with the usual relation, this can also be explicitly shown by computing the normalized conditional variance in magnitude space:

$$\sigma_p^2(m) \equiv \frac{\langle N(<m)^2\rangle_p - \langle N(<m)\rangle_p^2}{\langle N(<m)\rangle_p^2} = \frac{\mathcal{Q}_2}{\mathcal{Q}_1^2} > 0 , \tag{10.19}$$

which is independent of m (i.e. f). The value of the constant in this relation is determined by three-point properties of the fractal (see Sect. 4.4). This constancy translates in magnitude space the scale-invariant nature of the fluctuations characterizing every fractal in real space (see Fig. 10.4).

Thus, when normalized to the average, the fluctuations in the number counts in apparent magnitude are constant as a function of apparent magnitude, while in the Poisson case they decrease exponentially. (As mentioned in a fractal the behavior of such a quantity depends on three-point properties which control the rate of decrease).

10.3.3 Effect of Long-Ranged Correlations in Homogeneous Distributions

Let us now finally consider the case in which the system is homogeneous beyond a certain finite scale with average density $n_0 > 0$, but presents scale-invariant density fluctuations about the average (the "critical" case of our classification in Chaps. 2 and 3). This is the case where the correlation length r_c, which is usually defined as the scale beyond which $\xi(r)$ is exponentially damped (see Chap. 2), is divergent. That is, $\xi(r) \sim r^{-\alpha}$ where $0 < \alpha < 3$ and $\xi(r) < 1$ (the case $\xi(r) \gg 1$ can be included in the previous fractal behavior).

Recall (see Chap. 2) that in the definition of l-point conditional densities for any l the contribution of the origin of coordinates is not considered, so that

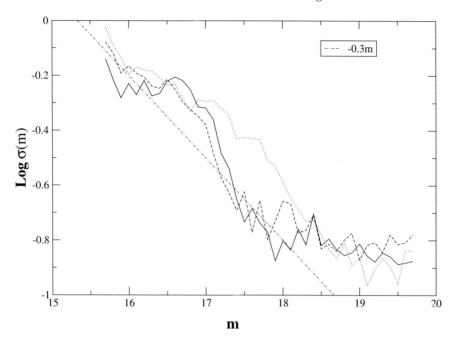

Fig. 10.4. Behavior of the square root of the normalized conditional variance $\sigma_p^2(m)$ as a function of apparent magnitude in a fractal distribution. The different lines correspond to different realizations of the same particle distribution with fractal dimension $D = 2.1$ in $d = 3$. At bright magnitudes there is the additional contribution from Poisson noise (i.e. shot noise) due to sparseness of points, while at faint (large) magnitudes one observes only the effect of the intrinsic scale-invariant fluctuations of a fractal making the variance almost constant in m

$$\langle n(\mathbf{r}) \rangle_p = n_0 [1 + \xi(r)] \ . \tag{10.20}$$

In this case the density fluctuations around the average are scale-invariant and a density field $\rho(\mathbf{r})$, defined by

$$\rho(\mathbf{r}) = [n(\mathbf{r}) - n_0] \ , \tag{10.21}$$

can be considered as like a fractal field whose 1-point conditional density is $n_0 \xi(r) \sim r^{-\gamma}$, and $D = 3 - \gamma$ is the fractal dimension. For the same reason the quantity

$$\langle (n(\mathbf{r_1}) - n_0)(n(\mathbf{r_2}) - n_0) \rangle_p = n_0^2 \left[\tilde{\xi}(r_{12}) + \zeta(r_1, r_2, r_{12}) \right] \tag{10.22}$$

is analogous to the two-point conditional density of the fractal case. Therefore one can impose the equivalent of (4.26)–(4.28) to the present case:

$$\xi(r_{12}) + \zeta(r_1, r_2, r_{12}) = \xi(r_1)\xi(r_2)\mathcal{L}(r_1/r_2, \theta) \ , \tag{10.23}$$

10.3 Number Counts as a Function of Apparent Magnitude

where θ is the angle between $\boldsymbol{r_1}$ and $\boldsymbol{r_2}$, and $\mathcal{L}(r_1/r_2,\theta)$ is the lacunarity function of the fractal fluctuation field (see Chap. 4). Thus we can write

$$\langle n(\boldsymbol{r_1})n(\boldsymbol{r_2})\rangle_p - \langle n(\boldsymbol{r_1})\rangle_p \langle n(\boldsymbol{r_2})\rangle_p = n_0\delta(\boldsymbol{r_1}-\boldsymbol{r_2}) + $$
$$n_0^2 \xi(r_1)\xi(r_2)g(r_1/r_2,\theta) \,, \qquad (10.24)$$

where the term in $\delta(\boldsymbol{r_1}-\boldsymbol{r_2})$ is due to the diagonal part of $\tilde{\xi}(r_{12})$ and $g(r_1/r_2,\theta)$ is related to the lacunarity function through (4.27). As shown below, this term is important only for $\gamma \geq 3/2$ (i.e. $D \leq 3/2$). In the case of a purely fractal point distribution this contribution, due to the constant and positive background density, was omitted because in that case it is always irrelevant at sufficiently large scales. At this point we can evaluate count fluctuations around the average both in a spatial VL sample (for large R) and a flux limited one (for small f).

First of all, we obtain (considering only the dominant contribution):

$$\langle N(<R;\, >L_{VL})\rangle_p \sim R^3 \qquad (10.25)$$

and

$$\langle N(>f)\rangle_p \sim f^{-\frac{3}{2}} \,. \qquad (10.26)$$

Moreover, if $\gamma < 3/2$, we have

$$\sigma_p^2(R) \sim R^{-2\gamma} \text{ and } \sigma_p^2(f) \sim f^\gamma \,, \qquad (10.27)$$

while if $\gamma \geq 3/2$ the contribution from $\delta(\boldsymbol{r_1}-\boldsymbol{r_2})$ in (10.24) dominates, and the same result as in the Poisson case is obtained.

In terms of magnitude one finds:

$$\begin{aligned}\sigma_p^2(m) &\sim 10^{-0.4\gamma m} \quad \text{for } \gamma < 3/2 \\ \sigma_p^2(m) &\sim 10^{-0.6m} \quad \text{for } \gamma \geq 3/2 \,.\end{aligned} \qquad (10.28)$$

Therefore, also in the case of a *homogeneous*, but long-range correlated, point distribution, the normalized fluctuations of the counts around the average decay exponentially with apparent magnitude. We have only a difference of the decay rate for the two cases $\gamma \geq 3/2$ and $\gamma < 3/2$. In the former case, because of the rapid decay of $\xi(r)$ at large scales, the behavior is the same as in the Poisson case without correlations. The same behavior is also found in the case of more rapid decay of $\xi(r)$. In the latter case the damping of normalized fluctuations is slower as correlations become important. *Clearly for both cases the average counts should behave as* $10^{0.6m}$. We conclude that the only case in which persistent and scale-invariant normalized counts fluctuations can be observed, is the case of a purely fractal point distribution extending to the probed scales [90].

10.4 Normalization of the Magnitude Counts to Real Space Properties in Euclidean Space

In this section we consider the normalization of counts of galaxies as a function of distance in a VL sample to the counts as a function of magnitude in a flux-limited catalog. We will treat explicitly the case of a pure fractal distribution with fractal dimension $0 < D < 3$ and average conditional density

$$\langle n(\mathbf{r})\rangle_p = \frac{DB}{4\pi} r^{D-3} .$$

With $D = 3$, and by using that

$$\frac{DB}{4\pi} = n_0 ,$$

we obtain the behavior of a Poisson particle distribution of average density n_0 (i.e. $\langle n(\mathbf{r})\rangle_p = n_0$). Moreover we will again make use of the approximations (8.2) and (8.3) consisting in the statistical separation of position and luminosity of all galaxies, and the assumption of the Schechter function for the luminosity PDF.

This is an interesting exercise which can be useful for the study of galaxy counts at bright apparent magnitudes and low redshifts. We derive in this case some useful formulas for the analysis of real samples. Note that for higher redshift ($z > 0.1$) and faint apparent magnitudes ($m > 16$) one has to take into account relativistic corrections which we have not considered here. In Appendix C we discuss both the distance redshift and magnitude redshift relations in FRW models.

10.4.1 Average Distance

By using (8.2) and (8.3) it is simple to show that the average distance $r(f)$ which can be associated to a galaxy of observed apparent flux f is (in the small redshift approximation $z \ll 1$)

$$r(f) = \frac{4\pi \int_0^\infty AL^\delta e^{-\frac{L}{L_*}} \int_0^\infty dr\, Br^D \delta\left(f - \frac{L}{4\pi r^2}\right)}{4\pi \int_0^\infty AL^\delta e^{-\frac{L}{L_*}} \int_0^\infty dr\, Br^{D-1} \delta\left(f - \frac{L}{4\pi r^2}\right)} , \qquad (10.29)$$

which gives

$$r(f) = \frac{\Gamma_e\left(\frac{D+3}{2} + \delta\right)}{\Gamma_e\left(\frac{D+2}{2} + \delta\right)} \left(\frac{L_*}{4\pi f}\right)^{\frac{1}{2}} , \qquad (10.30)$$

where Γ_e is the *Euler gamma function*. The relation between the apparent flux f and the apparent magnitude m is[1] given by [185]

---

[1] The relation between L_* and M_* is given by [185] $L_* = 10^{0.4(M_\odot - M_*)} L_\odot$ where $M_\odot = 4.58$ is the absolute magnitude of the Sun and $L_\odot$ is its intrinsic luminosity.

10.4 Normalization of the Magnitude Counts

$$f = \frac{L_*}{4\pi(10 \text{ pc})^2} 10^{0.4(M_*-m)} . \tag{10.31}$$

We can re-express (10.30) in terms of m by denoting, for simplicity, the average distance $r(m) = r(f(m))$

$$r(m) = \frac{h}{10^5} \frac{\Gamma_e\left(\frac{D+3}{2} + \delta\right)}{\Gamma_e\left(\frac{D+2}{2} + \delta\right)} 10^{-(0.2M_*)} 10^{(0.2m)} , \tag{10.32}$$

where, from now on, distances are expressed in Mpc/h. For example, by taking $M_* = -19.5$, $h = 1$, $\delta = -1$ and $D = 2.5$ we have

$$r(m) = 0.08 \cdot 10^{(0.2m)} . \tag{10.33}$$

Using instead $D = 3$ (Poisson case) the pre-factor in (10.32) changes, and we obtain

$$r(m) = 0.09 \cdot 10^{(0.2m)} . \tag{10.34}$$

which is of the same order of magnitude. Hence the pre-factor is a slowly varying function of the fractal dimension.

10.4.2 Normalization of Distance to Magnitude Counts

Let us now consider the normalization of the counts in space to the ones in apparent magnitude. By differentiating (10.14) with respect to f, and using (8.3), the average differential number counts of a system with fractal dimension D, per steradian, can be written as:

$$\langle n(f) \rangle_p = \frac{d\langle N(f) \rangle_p}{df} = \frac{DB}{4\pi} \frac{1}{2(4\pi)^{D/2}} L_*^{\delta + \frac{D+2}{2}} C_1 f^{-\frac{D+2}{2}} \tag{10.35}$$

where we have defined

$$C_1 = \Gamma_e \left(\delta + \frac{D+2}{2} \right) . \tag{10.36}$$

The average distance $r(f)$ associated to the apparent flux f is given by (10.30), which we can rewrite as

$$r(f) = \frac{C_2}{C_1} \left(\frac{L_*}{4\pi f} \right)^{\frac{1}{2}} . \tag{10.37}$$

where

$$C_2 = \Gamma_e \left(\frac{D+3}{2} + \delta \right) . \tag{10.38}$$

By differentiating (10.37) and inverting the relation, we obtain

$$\frac{df(r)}{dr} = \frac{-2L_*}{4\pi r^3}\left(\frac{C_1}{C_2}\right)^2. \tag{10.39}$$

Let us now define $\langle N(r)\rangle_p^{ML} = \langle N(f(r))\rangle_p$, where $f = f(r)$ is given by solving (10.39), and where we have used the superscript ML to refer to a magnitude limited sample. We can then write

$$\frac{d\langle N(r)\rangle_p^{ML}}{d\langle r\rangle} \equiv \frac{d\langle N(f)\rangle_p}{df}\frac{df}{d\langle r\rangle} = \frac{DB}{4\pi}L_*^{\delta+1}\left(\frac{C_1}{C_2}\right)^{-D}C_2\langle r\rangle^{D-1}. \tag{10.40}$$

Now, assuming (8.2) and (8.3), and considering the case that we observe all galaxies without any limit on the absolute magnitude (indicated with the superscript all), we have

$$\frac{1}{4\pi}\frac{d\langle N(r)\rangle_p^{all}}{dr} = \frac{B}{4\pi}r^{D-1}AL_*^{\delta+1}C_3, \tag{10.41}$$

where

$$C_3 = \Gamma_e(\delta+1). \tag{10.42}$$

In a real VL we have a cut-off in absolute magnitude and hence

$$\frac{1}{4\pi}\frac{d\langle N\rangle_p^{VL}}{dr} = \frac{C_4^{VL}}{C_3}\frac{1}{4\pi}\frac{d\langle N\rangle_p^{all}}{dr} \tag{10.43}$$

where

$$C_4^{VL} = \int_{L_{VL}}^{\infty}\phi(L)dL. \tag{10.44}$$

Hence from (10.43) and (10.40) we obtain that

$$\frac{d\langle N\rangle_p^{ML}}{dr}\left(\frac{1}{4\pi}\frac{d\langle N\rangle_p^{VL}}{dr}\right)^{-1} = \left(\frac{C_1}{C_2}\right)^{-D}C_2\left(\frac{1}{C_4^{VL}}\right) \approx \frac{1}{C_4^{VL}} \tag{10.45}$$

as $C_1 \approx C_2 \approx 1$. By using (10.45) we can transform the galaxy counts as a function of apparent magnitude into counts as a function of the (average) distance and then we may normalize them to the counts in a VL sample. This is a further test to check the consistency of the assumption in (8.2), and possibly, to extend the real space analysis to magnitude space.

10.5 Galaxy Counts in Real Catalogs

We now turn to real data, considering the counts in real space first before turning to the much more copious magnitude space data. Just as in the previous chapter we do not attempt to treat here exhaustively the data available which can be used to give observational constraints on these quantities. Rather we take some representative data sets which give an indication of the typical results which can obtained from current data.

10.5.1 Real Space Counts

For this case we consider the same survey – the CfA2 survey – which has been presented in some detail in Sect. 9.6 of the previous chapter. We have cut a given VL sample (limited by $M \geq -19.5$) in several angular slices, with similar solid angle, and we have considered the behavior of the integrated non-average radial counts in each of the slices. Then we have performed the average between the various determinations (each normalized to its solid angle). The results (see Fig. 10.5) show a highly fluctuating behavior. The best-fit slopes of the counts in the individual samples vary in the range from 2 to 3.5 (depending also on the range of scale considered) with an average slope $D \approx 2.6$ obtained through a best fit of the exponent of a single power law. The behavior of the fluctuations is quantified in Fig. 10.6, which shows the conditional variance as a function of scale, calculated using the same set of samples. The conditional variance shows an almost constant behavior up to about 20 Mpc/h. It then decreases up to around 50 Mpc/h with a scaling as a function of distance similar to that of a Poisson distribution,

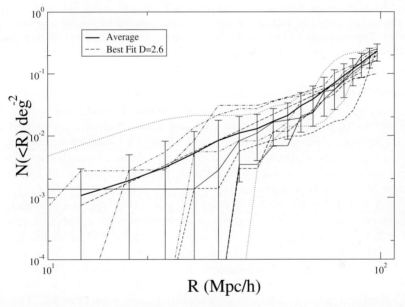

Fig. 10.5. Behavior of the radial counts of galaxies (10.2) per unit of solid angle (deg^{-2}) as a function of the radius (measured in Mpc/h) in a VL sample of the CfA2 survey (in this case $M_{VL} = -19.5$). The different lines correspond to different angular slices of the survey, while the line labeled with "average" represents the average behavior and its best fit with a single power law gives an effective fractal dimension $D = 2.6$. Note the highly fluctuating behavior of these counts over most of the range, with some suggestion of a decrease in their amplitude as the sample depth ($\sim$100 Mpc/h) is approached

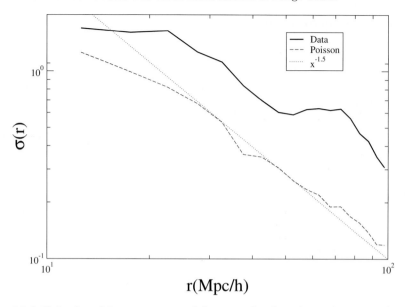

Fig. 10.6. Behavior of the square root of the normalized conditional variance (10.2) of the radial counts of galaxies in the VL sample shown in Fig. 10.5 of the CfA2 survey. For comparison the behavior is given for a Poisson distribution in a sample with the same geometry and the same number of points

possibly due to a sparseness effect in the observation from a single observer. It then stays constant between 50 and $70 \div 80$ Mpc/h, indicating possibly the persistence of some strong irregularity, and finally it shows further decrease in the last bins. Whether this decrease is a sign of a transition toward a more homogeneous and uniform behavior on larger scales, or whether it is just a local non averaged large scale fluctuation (e.g. a coherent wall) cannot be determined from this single sample.

In Chap. 9 we saw evidence from the analysis of real samples with the conditional density of scale invariance characterized by a fractal dimension $D \approx 2$, extending at least up to a scale ~ 20 Mpc/h. What we see here in the number counts probes scales which are slightly larger. What we find could be indicative of a continuing scale invariance up to ≈ 40 Mpc/h. The average number count shows, however, a best-fit with a single power law to a fractal dimension which is significantly larger, with $D \approx 2.5$. This can be indicative of either a slow transition to homogeneity or as a cross-over between two different fractal regimes. We will see that the same higher slope is also indicated by the number counts in magnitude space.

In interpreting these results it has to be remembered that the number counts are un-averaged over observers, in contrast to the conditional density estimations in VL samples (as explained in Chap. 9) which are effectively number counts averaged over many different observers (for different regions

10.5 Galaxy Counts in Real Catalogs

of the sky, for the regime in which our results are robust). Thus they are less reliable probes of the average behavior. Further they are more sensitive to many systematic errors: in particular the slope of the counts relative to the observer can be changed by any systematic effect which depends on the distance from the observer. For example Teerikorpi et al. [228] have studied, using the KLUN sample of 5171 spiral galaxies with Tully-Fisher distance moduli, the radial space distribution of galaxies out to a distance of about 100/h Mpc. They have used a method based on photometric Tully-Fisher distances, independent of redshift, to construct the number density distribution. The result is that at larger distances the radial distribution shows a dimension about $D = 2.2$.[2]

On the other hand, while such a systematic effect could be at the origin of the difference in the observed effective fractal dimension, it would not be expected to have any effect on the behavior of the fluctuations. It is not difficult to show that such a continuous rescaling of distances with respect to the origin should leave the qualitative behavior (constant or decreasing in m) of the conditional variance unchanged: what these behaviors depend on is how the fluctuations behave as a function of the number of points in the volume probed.

However, apart from the fact that the known effects of this kind (cosmological and k-corrections discussed in Appendix D) should not be important at the distance scales we are discussing, there is some further evidence that the increase of effective dimension found by fitting the counts through a single power-law may indeed be a real average behavior: In particular we note that, in two real samples which allow one to access with the average conditional density slightly larger scales than the CfA2 sample, we have used for our "robust" constraint in the previous chapter, one finds an indication for a similar flattening [223]. In Figs. 10.7–10.8 are shown results for averaged quantities in the LEDA database [176] and the PSCZ redshift survey [209]. In both cases a change in the behavior, consistent with what we see in the number counts, is observed around 20 Mpc/h.

Applying the same strict criteria as in the previous chapter to these last two samples – imposing that the behavior must be averaged over non-overlapping regions – reduces however the "robust" range to only slightly above the scale of ~20 Mpc/h. Our conclusion is thus that we need to see if this behavior is found in larger forthcoming surveys, in particular the SDSS survey over a significant portion of the sky.

---

[2] In [19] a method is developed to estimate the fractal dimension based on the consideration of the behavior of non-averaged galaxy counts in one-dimensional cylinders which has the advantage that it allows to probe very deep distances. The average, as for the angular case, can be performed over many cylinders. The fractal dimension is estimated to be $D = 2.1 \pm 0.1$ for a cylinder of length 100 Mpc/h.

282   10 Number Counts and Their Fluctuations at Large Scales

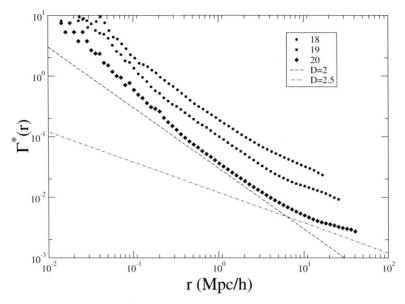

Fig. 10.7. The conditional average density $\Gamma^*(r)$ for samples constructed from the LEDA database [176] cut in apparent magnitude at $m = 14.44$. At scales larger than 20 Mpc/h, there is an apparent change of slope with a best fit fractal dimension $D \approx 2$ interpolating to $D \approx 2.5$ at larger scales. While the first slope at small scale is well defined for more than a decade, the range of scales is very limited at large distances and finite size effects – due to systematically poorer averaging at larger scales – may be important. Larger surveys, in particular the SDSS survey, will allow unambiguously to determine if this is the real average behavior or a cross-over to homogeneity. (Adapted from [67])

Let us remark finally on the evidence we see in Figs. 10.5–10.6 for a decrease of the fluctuations toward the sample size: all the subsamples appear to converge just at the limits of the sample to a common number density, with a suggestion also of a larger slope. This kind of behavior would be indicative of a greater degree of homogeneity at this scale. We note however that there is still a large variance – of order unity – at this scale. Further examination of the sample also shows that the notable increase in the counts at this depth corresponds in fact to a single feature in this sample – a large structure known as the Perseus-Pisces super-cluster. Thus what we observe could not appear to characterize an average behavior of the sample. Once again to conclude on the indications for a tendency to homogeneity we see in the range we can probe with these counts – up to ~ 100 Mpc/h, we believe it is appropriate to await the determination of truly averaged quantities at this scale from future data.

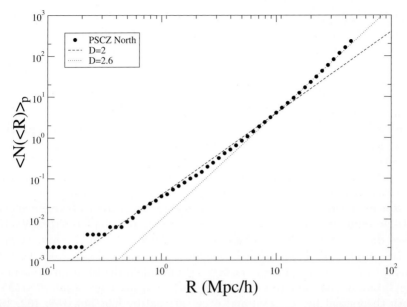

Fig. 10.8. The average conditional number of galaxies $\langle N(<R) \rangle_p$ in balls of radius R (which is equivalent to the integral of $\Gamma^*(r)$ in the same sphere) for a VL sample from the PSCZ redshift survey [209]. Again in this case there is evidence for a change in slope consistent with that seen in Fig. 10.7

10.5.2 Magnitude Space Counts

Counts of galaxies as a function of apparent magnitude are among the most time-honored observations in cosmology [117, 118]. For a long time their determination has been considered to be of great importance to cosmology, as a potentially powerful test to measure the global geometry of the universe: we discuss here only the Euclidean case, valid for the modest depths to which we limit ourselves, but in more general one can predict the expected behavior of the average counts in different cosmological models, which predict deviations from the Euclidean counts at very large scales. After many years of extensive programs of observation, this promise has not been fulfilled as many more fundamental problems with the interpretation of (average) number counts have emerged: the data cannot be explained to a first approximation, over any range of scales, by any purely cosmological model with a well defined mean density of galaxies, but rather the data must be interpreted in terms of an evolution – originally unanticipated – of the relevant galaxy populations. For example the mismatch between the bright (near) and the faint (far) counts is a long-standing problem. This leads to the hypothesized *faint blue galaxy excess* (e.g. [99]): if one normalizes the counts with the amplitude of the luminosity function found in nearby redshift samples, one must postulate the existence of a very large population of faint blue galaxies to explain the faint end data in the B-band. This problem is reflected in the uncertainties in the determination of the amplitude of the galaxy luminosity function parametrized as a power-law followed by an exponential cut-off $\phi(L) = \phi_* L^\delta \exp(-L/L_*)$ (see Chap. 11) which has three free parameters. Over the past two decades there have been numerous estimates of these parameters (see e.g. [146] for a recent discussion) with the result that the exponent δ is determined with good precision ($\Delta\delta/\delta \approx \pm 0.15$) while the typical luminosity shows a significant variation of about 40%, and of the normalization of about 50%. For this reason our knowledge of the average luminosity density has an uncertainty of about 60% (see discussion in Chap. 8).

In this analysis of number counts, just as in the $\xi(r)$ analysis of redshift surveys, the underlying assumption of homogeneity is always made, i.e. it is always assumed that the number count of galaxies, without taking evolution into account, should be that of a homogeneous population. When fluctuations due to large scale structure are considered, it is in the framework of distributions with a well defined mean density at the scales probed. Thus, for example, many attempts have been made to explain the large variation of the amplitude of the luminosity function, known as *the normalization problem*, both theoretical by [78, 131] and observational by [69, 146, 158, 162, 163], but always within the framework of an homogeneous (Poisson-like) galaxy distribution. The possibility is considered that local clustering could be at the origin of this problem, but then a normalization of the counts at an intermediate range of magnitude is sought where in turn it is assumed that the effect of galaxy large scale structures should be less important.

Alongside these problems in the average counts of galaxies, large fluctuations from field to field in the sky and from survey to survey, both at faint and bright counts, and in different spectral bands, have been reported (e.g. [7, 31, 51, 149, 189, 236]). These fluctuations can be as large as a factor of two. There has been controversy as to whether these fluctuations are due to real clustering or to differences in the magnitude zero point of the various surveys [7, 31, 189, 242]. It is, in fact, possible that discrepancies among these surveys are not due mostly to differences in photometric systems or in data reduction effects, but rather to real effects, i.e. large scale structures. This is the possibility which we consider explicitly here.

Before proceeding let us recall what we have found in Sect. 10.3 above: when relativistic correction, k-corrections and evolutionary effects are neglected (i.e. an approximation which should be valid for the bright end of the counts) we have the following relation between real-space and magnitude-space counts: if $N(< r) \sim r^D$ then

$$N(m) = A 10^{\alpha m} = A 10^{\frac{D}{5} m} \tag{10.46}$$

where the normalization constant A is related to the parameters of the luminosity function and of the average conditional density. In the case of an uniform distribution $D = 3$ and thus $\alpha = 3/5 = 0.6$. It is clear that if one considers the possibility that the exponent α may have a value other than 0.6, and even depends on scale, then the *normalization problem* takes on a completely different perspective: at small scales (bright end) one may fit the counts with a different pre-factor than that obtained by imposing $\alpha = 0.6$. This is the procedure that is naturally followed for the counts if one analyzes them in the broader framework which we have applied to the real space counts and the conditional average density. An observed value of α other than 0.6 then gives an indirect measure of the effective fractal dimension, with a normalization of the luminosity function as described in Sect. 10.3. On the other hand there may be subtle selection effects, related to errors in the measurement of apparent magnitudes, which can slightly change the counts slope, as for example the Eddington bias discussed by [229].

Let us now consider some data sets. The largest and best data sets are in fact those at the intermediate and fainter magnitudes, which we have implicitly excluded from our analysis. We concentrate on the bright magnitude data i.e. a range of magnitude which in real space corresponds roughly to the same range of distances [0,100] Mpc/h where there are complete redshift data. Unfortunately this data is relatively limited, and not excellent in terms of its photometric quality. More accurate CCD calibrated data will however become available in the near future in particular from SDSS.

We consider first the Catalogs of Groups and Cluster of Galaxies (CGCG) [82, 254] which contains 27,837 objects with $m_{Zw} \leq 15.7^m$, the limit where Zwicky estimated that this catalog was complete. This is the "target" catalog used for the measurements of the CfA redshift survey. The north galactic hemisphere ($b > 20°$) has a total solid angle of $\Omega \approx 3.6~sr$, and the south galactic hemisphere ($b < -20°$) $\Omega \approx 1.6~sr$. The magnitudes are given in the Zwicky system and the error is estimated to be 0.3^m up to 15.0^m, and then increasing up to $\sim 1^m$ at the very faint end [23]. We have computed counts as a function of apparent magnitude in 14 angular fields, which cover the total solid angle of the survey, and have then calculated their average and variance, shown in Fig. 10.9. The best fit to the slope up to $m_{Zw} \leq 15.0$ gives $\alpha = 0.50 \pm 0.02$ which, interpreted as a measure of the fractal dimension of the distribution, gives $D \approx 2.5$. In the figure we also note that the variance as a function of apparent magnitude is approximately constant. This result, interpreted in terms only of intrinsic large fluctuations, as discussed above, is the behavior expected in strongly irregular and scale-invariant distributions, and thus gives further grounds for the interpretation of the slope in terms of the fractal dimension. Further these results are in line with the behavior we saw in the real space counts in the previous section. In this case, however, we see no evidence for a decrease in the fluctuations at any scale.

Next we consider the Lyon-Meudon extragalactic database (LEDA) (see [175, 176, 177, 204]) for which we also gave some redshift space results above. This survey has a high level of completeness up to an apparent magnitude $B_T = 15.0^m$, where B_T is the B-magnitudes reduced in the RC3-system [176].

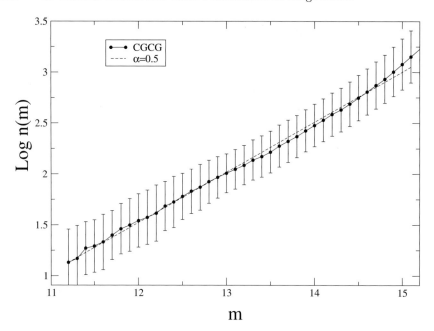

Fig. 10.9. Differential number counts $n(m)$ (in unit of $magn^{-1}deg^{-2}$) in the CGCG. The slope and error bars have been computed by averaging the counts in the 14 angular slices considered in this analysis. The best fit for $m_{Zw} \leq 15.0$ with slope $\alpha = 0.50 \pm 0.02$ is also reported

Up to this limit the sample contains ≈60000 objects. Paturel and coworkers have pointed out that the counts do not agree with the "Euclidean" value of $\alpha = 0.6$ in a series of papers (see e.g. [175, 204] and references therein) and they have related such a departure from homogeneity to the presence of a two dimensional large scale structure ("local super-cluster") in the local universe. This point of view qualitatively coincides with the connection of the counts slope with real space structures, which we may characterize in terms of the fractal dimension. We have done the same tests as for CGCG, in this case dividing the total solid angle of ≈10 sr in 20 angular slices [3] of solid angle $\Omega \approx 0.5$ sr each. The results of the calculation of the average and variance of the counts in these regions is shown in Fig. 10.9 for the range $11.5^m \leq B_t \leq 14.5^m$. The best-fit slope is $\alpha = 0.50 \pm 0.02$, the same as found by [204]. This again is consistent with our previous findings, corresponding to a fractal dimension $D \approx 2.5$.

---

[3] We have considered only the sample cut at $|b| > 20°$ in order to avoid the region with high galactic extinction, and we have limited ourselves to $B_T \leq 14.4^m$, in order to avoid possible systematic errors which may be important at the faint end (see discussion in [23]).

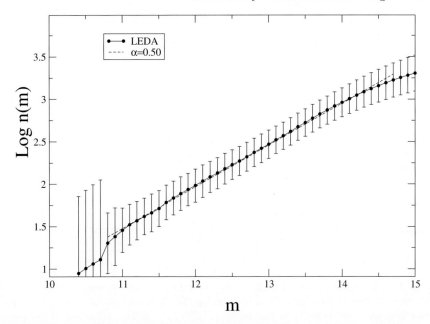

Fig. 10.10. Differential number counts (in unit of $magn^{-1}deg^{-2}$) as a function of apparent magnitude in LEDA. The average has been performed between 20 angular slices. The best fit in the range $[12^m, 14.5^m]$ is also reported. The tendency to a flattening at faint magnitudes $\gtrsim 15.0^m$ is due to the incompleteness of the sample

Finally we consider another smaller data set, a survey of bright galaxies by [31] in the blue ($16^m < B_J < 21^m$) and red ($15^m < R < 19.5^m$) pass-bands performed over $145\ deg^2$. This is a survey which has much better calibration properties, as it has been calibrated against a fair density of CCD standards. The photometric systematic error is $\leq 0.1^m$ over the whole magnitude domain. Given these good characteristics the data allows a more reliable study of the field to field fluctuations. The survey consists of seven fields, four in the southern galactic cap (SGC) and three in the northern one (NGC). The southern fields are 90% complete for $B_j \leq 21^m$ and the northern fields are 90% complete for $B_j \leq 20^m$, and so we have limited the analysis at $B_j \leq 20.0^m$ in both the SGC and NGC: up to this limit the survey contains 46663 galaxies. We have computed the integral number counts in each of the seven fields, and then we have made an average over the different fields (see Fig. 10.11). The average slope of the counts we find is about $\alpha = 0.51 \pm 0.02$ in the range $15.2^m \leq B \leq 20^m$, in agreement with that found by the authors of the survey in [31]. The fluctuations in Fig. 10.11 show a behavior which is quite consistent with that we have seen in the previous samples: from magnitude $\approx 18^m$ to the limit at $\approx 20^m$ the amplitude of fluctuations is roughly constant, with no detectable sign of decrease. The decreasing larger amplitude

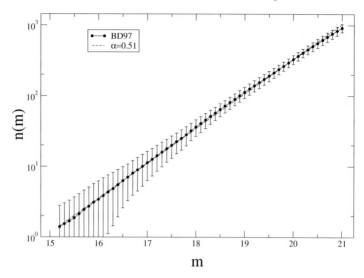

Fig. 10.11. Differential average number counts (in units of $magn^{-1} deg^{-2}$) as a function of apparent magnitude in the catalog [31]. The average has been performed over 7 fields. The best fit in the range $[15.2^m, 20^m]$ is also reported: its slope is $\alpha = 0.51 \pm 0.02$

at the very bright end is due to the additional shot noise term (as the number of points in the field is very small).

All these data sets thus give a possibly consistent picture when interpreted in terms of scale-invariant distributions, both in terms of the apparent persistence of fluctuations at all scales with an amplitude which is independent of scale, and in terms of their average behavior which is well fitted by a single fractal dimension $D \approx 2.5$. Many similar results have been published in the literature and we refer to [146] for a recent review of the subject. It is interesting to note – but beyond our scope to treat in detail here as we have limited ourselves to the bright end where the interpretation of the counts should be relatively simple – that at faint end the exponent α deviates from 0.6, but not in a way which can be explained by cosmological corrections (i.e. by the non-Euclidean geometry of the FRW models). In fact the whole range from the bright to the faint end can be fit very well by the same value $\alpha \approx 0.5$. The interpretation of such a behavior in terms of a continuing fractal behavior is however more delicate, as at faint magnitudes one must necessarily consider the possible effects of space-geometry, k-corrections and galaxy evolution.

10.6 Summary and Discussion

We have seen in this chapter that number counts, in real and mganitude space, provide, in principle, a large scale probe of the nature of the underlying distribution of galaxies. We have focused on the case that the distances probed are relatively small (up to $\sim 100\,\mathrm{Mpc/h}$), as in this case we can neglect effects of the non-trivial geometry of cosmological models, k-corrections and evolutionary effects in the galaxy distribution. The interpretation of the counts in this range is then straightforward, probing simply the distribution of galaxies on this scale. The slope of the counts give a direct measure of the effective fractal dimension of the distribution, while the fluctuations in these counts (measured in different fields seen by the observer) give a direct probe of the fractal or homogeneous nature of the distribution depending on their behavior as a function of scale or the apparent magnitude: in a pure fractal particle distribution these fluctuations are independent of scale (also in apparent magnitude space) while, in any distribution with small fluctuations about a mean density well defined at the scales probed, they decay (exponentially when expressed in terms of apparent magnitude).

Examining some representative real data we find that a possible consistent picture emerges: a best-fit fractal dimension of $D \approx 2.5$, with the behavior of the fluctuations in the counts that can be ascribed in principal (and with weak evidence at the moment) to a fractal-like nature of fluctuations on this scale. This behavior describes the galaxy distribution at scales starting from a little larger than those probed by the redshift data we discussed in the previous chapter, and extends up to $\sim 100\,\mathrm{Mpc/h}$. It is qualitatively consistent with that observed in these samples, but the measured fractal dimension is larger, meaning either a cross-over to a different fractal regime or a slow transition to homogeneity. Several caveats, however, need to be given: because counts are not averaged over different observers (in contrast to the redshift data analyzed with the conditional density) it is quite conceivable that this is not a true average behavior, but represents a local fluctuation around us. In this case, we emphasize, the conclusion is still that the galaxy distribution is characterized by large fluctuations at this scale ($\sim 100\,\mathrm{Mpc/h}$) and thus that it is essential to use the statistical instruments (which do not build in the assumption of homogeneity) we have introduced to characterize it in this and the preceeding chapters. Further, as mentioned, it must be born in mind that there may be other, purely observational, systematic effects which could be at the origin of such observed large fluctuations e.g. magnitude calibration problems (see also [229]). This is an issue about which there has been a certain amount of controversy [7, 31, 189, 242]. Observations forthcoming in the next years will allow us to determine with much greater reliability in a well averaged way the correlation properties of the galaxy distribution up to several hundred Mpc/h, and these questions will decisively be answered. What is learned from this about the relation between number counts and averaged quantities will be of great use also in attempts to use number counts to extend constraints on the galaxy distribution to even larger scales.

11 Luminosity in Galaxy Correlations

11.1 Introduction

Galaxy luminosities vary over many orders of magnitude, with typical surveys sampling galaxies with absolute magnitude in the range $[-12, -23]$. In Chaps. 8 and 9 we have discussed solely the spatial correlation properties of galaxies, without considering explicitly their luminosity (other than in the construction of volume limited samples). In Chap. 10 we have analyzed also the problem of number counts in apparent magnitude space. However in all three chapters we have made the assumption that spatial position and luminosity are uncorrelated i.e. we have made the assumption given in (8.2) of Chap. 8 of separability of the dependence on space position and luminosity. We mentioned that such an assumption is in fact, for what we have done so far, a good approximation. It is well known, however, that it is not valid absolutely. That luminosity – and other galaxy properties such as morphology – are correlated with position is in fact well known: for example larger (brighter) galaxies are found preferentially in clusters, while smaller (fainter) galaxies do not show such a tendency.

In this chapter we discuss a more general framework which allows an analysis of galaxy correlations without this assumption. This approach is based on the concept of multifractality discussed in Chap. 5. This is a mathematical framework which allows one to detect and characterize how luminosity and space correlation properties are intertwined. In particular it links together in a broader picture the two principal quantities we have considered for space and luminosity properties, the conditional average density and the luminosity function (LF). As in Chap. 8 we focus on the low redshift range, and we do not discuss the complexities of higher redshift where effects such as evolution and relativistic ones are expected to become important.

The important point about this formalism is that it allows, as in the previous chapters, the study of correlation properties without the prior assumption of homogeneity. We can characterize with it the luminosity dependence of clustering, even when this clustering is strong, associated with large fluctuations at the scale of the sample. The crucial point in this respect is that in this case the quantities of interest are scaling exponents rather than amplitudes. Luminosity dependence of clustering is described by a spectrum of (multifractal – MF) exponents, and not by a different amplitude of the reduced two-point correlation function. As we discussed in Chap. 9 the standard analysis with the reduced two-point correlation function gives the latter kind of description of the dependence on luminosity. In particular the phenomenon of "luminosity segregation" or "luminosity bias" corresponds to the observation that r_0 (which is directly related to the amplitude of the reduced two-point correlation function) varies for different luminosities. As we have discussed at the end of Chap. 9 such a variation in amplitude can be interpreted also as a possible manifestation of the fact that the different samples have different densities, because the underlying distribution still has large fluctuations at the corresponding scales. We will return to this question in the next chapter, where we discuss the interpretation of a very similar phenomenon in catalogs of galaxy clusters.

11.2 Standard Methods for the Estimation of the Luminosity Function

Before discussing the MF description let us briefly review the standard methods usually adopted for the estimation of the galaxy luminosity function. There are several methods to determine the LF, defined as the number of galaxies of luminosity L per unit of volume and of luminosity. [32]. In some cases, special emphasis is placed on the systematic differences in the LF for the various Hubble types[1]. Here we are interested in the determination of the *general* LF defined as the sum over all Hubble types. All the methods we are going to describe in this first section are based on the assumption that the luminosity and space density are not correlated.

The so-called *classical method* to determine the LF is based on the a-priori hypothesis of homogeneity (spatial uniformity) of the galaxy distribution inside a given sample: the average density n_0 of galaxies in space is constant and well defined and correlations can be neglected, i.e. one assumes (8.4). Under this assumption one just makes a histogram normalized to the observed number of galaxies with absolute magnitude in a given range and from it derives the LF. This method is clearly highly sensitive to spatial inhomogeneities and correlations in the distribution of galaxies which may distort the shape of the LF and cause n_0 to fluctuate from sample to sample. For this reason many authors in the past excluded a region of a given catalog containing strong "inhomogeneities" in the galaxy distribution (e.g. the Virgo cluster [85]).

More refined methods to measure the LF aim to separate the determination of the shape from that of the amplitude, i.e. one uses the less stringent condition given by (8.2) rather than (8.4), so that $\phi(L)$ is a normalized PDF, i.e. with unit integral over all luminosities. The so-called *inhomogeneity-independent* methods have been developed with the aim of determining only the shape of the LF. The basic idea is to consider the ratio of galaxies having

---

[1] This refers to Hubble's classification of galaxy types.

intrinsic luminosity between L and $L + dL$ to the total number of galaxies brighter than L. If one assumes the statistical independence of space and luminosity distributions (8.2) then one can derive that

$$\frac{dV \, dL \langle \nu(L, \boldsymbol{r}) \rangle_p}{dV \int_L^\infty dL' \langle \nu(L', \boldsymbol{r}) \rangle_p} = \frac{dL \, \langle n(\boldsymbol{r}) \rangle_p \phi(L)}{\langle n(\boldsymbol{r}) \rangle_p \int_L^\infty dL' \phi(L')} =$$

$$\frac{dL \phi(L)}{\Phi(L)} = -d \log \Phi(L) \,, \qquad (11.1)$$

where now $\phi(L)$ is a normalized PDF with amplitude independent of the average density of galaxies and where we have called

$$\Phi(L) = \int_L^\infty dL' \phi(L') \,.$$

By measuring the first expression of (11.1), with the hypothesis of statistical independence between spatial coordinates and luminosity of galaxies, one can obtain directly $\Phi(L)$, and by simple differentiation $\phi(L)$. Note that this method is valid also if the spatial distribution is fractal (i.e., $\langle n(\boldsymbol{r}) \rangle_p \sim r^{D-3}$)) provided the correlation between space and luminosity distributions can be neglected.

Usually this LF is then fit by an analytic function. The most popular fit is the one proposed by Schechter given by (8.3) [211] where the experimentally determined parameters are L^*, the cut-off, ϕ^*, the normalization constant[2] and δ the exponent. The LF has been measured by several authors in different redshift surveys and the agreement between the various determinations in very different volumes is excellent. Typical current best fit parameters are $\delta = -1.23$ and $M_{bj}^* + 5 \log h = -19.72$ (see [146] for an up to date review).

Other more sophisticated methods have been introduced more recently to fit the observations of the luminous distribution of galaxies. For example a maximum likelihood method which allows for a generic LF shape and which can take into account also evolutionary effects, and measurements uncertainties [33]. The overall normalization cannot however be determined by this procedure. For this purpose a method called minimum variance estimation is often used [54]: this method requires in fact both the assumption of independence between space and luminosity distributions, given by (8.2), and the assumption that the average space density is a well defined quantity well inside the samples considered.

11.3 Multifractality, Luminosity and Space Distributions

Observationally, there are many indications that the location of certain galaxies is correlated with their intrinsic luminosity: the different distribution of

---

[2] When the assumption of purely Poisson distribution is made – i.e. when (8.2) is replaced by (8.4). In the more general case of (8.2) the amplitude is just a normalization factor such that the LF becomes a PDF.

bright elliptical and faint spiral galaxies mentioned in the introduction being one of the most evident cases. However from a statistical point of view one usually does not take into account the possible correlation between space and luminosity distributions. This case can be studied quantitatively with the MF formalism. The MF analysis proceeds now along the lines explained in Chap. 5: the microscopic density function can be written as

$$\mu(\boldsymbol{r}) = \sum_i^N \mu_i \delta(\boldsymbol{r} - \boldsymbol{r}_i) \qquad (11.2)$$

where the sum is extended to the N objects (galaxies) contained in a given volume and μ_i is the weight of the ith point located at $\boldsymbol{r}_i$. The weights are thus taken to be proportional to the intrinsic luminosities of the galaxies and normalized so that $\sum_i^N \mu_i = 1$. By using the MF technique one can then measure the spectrum of exponents $f(\alpha)$ and the generalized MF dimensions $D(q)$. This approach has been applied in some galaxy catalogs [47, 223] and the result is that there is evidence for MF, albeit weak due to the small number of points in the samples analyzed. Let us now see the usefulness of the MF in relation to the various statistical and morphological properties of galaxy clustering: in this context the strongest (weakest) singularities correspond to the most (less) luminous galaxies.

The relation between the properties of the $f(\alpha)$ spectrum and real space correlations is straightforward. The exponent that describes the (power-law) behavior of the number density (i.e. the conditional average density discussed in Chap. 9 of the microscopic density (11.2) with all $\mu_i = 1$) is given by $D - 3$ where D is the fractal dimension, of the *support* (see Chap. 5) of the MF distribution. Therefore, in terms of the multifractal spectrum of dimensions, it is given by $D(0) - 3$ which is related to the maximum of $f(\alpha)$: clearly this includes the possibility of having $D(0) = 3$, i.e. an homogeneous and uniform support. Varying α one may explore the different singularities of the distribution, i.e. the correlation properties of different types of galaxies.

A simple way to understand the MF behavior is shown in Fig. 11.1: by defining a threshold in the measure and considering only those singularities which lie above it, we select only the largest peaks in the measure distribution: the set defined by these peaks has a different fractal dimension to that of the set defined by the entire distribution, or any other set defined by a different threshold. If the distribution is MF the fractal dimension decreases as the threshold increases[3]: in fact, given that the generalized MF spectrum is, in

---

[3] We note that, strictly speaking, the presence of the cut-off in the threshold can lead (for a certain well defined value of the cut-off itself) to the so-called *multiscaling* behavior of the MF measure [135]. In fact, the presence of a lower cut-off in the calculation of the generalized correlation function affects the single-scaling regime of $\chi(\epsilon, q)$ for a well determined value of the cut-off $\alpha_{cut-off}$ such that $\alpha_{cut-off} < \alpha_c$, and this function exhibits a slowing varying exponent proportional to the logarithm of the scale ϵ.

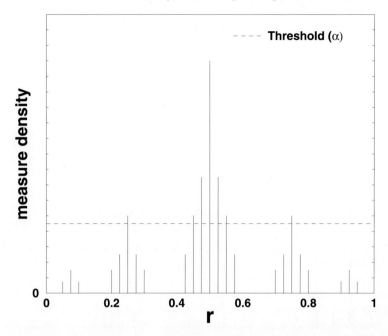

Fig. 11.1. Simple example of a multifractal measure in one dimension. The strongest singularities lie in the largest clusters. The multifractal spectrum is computed in a way substantially equivalent to changing the measure (i.e. luminosity) threshold, i.e. by selecting peaks of different height

general, convex with the maximum at the dimension of the support, one expects that the stronger are the singularities of the field the smaller is their fractal dimension.

As discussed in Chap. 4 (see Fig. 4.17) the fractal dimension is a measure of the "intensity" of clustering for fractal objects: The clustering increases when the fractal dimension decreases. It is in this sense that the clustering properties of the most massive and luminous elliptical galaxies are different. In fact, massive galaxies are mostly found in rich clusters while field galaxies are usually spirals or gas rich dwarfs. These observational properties are consistent with a MF picture of the galaxy distribution, i.e. with a self-similar behavior of the whole visible matter distribution. The largest peaks are located in the largest clusters and they should show correspondingly a smaller fractal dimension (see Fig. 11.2). This behavior can be related to the different correlation exponent found by correlation analysis for the elliptical, lenticular and spiral galaxies: In fact the observational evidence is that the correlation exponent is higher for elliptical than for spiral galaxies (see discussion in [220]).

In the MF description of galaxy luminosity-space correlation some well-known observational facts, e.g. the giant-to-dwarf ratio depends on the envi-

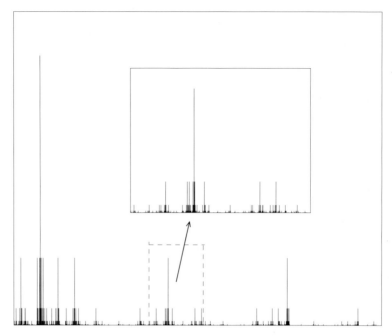

Fig. 11.2. One-dimensional generalized multiplicative random Cantor set with compact support (see Chap. 5). One the y-axis it is reported the measure density, while the x-axis corresponds to the spatial coordinate. Largest fluctuations belong to large "clusters" and the smaller ones are more uniformly distributed. In the *inset panel* it is shown a "zoom" of a small region of the distribution

ronmental density, are naturally described. In fact, dwarf galaxies can belong to the rich clusters where giants lie, but they can also be in small groups. Hence what is known in the literature as "morphological-segregation", i.e. the fact that galaxies of different type are distributed differently in space, can be understood precisely in terms of the self-similar character of the whole matter distribution.

The relationship between $f(\alpha)$ and the LF (measure distribution) has been discussed in Sect. 5.8: indeed it has been shown that, if the underlying distribution has a MF nature with a certain $f(\alpha)$ spectrum, the probability, once the spatial scale is fixed, of finding an object with singularity μ is proportional to a power-law function with an exponential cut-off, i.e. whose shape is very well represented by the Schechter functional behavior: this result is very weakly dependent on the specific shape of $f(\alpha)$. In this case the galaxy LF can be interpreted as a complementary aspect of the MF nature of the galaxy distribution.

Finally it is interesting to note that for a more physical interpretation one would like to express these properties in terms of galaxy mass rather than magnitude or luminosity. The mass of each galaxy is generically assumed to be related to its absolute luminosity through a relation

$$M = k(i)L^\beta \tag{11.3}$$

where k is the "mass to light ratio" which depends on the galaxy morphological type i. In relation to MF properties what is important is that there is a large range of galaxy masses (varying by a factor of order 10^6 or more). The variation of k produces only small effects on a logarithmic scale. The exponent β, which typically has a value $\beta \approx 1$ [81], is more important. A different value of β does not change the MF nature of the mass distribution, but it does modify the parameters of the spectrum [47, 190, 223].

11.4 Summary and Discussion

In this chapter we have applied the MF analysis introduced in Chap. 5 to the galaxy space-luminosity distribution. Such a scheme gives us a mathematical framework to treat the full space-luminosity distribution in a unified way without assuming a-priori that the luminosity and position of galaxies are uncorrelated, and further without assuming a-priori the spatial homogeneity of galaxy distribution. The galaxy luminosity function and space density are then linked to the MF spectrum of generalized exponents. Hence the MF picture provides a mathematical framework to describe the different *strong* clustering of brighter and fainter galaxies: in this context it is natural to expect that the massive galaxies are mostly located in large clusters, as well as a number of other properties compatible with observations (e.g. morphological segregation). The quantitative characterization of the different clustering properties of galaxies of different luminosity is therefore expressed in terms of a spectrum of *scaling exponents* of the measure density one can associate to the available set of galaxies rather than in terms of *amplitudes* of the reduced two-point correlation function (see discussion in Chap. 4 and Fig. 4.17). In particular brighter galaxies should have a larger correlation exponent (i.e. smaller fractal dimension) than fainter ones. This is a new and different perspective for the description of the statistical properties of different "peaks" of the mass distribution with respect to the canonical one where peaks are given by a selection on a Gaussian field (see Chap. 13). Whether very faint galaxies, or low surface brightness ones, which are difficult to observe, may fill the voids of brighter galaxies so that the whole distribution is homogeneous, is an open problem. This can very well be the case and it would correspond to the MF measure having a compact support[4]. However there is observational evidence [37] that low-surface brightness galaxies are located around the same structures as bright galaxies.

---

[4] In [71] a simple model of this type is described.

12 The Distribution of Galaxy Clusters

12.1 Introduction

In this chapter we turn to catalogs of galaxy clusters, in which there has been much observational and theoretical interest in parallel to that in galaxies. We consider what the relation is between the correlation properties of clusters and those of the galaxy distribution from which they are selected. In particular we discuss in a simplified way both the case of an underlying galaxy distribution which is fractal-like up to the scales probed by the sample, and the case where there is a cross-over to homogeneity. We then discuss the "galaxy-cluster" mismatch problem, which is very analogous to the "luminosity bias" we have discussed previously: when the correlation properties of clusters are characterized with the reduced two-point correlation function, one finds that its form is very similar to that of galaxies – with approximately the same exponent – but that its amplitude is higher. Thus the value of r_0 – interpreted as a real physical scale characterizing the cluster distribution – is larger than its value for galaxies. Just as for the case of "luminosity bias", in the fractal framework, this behavior can be simply understood as giving evidence for the absence of a well defined mean density at least up to the sample scale: clusters, as they are brighter objects, probe greater depths than available single galaxy distributions, and an increase in r_0 would also be associated to a decrease in the mean density. Further the fact that the exponent remains the same as that in the galaxy distribution suggests a possible simple interpretation in terms of a continuing fractal to scales larger than those on which we could place a good constraint (see Chap. 9) from galaxy catalogs alone.

Just as for galaxies both angular and redshift catalogs of clusters have been constructed from observations. In the case of clusters the redshift one considers is an average redshift. This latter is obtained by measuring the redshift of several galaxies which are members of the same "cluster". It is important to note that numerous different observational strategies for the definition of a galaxy cluster have been used: sometimes they have been identified in optical angular catalogs, in other cases they are selected by a criterion based, for example, on X-ray observations [208]. Theoretically clusters are believed to be gravitationally bound (and approximately virialized) systems of galaxies and dark matter, but it is not feasible observationally to use such a definition by probing the relation between kinetic and potential energy. We do not discuss here the details of different methods used to define a cluster of galaxies. We treat the problem from a purely statistical point of view, pointing out the principal statistical properties of the new set of "points" defined by galaxy clusters.

12.2 Cluster Correlations and Multifractality

Given a point distribution one can identify a "cluster" as a group of points in the simple visual way shown in Fig. 12.1: this is the empirical method used in the construction of early cluster catalogs. This means identifying high local density regions in the point distribution, considering volumes of a certain size centered on the density "peak" and selecting a cluster when the number of points in the volume exceeds a certain defined value. The question is then to understand the relation between this new set, and in particular its correlation properties, and the original one.

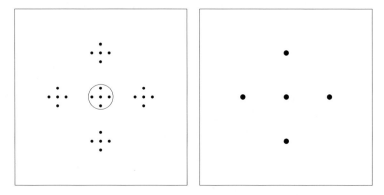

Fig. 12.1. Empirical construction of a catalog of clusters (*right panel*) from one of galaxies (*left panel*)

To define the problem more precisely from a statistical point of view, the simple sampling shown in Fig. 12.1 must be refined. For simplicity we consider that the sample volume is a box of size L. We may cover such a volume with a regular grid with spacing $\ell \ll L$, which gives $(L/\ell)^3$ cells in three-dimensional space. The set "clusters" is then defined as the set of cells in which the number of points exceeds a given lower limit: clearly the grid spacing ℓ must be appropriately chosen as this gives the physical dimension of a cluster. Such a procedure corresponds to coarse-graining the galaxy distribution (sometimes called also "point decimation"), and then considering a threshold in the number of points contained in each coarse-grained cell. In this way we have identified a set of points (with coordinates given by the centers of the cells) and to each point we can associate a measure given by

12.2 Cluster Correlations and Multifractality

the number of points contained in the cell. Thus, we can study the complete correlations of this measure distribution by applying the multifractal (MF) formalism discussed in Chap. 5. In this context, we define the normalized density function for the point (galaxy) distribution as

$$\mu(\boldsymbol{r}) = \sum_{i=1}^{N} \mu_i \delta(\boldsymbol{r} - \boldsymbol{r}_i) \tag{12.1}$$

with $\mu_i = 1/N$, where N is the number of points (galaxies) of the original set contained in the sample volume. We label each box of the superimposed grid of spacing ℓ by the index i and construct for each box the function

$$\mu_i(\epsilon) = \int_{ith\ box} \mu(r)dr , \tag{12.2}$$

where clearly $0 < \mu_i < 1$ and $\epsilon = \ell/L$. This procedure defines a discretized measure density suitable for the MF analysis presented in Chap. 5. The fractal dimension of the cluster distribution with a given richness corresponds approximately to a certain value of the moment $q > 0$, and then essentially to a certain threshold in the number of points. Its scaling behavior is described by the corresponding value of the $f(\alpha)$ spectrum. Changing the value of $q > 0$ one considers a sampling strictly related to that consisting of selecting clusters of different richness.

As explained below, the fact that the exponent characterizing the decay of the two-point correlation function of galaxies and clusters is nearly the same is a clear sign that there is a very weak MF behavior in the sense explained above, i.e that at least to a first approximation we may write, in the MF analysis $f(\alpha) \approx D$ for any value of α, where D is the fractal dimension of the support observed in the analyzed sample. However, on the other hand, the simple fractal picture can give us a natural and simple explanation of the observation that the exponent of the two-point correlation function of galaxies and clusters is the same as clusters represent a coarse-grained view of galaxy distribution.

Let us now consider two simple cases which may clarify some of these points:

(i) Suppose that the original point distribution is a simple fractal with dimension $0 < D < 3$ up to a scale larger than the sample size L (see Fig. 12.2). It is clear that the selection of the set "clusters" defined above will change only the lower cut-off of the distribution: that is, the average distance between first neighbor "clusters" is of the order of the cell size ℓ. As ℓ changes only this cut-off can vary, as no other length scales is introduced in the system. This means that in such a situation the point distribution identified by "clusters" will show the same fractal properties as the original set, i.e. with the same dimension D, except for the fact that the average conditional density will have a smaller amplitude (i.e. the number density of

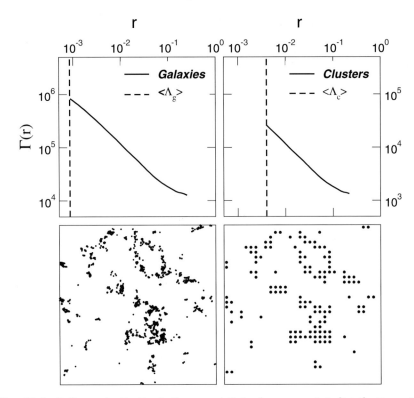

Fig. 12.2. *Left panels*: In the bottom part it is shown a point distribution with fractal correlation, while in the upper part the corresponding conditional average density. *Right panels*: In the bottom part it is shown the "cluster" distribution constructed as explained in the text, while in the upper part it is shown the corresponding conditional average density. One may note that the lower cut-off and the amplitude are different from the one on the left, but the exponent is the same

richest clusters is smaller than the number density of the poorest ones) and a larger lower cut-off. Both the amplitude and the lower cut-off will depend on the parameter ℓ. The reduced two-point correlation function (which we recall is sample size dependent in the case of a fractal distribution – see Chap. 9), instead, will have the same amplitude for both sets because we have assumed that the sample size is unchanged by the coarse graining procedure.

(ii) As a second simple example, we suppose that the original point distribution has fractal correlations up to a scale $\lambda_0 \ll L$ and then that it becomes purely Poisson-like, i.e. without any residual correlation (this set can be constructed, for example, by making a grid of $(L/\lambda_0)^3$ independent fractal structures with the same fractal dimension, each generated in a sample of size λ_0. Even in this case, the selection procedure defined above will not change the unique length scale in the distribution, which is λ_0: it will therefore not

change the functional behavior of $\Gamma(r)$, but only its amplitude along the lines discussed for the pure fractal case. Thus $\Gamma(r)$ for the cluster set will show the same slope up to λ_0 and then the same flattening for scales $r > \lambda_0$ (see Fig. 12.3). The amplitude of the reduced two-point correlation function *does not* change as it is related to the unique length scale in the distribution given by λ_0.

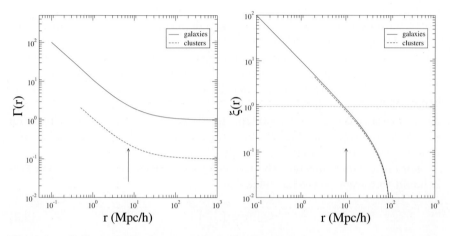

Fig. 12.3. *Left panel*: If galaxies are fractally distributed up to λ_0 and Poisson-like for $R > \lambda_0$ (the figure shows results for an artificial distribution with these features), galaxy clusters, identified by the procedure shown in Fig. 12.2, have a similar $\Gamma(r)$, but with a lower amplitude. The reduced two-point correlation correlation (*right panel*) function is, on the other hand, unchanged between galaxies and clusters

There remains the question of whether the selection procedure discussed here can change the "homogeneity" scale of the original set, while leaving the exponent of the conditional density the same. The answer to this question will in fact depend on details of the correlations in the original set. We will return to this point in Sect. 12.4.

12.3 Galaxy Cluster Correlations

With respect to galaxy catalogs, cluster surveys offer the possibility of studying the visible matter distribution in much larger volumes, reaching depths beyond the redshift $z \approx 0.2$, and extending over the whole sky. Using clusters, one can trace the matter distribution with a smaller number of objects, in a given volume, with respect to the galaxies. For example, in the northern hemisphere there are $\approx$500 rich clusters up to $z \approx 0.15$ which correspond to $\approx 10^6$ galaxies. The main problem of cluster catalogs is their incompleteness, since clusters are usually identified as density enhancements in angular galaxy surveys (through an empirical procedure very similar to the one shown in Fig. 12.1), and their distance is usually determined through the redshift of one or two galaxy members. It is in fact clear that only the measurements of the magnitude and the redshift of every cluster (and its member galaxies) allow the construction of truly volume limited samples.

As for the galaxy distribution, the observations of large scale structures defined by galaxy clusters, like super-clusters and voids, which extend up to the limits of the samples investigated, raise a question mark about the existence of a well-defined and sample-independent average density at the scales of these samples. When these large scale inhomogeneities have been attributed to some incompleteness in the surveys, better and more extensive studies have usually confirmed the reality of these structures, making the observed agglomerations sharper and leaving the voids empty. For instance, Tully [234, 235], by investigating the spatial distribution of the Abell and ACO clusters catalogs up to 300 Mpc/h, stressed that there are structures on a scale of order of $\sim$300 Mpc/h lying in the plane of the local super-cluster. Other authors found similar results [9, 21, 75, 77, 124, 145, 196].

For cluster distributions, the reduced two-point correlation function $\xi(r)$ is found to have a power-law behavior

$$\xi(r) \approx \left(\frac{r_0}{r}\right)^\gamma . \qquad (12.3)$$

The exponent γ appears, generally, consistent with the value observed for galaxies, i.e. $\gamma \approx 1.8$. The value of r_0 is higher than for galaxies, and is not stable in different samples. In general, high values for $r_0 \approx 20 \div 25$ Mpc/h are obtained for samples which contain the richer Abell/ACO clusters [9, 196] (see Fig. 12.4), although some authors claim these are overestimates (finding $r_0 \approx 14$ Mpc/h, e.g. [61]), produced by systematic biases present in the Abell/ACO catalog. Lower values of $r_0 \approx 13 \div 15$ Mpc/h are obtained also from the analysis of automated cluster catalogs (APM, EDCC) and from the cluster catalogs selected from X-ray galaxy surveys. In conclusion, for the samples analyzed so far we have 14 Mpc/h $\lesssim r_0 \lesssim$ 25 Mpc/h.

Clusters are then said to be *more clustered than galaxies*, for which $r_0 \approx 5$ Mpc/h. This *mismatch* between galaxy and cluster correlations is another puzzling feature of the usual analysis; clusters, in fact, are made by galaxies and many of these are included in the galaxy catalogs for which the smaller value of r_0 is derived. To explain the mismatch it is necessary to assume that fundamental differences between the cluster galaxies and the galaxies not belonging to clusters. This concept has given rise to the so-called *richness-clustering relation*[1] [9] according to which, objects with different mass or morphology segregate from each other and give rise to different correlation properties. Such a relation is expressed as

---

[1] The richness quantifies the number of galaxies contained, for example, in a sphere of radius 1 Mpc around the cluster's center.

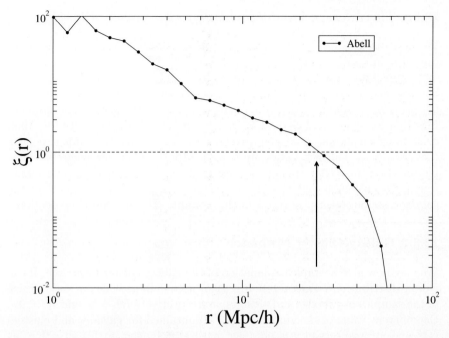

Fig. 12.4. The reduced two-point correlation function $\xi(r)$ for the Abell clusters: note that $\xi(r_0) = 1$ for $r_0 \simeq 25\,\text{Mpc/h}$. (Adapted from [223])

$$r_{0i} \approx 0.4\bar{n}_i^{-1/3}$$

where the index i refers to the system being considered, and n_i is the estimation of its mean spatial density (see e.g. [10]). The richness dependent amplitude of the correlation function increases monotonically from $10\,\text{Mpc/h}$ for poor clusters to 20–$25\,\text{Mpc/h}$ for rich clusters. This proposition leads eventually to the peculiar feature that every object is slightly different from any other and forms, on its own, a morphological class, which makes useless the concept of correlation between objects without specification of the type of objects considered.

In standard theoretical models, the fact that the most massive clusters, which correspond to the rarest and highest fluctuations of the matter field, exhibit a larger amplitude of the correlation function is interpreted in the framework of the (luminosity) bias selection introduced by [123] and discussed in Chap. 13. Here we consider a different explanation: in the next two subsections we describe the results of an analysis without the a-priori assumption homogeneity. This allows us to present a simple interpretation of observed cluster correlation based on the concept of coarse-graining.

12.3.1 The Average Conditional Density for Galaxy Clusters

We now discuss the same analysis of galaxy clusters without the assumption of homogeneity performed for galaxies (see discussion in Chap. 9). The motivations are, also in this case, related to the fact that the assumption of homogeneity within the sample size, on which the analysis leading to (12.3) is based, is questionable in the samples observed in the last two decades. In Fig. 12.5 we show the behavior of the conditional average density for Abell clusters which shows power-law correlations, with almost the same exponent (i.e. $D \approx 2$) as that of galaxies, up to $\sim 50\,\mathrm{Mpc/h}$. This fact suggests a simple interpretation that the galaxy cluster distribution, which has the same fractal dimension as that of galaxies, can be seen as a "zooming out" of the galaxy distribution, i.e. in terms of the example *(i)* of the previous section.

12.3.2 Galaxy-Cluster Mismatch

We may now give a simple explanation of the galaxy-cluster mismatch. If the galaxy distribution is fractal (see Fig. 12.6) up to a large scale λ_0, and the size of the samples of galaxies and clusters are respectively $R_s^g < \lambda_0$ and $R_s^c < \lambda_0$, the different value of the amplitude of $\xi(r)$ obtained for galaxies and clusters (r_0^{gg} and r_0^{cc} respectively) is obtained naturally. Indeed in this situation, as discussed in Chap. 9, the amplitude of the reduced two-point correlation

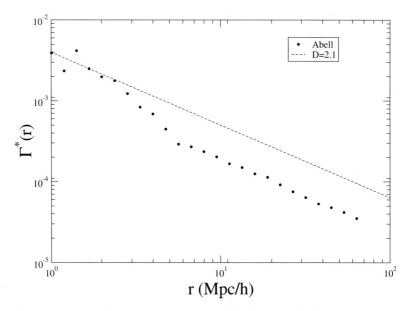

Fig. 12.5. The conditional average density $\Gamma^*(r)$ for the Abell clusters. (Adapted from [223])

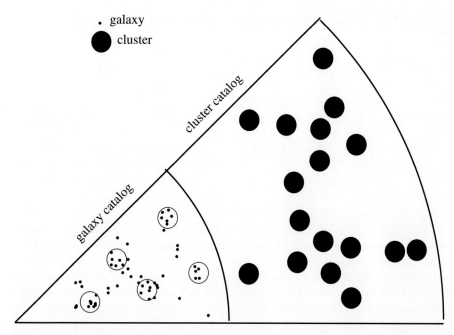

Fig. 12.6. Clusters are brighter than galaxies so that cluster catalogs sample volumes of space larger than those of galaxy surveys. In the interpretation discussed in the text, based on fractal geometry, the difference between the "correlation lengths", as r_0 is called in the cosmological literature (in reality it is a measure of the apparent homogeneity scale) of galaxies and clusters (the galaxy-cluster mismatch) can be simply ascribed to a difference in the depth of the respective catalogs. It is thus understood as finite size effect

function is linearly (on average) related to the sample size R_s: given that, on average, $r_0 \propto R_s$, it is simple to infer that the following relation should hold:

$$\frac{r_0^{cc}}{r_0^{gg}} = \frac{R_s^c}{R_s^g}. \tag{12.4}$$

The reasoning leading to (12.4) shows that the difference in the "correlation lengths" (as r_0 is widely referred to in the cosmological literature see Chap. 2) of galaxies and galaxy clusters can be due to the different sample sizes of the corresponding catalogs in the case in which the real homogeneity scale is at least as large as the larger (cluster) sample size: this accounts well for the results from the analysis of the Abell and ACO catalogs for clusters and CfA for galaxies [47, 223].

12.4 Luminosity Bias and the Richness-Clustering Relation

In this section we clarify a closely related point concerning the interpretation of the correlation properties of galaxies via the standard $\xi(r)$ analysis: the question we discuss concerns the kind of sampling which can give rise to a linear amplification of the two-point correlation function in the regime of strong clustering (i.e., $\xi(r) \gg 1$).

As already mentioned, it has been observed that in many observations the amplitude of $\xi(r)$ changes when different VL samples are considered [56, 28]. A VL sample is characterized by two related cuts in distance and in absolute magnitude (i.e. intrinsic luminosity) with a systematic trend: the larger the cut in distance the brighter are the galaxies in the sample. Let us call $R_1 > R_2$ the two cuts in distance of two VL samples corresponding to cuts in absolute magnitude $M_1 < M_2$, in the same magnitude limited survey. It is thus observed[2] that in a certain range of scales one has $\xi_1(r) > \xi_2(r)$ (e.g. [168, 250]), and that in the same range of scales the functional behavior of $\xi(r)$ is a power-law with the same exponent in both samples. We can envisage two possible explanations: either sampling or finite size effects. Namely that the brighter are the galaxies considered the larger is the amplitude of $\xi(r)$, or that the larger is the sample size then the larger is the amplitude of $\xi(r)$ because of a finite-size effect due to the variation of the sample depth. While the latter is a typical finite-size effect present in the case of a fractal distribution, as we have discussed at length, the first explanation has been called "luminosity bias". Let us consider further what this latter hypothesis means in terms of correlation properties.

Consider first the following example. Suppose we have two kind of galaxies of different mass, one of type A and the other of type B. Suppose further that the mass of the galaxies of type A is twice that of type B. The proposition "galaxies of different luminosity (masses) correlate in different ways" (in the sense that they have different correlation amplitudes of the $\xi(r)$ function) implies that the physics of the origin of these correlations distinguishes between a situation in which there is, in a region, a galaxy of type A or two galaxies of type B close to one another. This would be a paradox if only gravity is considered to form such correlations, as the gravitational interaction depends only on the total mass.

A more specific meaning can be given as follows. Suppose that the galaxies of type A have a smaller real homogeneity scale than that of the galaxies of type B. The galaxies of type B are still then strongly correlated when the galaxies of type A are almost uniformly distributed. This means that the galaxies of type A should fill the voids of galaxies of type B.

---

[2] Using $\xi(r)$ without having previously tested that a crossover of the average conditional density, as it should be as explained in Chap. 9, toward homogeneity is present within the samples considered.

12.4 Luminosity Bias and the Richness-Clustering Relation

In terms of the average conditional density, and by using a simple approximation for the crossover to homogeneity (Chap. 4), we obtain (see Fig. 12.7):

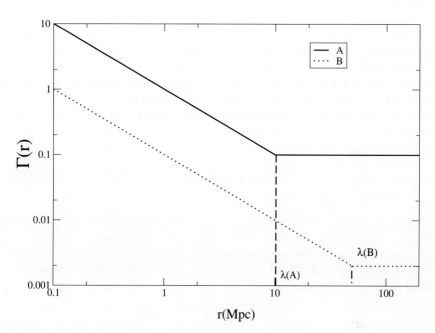

Fig. 12.7. Luminosity bias: in this example we have two kind of objects. Galaxies A have power-law correlation up to $\lambda(A)$ and then became uniformly distributed in space. Galaxies B have the same behavior, but the power-law behavior is up to $\lambda(B) > \lambda(A)$. In this case the voids of B galaxies should be full of galaxies A

$$\begin{cases} \Gamma^A(r) = B^A r^{D-3} & \text{for} \langle \Lambda \rangle \leq r \ll \lambda_0^A \\ \Gamma^A(r) = B^A (\lambda_0^A)^{D-3} & \text{for} \lambda_0^A \ll r \leq R_s \end{cases}, \quad (12.5)$$

where $\langle \Lambda \rangle$ is a measure of the average first neighbor distance. Moreover, analogously, we can write

$$\begin{cases} \Gamma^B(r) = B^B r^{D-3} & \text{for } \langle \Lambda \rangle \leq r \ll \lambda_0^B \\ \Gamma^B(r) = B^B (\lambda_0^B)^{D-3} & \text{for } \lambda_0^B \ll r \leq R_s \end{cases}. \quad (12.6)$$

If $R_s \gg \lambda_0^B > \lambda_0^A$ then we simply have that the amplitude of $\xi(r)$ (for example at the scale $\xi(r_0) = 1$) has the following property

$$\frac{r_0^A}{r_0^B} = \frac{\lambda_0^A}{\lambda_0^B}. \quad (12.7)$$

In this case there is indeed an amplification of $\xi(r)$ which is simply due to the difference in the homogeneity scale of the two galaxy types [223]. Physically as we have noted, this model seems quite unrealistic (if gravity plays a central role in the developments of correlations). Note that, in any case, the simplest way to detect such a behavior is to measure the conditional density (for the different galaxies).

Let us consider now "luminosity segregation" more generically. A simple selection of a point set which describes more realistically the case of galaxy samples is the following. Let us characterize each galaxy by its three space coordinates $\boldsymbol{x}_i$ and its luminosity, which we call hereafter measure μ_i. We then have a point set $\{\boldsymbol{x}_i, \mu_i\}$ with, in general, a non-trivial measure distribution and a given reduced two-point correlation function $\xi(r)$. We now consider the selection of points by using the measure. For example we put a threshold in the measure $\bar{\mu}$ and we select only points with $\mu_i > \bar{\mu}$ in a way such that the selected set has a correlation function

$$\xi_S(r) = A\xi(r) \gg 1 \qquad (12.8)$$

where A is a constant.

Let us consider how the selection affects the conditional density, which simply measures the number density of points around an average particle. We can make the following hypothesis for its behavior in the full set (i.e. a volume-limited sample cut at faint absolute magnitude, and hence containing both faint and bright galaxies):

$$\begin{aligned} \langle n(r) \rangle_p &\gg n_0 \quad \text{for} \quad r \leq \lambda_0 \\ \langle n(r) \rangle_p &\approx n_0 \quad \text{for} \quad r \geq \lambda_0 \end{aligned} \qquad (12.9)$$

(making the hypothesis that the homogeneity scale is well inside the sample volume). Thus our distribution is made of non-linear structures of size λ_0 which we denominate "clusters". Hence let us suppose we make a sampling such that at small scales ($r \leq \lambda_0$) an average point continues to see the same number density while at large scale the average density decreases. This means that each particle inside a cluster sees the same number density of particles, but that the number density of clusters has been decreased (in the infinite volume limit). Doing this we have that for the selected set $\langle n(r) \rangle_p^S \approx \langle n(r) \rangle_p$ if $r \ll \lambda_0$, while the average density at large scales in the selected set is $n_0^S \ll n_0$. In this way a relation of the type (12.8) can be obtained. Such a selection has to be made using the measure distribution and there must be a correlation between space positions and measure. For example if we have "clusters" of particles where $\mu_i \ll \bar{\mu}$, and clusters with $\mu_i \gg \bar{\mu}$ only the latter survive when we select using the threshold $\bar{\mu}$. Hence the linear amplification in the strong clustering regime may be obtained with this very particular sampling. Note this kind of sampling changes an intrinsic length scale of the system: while the scale λ_0 (which is interpreted as the typical cluster size) is not changed by the selection, the original objects are typically separated by a distance of order $n_0^{-1/3}$ which increases to $(n_0^S)^{-1/3}$ after sampling.

12.5 Summary and Discussion

The observational galaxy-cluster mismatch is at the origin of the concept of bias which is a crucial ingredient in the theoretical and phenomenological interpretation of matter density fields in cosmology [123]. We will discuss this concept further in the next chapter. Here we have presented a different interpretation of the relevant observations. The essential idea is that clusters correspond to a distribution obtained by a coarse-graining of the galaxy distribution. We have explained that the coincidence of the exponents of the power-law correlation function of galaxies and galaxy clusters can be easily understood in this way, as well as the different amplitudes of the reduced correlation function. In this explanation the observed mismatch of the correlation functions is a finite size effect.

13 Biasing a Gaussian Random Field and the Problem of Galaxy Correlations

13.1 Introduction

We have discussed at some length the problems of "luminosity bias" and the related "galaxy-cluster mismatch" which arise from what is found observationally in the analysis of galaxies and galaxy clusters with the reduced two-point correlation function $\xi(r)$: a considerable variation of the amplitude of this function measured on different kind of objects. We have proposed a possible resolution which ascribes these observations to the sample dependence of this statistical quantity, and not to an intrinsic difference between the correlation properties of the different objects. This explanation posits that the underlying distribution is still strongly fluctuating at the scales probed by these samples, and that it is simply the sample dependence of the mean density which is giving rise to the observed variations of the amplitude of $\xi(r)$. In this chapter we go in an orthogonal direction and examine aspects of the conceptual framework used predominantly in the cosmological community to explain these observations: this is what is known as "bias", which generically covers the notion that the correlation properties of different kinds of objects (in particular galaxies of differing luminosity, and galaxy clusters) are really different. The core theoretical idea is that different kinds of objects correspond to different kinds of selections on the underlying dark matter field dominating the gravitational evolution of the universe. Our aim here is to analyze just the simplest canonical model of bias, that introduced by Kaiser [123] to explain the galaxy-cluster mismatch. Our main conclusion is that it is very problematic in terms of its capacity to explain the relevant observations, as it does not give rise to a scale-independent renormalization of either the reduced two-point correlation function or the PS of the different objects. We explain that these results should be valid for a generic model of bias, and not just the specific one we analyze, and we consider the implications for models of structure formation.

In the original model of bias [123] the underlying distribution of dark matter is treated as a correlated Gaussian density field with a given two-point correlation function $\tilde{\xi}(r)$: the various visible objects, such as galaxies of different luminosities or galaxy clusters, are interpreted as the peaks of the matter distribution, which have collapsed by gravitational clustering. Different kinds of objects are selected as peaks above a given threshold ν. A change in the threshold selects different regions of the underlying Gaussian field, corresponding to fluctuations of differing amplitudes. The reduced two-point correlation function of the selected objects is then that of the peaks $\tilde{\xi}_\nu(r)$. It is enhanced with respect to that of the underlying density field $\tilde{\xi}(r)$ and such an amplification goes in the direction of the observational results but only in a qualitative way. Indeed we discuss some problematic aspects of this mechanism: in particular the fact that the amplification of the correlation function is in fact only linear in the regime in which $\tilde{\xi}_\nu(r) \ll 1$ [89]. In the region of most observational relevance (where $\tilde{\xi}_\nu(r) \gg 1$) there is actually at least an exponential and scale dependent distortion of the correlation function. Furthermore we explain that the amplification of the correlation function by biasing, i.e. by increasing the threshold, reflects simply that the distribution of peaks is *more clustered* because peaks are exponentially *sparser* while their size is only slightly modified.

We then discuss other subtle properties introduced by this selection mechanism for what concerns the power spectrum (PS). While threshold biasing of a Gaussian random field does give rise to a linear amplification of the reduced two-point correlation function at large distances, i.e. in the regime where correlations are very weak, we show that, for standard cosmological models, this *does not* translate directly into a linear amplification of the power spectrum (PS) at small k [72]. For example, for standard cold dark matter (CDM) type models (see Chap. 6) this means that the "turn-over" at small k of the original PS disappears in the PS of the biased field for the physically relevant range of the threshold parameter ν. In real space this difference is manifest in the asymptotic behavior of the normalized mass variance in spheres of radius R, which changes from the super-homogeneous behavior $\sigma^2(R) \sim R^{-4}$, typical of standard CDM models, to a Poisson-like behavior $\sigma_\nu^2(R) \sim R^{-3}$ (Chap. 3). This qualitative change results from the extreme sensitivity of the condition of *super-homogeneity* $P(0) = 0$ for the PS typical of all standards cosmological models. While our quantitative results are specific to the simplest threshold biasing model, we argue that our conclusions should be valid qualitatively for a generic biasing mechanism involving a scale-dependent amplification of the correlation function. At the end of the chapter we return then to the problem of the relative normalization of fluctuations at early times (i.e. CMBR anisotropies) to fluctuations observed today (e.g. the galaxy distribution).

13.2 Biasing of Gaussian Random Fields

In this section the notion of biasing for a Gaussian random field is given in mathematical terms. We then calculate biasing for some examples and clarify the physical meaning of bias in the context of the cosmological literature [15, 123]. We consider a statistically homogeneous, isotropic and correlated continuous Gaussian random field, $\delta_\rho(\mathbf{x})$, with zero mean and variance $\sigma^2 = \langle \delta_\rho(\mathbf{x})^2 \rangle$ in a volume V as defined in Chap. 2. The marginal one-point

13.2 Biasing of Gaussian Random Fields

probability density function of δ_ρ (i.e. the a priori probability density that the field has the value δ_ρ in a randomly chosen point of space) is

$$p(\delta_\rho) = \frac{1}{\sqrt{2\pi}\sigma} e^{-\frac{\delta_\rho^2}{2\sigma^2}}.$$

The fraction of the volume V with $\delta_\rho(\mathbf{x}) \geq \nu\sigma$, is then

$$Q_1(\nu) = \int_{\nu\sigma}^{\infty} p(\delta_\rho) d\delta_\rho. \tag{13.1}$$

The correlation function between two values of $\delta_\rho(\mathbf{x})$ at two points separated by a distance r is given by $\tilde{\xi}(r) = \langle \delta_\rho(\mathbf{x})\delta_\rho(\mathbf{x}+r\mathbf{n})\rangle$. By definition, $\tilde{\xi}(0) = \sigma^2$, is the one-point variance of the field. In this context, statistical homogeneity implies that the mean ρ_0, the variance σ^2, and the correlation function, $\tilde{\xi}(r)$, do not depend on $\mathbf{x}$. Isotropy means that $\tilde{\xi}(r)$ does not depend on the direction $\mathbf{n}$[1]. An important application consists in the application to cosmological density fluctuations,

$$\delta_\rho(\mathbf{x}) = \frac{\rho(\mathbf{x}) - \rho_0}{\rho_0},$$

where $\rho_0 = \langle \rho \rangle$ is the average density; but the following arguments are completely general.[2]

Our goal is to determine the correlation function of local maxima from the correlation function of the underlying density field. We simplify the problem by computing the correlations of *regions* above a certain threshold $\nu\sigma$ instead of the correlations of *maxima*. However, these quantities are closely related for values of ν significantly larger than unity [123]. We define the threshold density, $\theta_\nu(\mathbf{x})$ by

$$\theta_\nu(\mathbf{x}) \equiv \theta(\delta_\rho(\mathbf{x}) - \nu\sigma) = \begin{cases} 1 \text{ if } & \delta_\rho(\mathbf{x}) \geq \nu\sigma \\ 0 \text{ else.} \end{cases} \tag{13.2}$$

Note the qualitative difference between δ_ρ which is a weighted density field, and θ_ν which just defines a set of regions having equal weight (see Fig. 13.1). The underlying idea is that one would like to relate the continuous density field (for example the CDM one) to the distribution of galaxies: on sufficiently large scales the galaxy distribution should represent a selection of a discrete set from the underlying CDM Gaussian field. However the new density field $\theta_\nu(\mathbf{x})$ does not correspond to a point distribution and a further discretization of this new field is then necessary (see below). We note the following

---

[1] In other words, we assume $\delta_\rho(\mathbf{x})$ to be an isotropic Gaussian SSP as defined in Chap. 2.

[2] Clearly, cosmological density fluctuations can never be perfectly Gaussian since $\rho(\mathbf{x}) \geq 0$ and thus $\delta_\rho(\mathbf{x}) \geq -1$, but, for small fluctuations, a Gaussian can be a good approximation. Furthermore, our results remain at least qualitatively correct also in the non-Gaussian case.

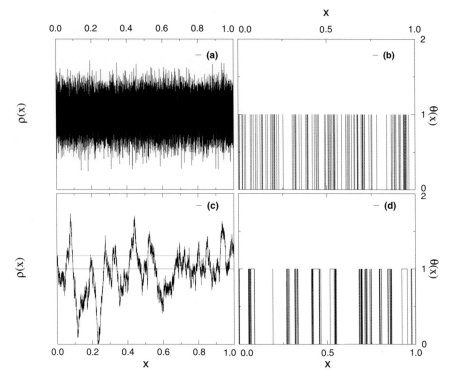

Fig. 13.1. Note the qualitative difference between $\delta_\rho(x)$ which is a weighted density field (*left panels*), and the biased field $\theta_\nu(x)$ (*right panels*) which just defines a new field having equal weight. In the upper part is shown the case of pure white noise, while in the lower part is shown the case of a Gaussian field with small scale correlations

simple facts concerning the threshold density, θ_ν, which follow directly from its definition, independently of the correlation properties of $\delta_\rho(\mathbf{x})$:

$$\langle \theta_\nu \rangle \equiv Q_1(\nu) \leq 1 \;,\quad (\theta_\nu(\mathbf{x}))^n = \theta_\nu(\mathbf{x}) \;, \tag{13.3}$$
$$\langle \theta_\nu(\mathbf{x})\theta_\nu(\mathbf{x}+r\mathbf{n}) \rangle \leq Q_1(\nu) \;,$$
$$\frac{\langle \theta_\nu(\mathbf{x})\theta_\nu(\mathbf{x}+r\mathbf{n}) \rangle}{Q_1(\nu)^2} - 1 \equiv \tilde{\xi}_\nu(r) \leq \tilde{\xi}_\nu(0) = \frac{1}{Q_1(\nu)} - 1 \;,$$
$$\theta_{\nu'}(\mathbf{x}) < \theta_\nu(\mathbf{x}) \;,\quad Q_1(\nu') < Q_1(\nu) \quad \text{for} \quad \nu' > \nu \;,$$
$$\tilde{\xi}_{\nu'}(0) > \tilde{\xi}_\nu(0) \quad \text{for} \quad \nu' > \nu \;. \tag{13.4}$$

The difference between the sets θ_ν for different values of ν is called *biasing*. The enhancement of $\tilde{\xi}_\nu(0)$ for higher thresholds has clearly nothing to do with how "strongly clustered" the peaks are but is entirely due to the fact that the larger is ν the lower is the fraction of points above the threshold (i.e. $Q_1(\nu') < Q_1(\nu)$ for $\nu' > \nu$). If we consider the trivial case of white Gaussian

13.2 Biasing of Gaussian Random Fields

noise ($\tilde{\xi}(r) = 0$ for $r > 0$) the peaks are just spikes (see Fig. 13.1). When a threshold $\nu\sigma$ is considered the number of spikes decreases and hence $\tilde{\xi}_\nu(0)$ is amplified simply because they are much sparser and not because they are "more strongly clustered": we show in the following that also in the case of a correlated field ($\tilde{\xi}(r) \neq 0$ for $r > 0$) the importance of sparseness is crucial in order to explain the amplification of $\tilde{\xi}_\nu(r)$, even though, as shown below, in this case it is somewhat related to clustering.

The joint two-point probability density $\mathcal{P}_2(\delta_\rho, \delta'_\rho; r)$ depends on the distance r between $\mathbf{x}$ and $\mathbf{x}'$, where $\delta_\rho = \delta_\rho(\mathbf{x})$ and $\delta'_\rho = \delta_\rho(\mathbf{x}')$. For Gaussian fields $\mathcal{P}_2$ is entirely determined by the mean of $\delta_\rho(\mathbf{x})$, and the two-point correlation function $\tilde{\xi}(r)$ [84, 203]:

$$\mathcal{P}_2(\delta_\rho, \delta'_\rho; r) = $$
$$= \frac{1}{2\pi\sqrt{\sigma^4 - \tilde{\xi}(r)^2}} \exp\left(-\frac{\sigma^2(\delta_\rho^2 + \delta_\rho'^2) - 2\tilde{\xi}(r)\delta_\rho\delta'_\rho}{2(\sigma^4 - \tilde{\xi}^2(r))}\right). \quad (13.5)$$

By definition

$$\tilde{\xi}(r) \equiv \langle \delta_\rho(\mathbf{x} + r\mathbf{n})\delta_\rho(\mathbf{x})\rangle = \int_{-\infty}^{\infty}\int_{-\infty}^{\infty} d\delta_\rho d\delta'_\rho \delta_\rho \delta'_\rho \mathcal{P}_2(\delta_\rho, \delta'_\rho; r). \quad (13.6)$$

The probability that both δ_ρ and δ'_ρ are larger than $\nu\sigma$ is

$$Q_2(\nu, r) = \int_{\nu\sigma}^{\infty}\int_{\nu\sigma}^{\infty} \mathcal{P}_2(\delta_\rho, \delta'_\rho, r) d\delta_\rho d\delta'_\rho \equiv \langle \theta_\nu(\mathbf{x})\theta_\nu(\mathbf{x} + r\mathbf{n})\rangle. \quad (13.7)$$

The conditional probability that $\delta_\rho(\mathbf{y}) \geq \nu\sigma$, given $\delta_\rho(\mathbf{x}) \geq \nu\sigma$, where $|\mathbf{x} - \mathbf{y}| = r$, is then just $Q_2(\nu, r)/Q_1(\nu)$. The reduced two-point correlation function for the stochastic variable $\theta_\nu(\mathbf{x})$, introduced above can be expressed in terms of Q_1 and Q_2 by (see Chap. 2)

$$\tilde{\xi}_\nu(r) = \frac{Q_2(\nu, r)}{Q_1^2(\nu)} - 1. \quad (13.8)$$

Defining

$$\tilde{\xi}_c(r) = \frac{\tilde{\xi}(r)}{\sigma^2}$$

as the normalized reduced two-point correlation function (i.e. $\tilde{\xi}_c(0) = 1$), we obtain

$$Q_1(\nu)^2(\tilde{\xi}_\nu(r) + 1) = \frac{1}{2\pi\sqrt{1 - \tilde{\xi}_c^2}}\int_\nu^\infty\int_\nu^\infty dx\,dx'$$
$$\times \exp\left(-\frac{(x^2 + x'^2) - 2\tilde{\xi}_c(r)xx'}{2(1 - \tilde{\xi}_c^2(r))}\right). \quad (13.9)$$

We note that the amplitude of $\tilde{\xi}_\nu(r)$ does not give information about how large the fluctuations are with respect to ρ_0, but it rather describes the correlation properties of the "fluctuations of the fluctuations", that is the fluctuations of the new field $\theta_\nu(\mathbf{x})$ around its average $Q_1(\nu)$. Similar arguments to those introduced for the original field can now be developed to characterize the typical scales of the new set defined by $\theta_\nu(\mathbf{x})$. In particular, one can define a correlation length $r_c(\nu)$ following the discussion in Chap. 2, by replacing $\tilde{\xi}(r)$ with $\tilde{\xi}_\nu(r)$. Like r_c, $r_c(\nu)$ does not depend on any multiplicative constant in $\tilde{\xi}_\nu(r)$, i.e. it does not depend on the amplitude of $\tilde{\xi}_\nu(r)$. Moreover a "homogeneity scale" $r_0(\nu)$ can be defined as the scale such that $\tilde{\xi}_\nu(r_0) = 1$ (see Chap. 2). The value of $r_0(\nu)$ is given by the amplitude of $\tilde{\xi}_\nu(r)$ and represents the minimal system size giving meaningful estimates of the average density $Q_1(\nu)$ and of $r_0(\nu)$ itself; $r_0(\nu)$ is the distance at which the conditional density $Q_2(\nu,r)/Q_1(\nu)$ begins to flatten toward $Q_1(\nu)$. We show below that while $r_0(\nu)$ depends strongly on ν due to a sparseness effect, $r_c(\nu)$ is almost constant and equal to r_c of the original field, i.e. the maximal size of the structures does not depend on the threshold. We recall in this context the fact that for any SSP the homogeneity scale r_0 is related to the average value ρ_0 of the field, while the correlation length r_c is independent (see Fig. 13.2) of it and related only to the correlation properties of the normalized fluctuation field $\delta_\rho(\mathbf{x})$.

13.3 Biasing and Real Space Correlation Properties

Let us now study in detail the behavior of $\tilde{\xi}_\nu(r)$ in terms of the properties of $\tilde{\xi}_c(r)$. We start from (13.9) which defines $\tilde{\xi}_\nu(r)$ as a function of $\tilde{\xi}_c(r)$. An important step is to find a simple approximation for this integral expression. For $\nu \gg 1$ and for sufficiently large r, such that $\tilde{\xi}_c(r) \ll 1$, it has been found [195] that (13.9) can be approximated by

$$\tilde{\xi}_\nu(r) \simeq \exp\left(\nu^2 \tilde{\xi}_c(r)\right) - 1. \tag{13.10}$$

If, in addition, $\nu^2 \tilde{\xi}_c(r) \ll 1$ we find [195, 123]

$$\tilde{\xi}_\nu(r) \simeq \nu^2 \tilde{\xi}_c(r). \tag{13.11}$$

Note that this last result is not obtained by the condition $\tilde{\xi}_c(r) \ll 1$ and separately $\nu \gg 1$. These conditions are significantly weaker than the required $\nu^2 \tilde{\xi}_c(r) \simeq \tilde{\xi}_\nu(r) \ll 1$. A better approximation for $\tilde{\xi}_\nu(r)$ is given [72] by

$$\tilde{\xi}_\nu(r) = \left[\sqrt{\frac{1+\tilde{\xi}_c(r)}{1-\tilde{\xi}_c(r)}} \exp\left(\nu^2 \frac{\tilde{\xi}_c}{1+\tilde{\xi}_c}\right) - 1\right](1 + o(\nu^{-1})). \tag{13.12}$$

This approximation is obtained by expanding the full expression for $\tilde{\xi}_\nu(r)$ given in (13.9) in $1/\nu$, and further assuming only that

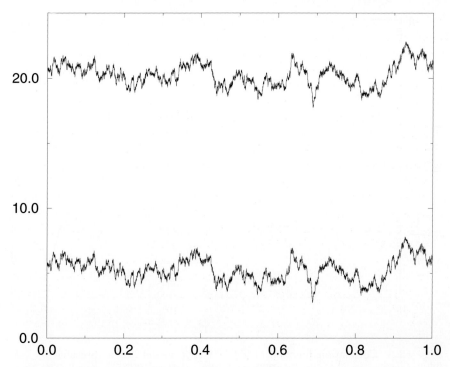

Fig. 13.2. Gaussian fluctuation field in $d=1$ with correlation up to a scale $r_c \approx 0.1$ super-imposed on a uniform background. The constant background density ρ_0 (i.e the average density) is smaller for the lower density field ($\rho_0 = 5$) than for the upper one ($\rho_0 = 20$), but the correlation length is the same for the two distributions. The amplitude of $\tilde{\xi}(r)$ at the same r is clearly larger for the lower distribution than for upper one: this is because the amplitude of the fluctuations with respect to the average density is larger. The correlation length r_c is finite and is related to the greatest spatial extension of the structures (see Chap. 3). Beyond r_c fluctuations from the average density can be considered substantially Poisson and mutually uncorrelated

$$\nu\sqrt{(1-\tilde{\xi}_c)/(1+\tilde{\xi}_c)} \gg 1\ .$$

It is a much better approximation than that of [195] both at small and larger values of $\tilde{\xi}_c$. In particular it gives an asymptotic behavior $\tilde{\xi}_\nu \approx (\nu^2 + 1)\tilde{\xi}_c$ for $\nu^2 \tilde{\xi}_c \ll 1$ which is a much closer to the exact behavior at typically relevant values of ν (see Fig. 13.3).

Let us come back to the exact expression of $\tilde{\xi}_\nu$ as a function of ν. Using the expression for $Q_1(\nu)$ given above, one can recast this after a simple change of variables into the form

$$\tilde{\xi}_\nu(r) = \frac{\int_\nu^\infty dx e^{-x^2/2} \int_\mu^\nu dy e^{-y^2/2}}{[\int_\nu^\infty dx e^{-x^2/2}]^2}\ . \tag{13.13}$$

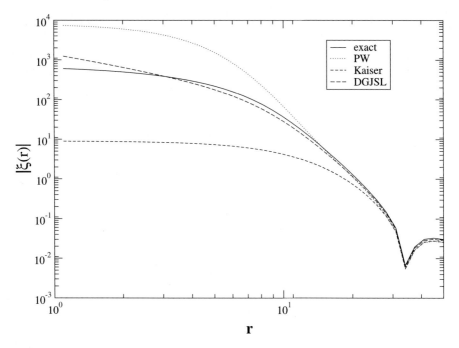

Fig. 13.3. Correlation function of a CDM type model, for a threshold $\nu = 3$. The approximations by Kaiser, Politzer & Wise (PW) and Durrer et al. (DGJSL – (13.12)) together with the exact numerical integration are reported

where $\mu = (\nu - \tilde{\xi}_c x)/\sqrt{1 - \tilde{\xi}_c^2}$. In this form it is evident that $\tilde{\xi}_c(r) = 0$ *if and only if* $\tilde{\xi}_\nu(r) = 0$ and that $\text{sign}[\tilde{\xi}_\nu(r)] = \text{sign}[\tilde{\xi}_c(r)]$. Taylor expanding this expression about $\tilde{\xi}_c = 0$, we find[3]

$$\tilde{\xi}_\nu(r) = b_1(\nu)\tilde{\xi}_c(r) + b_2(\nu)\tilde{\xi}_c^2(r) + \ldots \quad (13.14)$$

with

$$b_1(\nu) = e^{-\nu^2/2} \frac{\int_\nu^\infty dx\, x e^{-x^2/2}}{[\int_\nu^\infty dx\, e^{-x^2/2}]^2} \quad (13.15)$$

$$b_2(\nu) = \frac{1}{2}\nu e^{-\nu^2/2} \frac{\int_\nu^\infty dx(x^2 - 1)e^{-x^2/2}}{[\int_\nu^\infty dx\, e^{-x^2/2}]^2} . \quad (13.16)$$

The first term gives the linear relation obtained by [123] as $b_1(\nu) \approx \nu^2$ for $\nu \gg 1$, valid in the regime $|\tilde{\xi}_c| \ll 1$ and $|\tilde{\xi}_\nu| \ll 1$. Note that the fact that (13.12) is a much better approximation to (13.13) than (13.10) is made clear by the fact that $b_1(1) \approx 2.4$. It is important to note that in the regime

---

[3] For the expansion to all orders see [134].

13.3 Biasing and Real Space Correlation Properties

which is most relevant to observations, $\tilde{\xi}_\nu \gtrsim 1$, (13.11) does not apply and $\tilde{\xi}_\nu$ is actually exponentially enhanced. If this mechanism is the correct one to explain the observed so-called galaxy-cluster mismatch we expect that the connected two-point correlation function for clusters of galaxies to be thus exponentially enhanced on scales where the correlation function of galaxies $\tilde{\xi}_{cc} \gtrsim 1$, i.e. $r \lesssim 20h^{-1}$Mpc. However this is not the case, as the observed amplification is only linear. Therefore this model of bias does not appear capable of explaining the simple amplification observed. It cannot be excluded that some variant of this model might do so, but there is no evident reason why one would expect ever to produce the scale-dependent effect required. We note that, in contrast, the explanation proposed in Chap. 12 naturally produces such an amplification.

Now we discuss some simple examples through which a better characterization of the meaning of $r_0(\nu)$ and $r_c(\nu)$ in the context of biasing can be given. Let us start with a simple consideration: if the two-point correlation function $\tilde{\xi}(r)$ of the Gaussian field has a power-law tail and homogeneity scale ρ_0, we can write:

$$\tilde{\xi}(r) = \left(\frac{r}{r_0}\right)^{-\gamma} , \text{ for } r \gg r_0.$$

If, moreover, the threshold ν and the distance r are such that (13.11) holds (i.e. such that $\tilde{\xi}_\nu \ll 1$) for the biased field $\theta_\nu(\boldsymbol{x})$, we have

$$\tilde{\xi}_\nu(r) = \left(\frac{r}{r_0(\nu)}\right)^{-\gamma} .$$

The scales $r_0(\nu)$ for different biases all satisfying the conditions for the validity of (13.11) are related by

$$r_0(\nu') = r_0(\nu) \left(\frac{\nu'}{\nu}\right)^{2/\gamma} .$$

Therefore we see simply that the homogeneity scale $r_0(\nu)$ is very sensitive to the value of the threshold ν.

In order to clarify further the behavior of the two length scales $r_c(\nu)$ and $r_0(\nu)$ of the biased field $\theta_\nu(\boldsymbol{x})$ by changing the threshold ν, we first study an example of a Gaussian density field with finite correlation length r_c, and which is well approximated by a power-law over a certain range of scales. For instance we can take

$$\tilde{\xi}(r) = \frac{\sigma^2 \exp(-r/r_c)}{[1 + (k_s r)^\gamma]} ,$$

with $k_s^{-1} \ll r_c$. The distance r_c is approximately the correlation length as defined as in Chap. 2. The length k_s^{-1} can be seen as the smoothing scale of the continuous field, in the sense that on scales $r \gg k_s^{-1}$ the field is weakly fluctuating. In the region $k_s^{-1} \ll r \ll r_c$, the function $\tilde{\xi}(r)$ is well approximated by the power-law $(k_s r)^{-\gamma}$. The correlation length, $r_c(\nu)$ of the biased field $\theta_\nu(\mathbf{x})$ for any value of ν, is given by the slope of $\log \tilde{\xi}_\nu(r)$ which is clearly independent of the bias ν (see Fig. 13.4). This can also be obtained from (13.9)–(13.10). For relatively small values of the threshold, $\nu \ll \nu_c \approx$

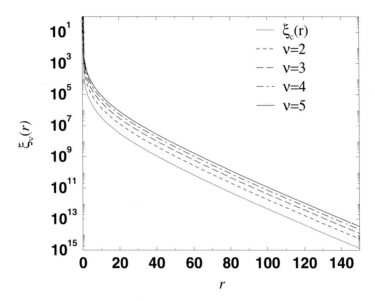

Fig. 13.4. The behaviors of $\tilde{\xi}(r) = \sigma^2/[1+(k_s r)^\gamma \exp(-r/r_c)]$ (where $\gamma = -2$, $k_s^{-1} = 0.01$ and $r_c = 10$) and $\tilde{\xi}_\nu(r)$ are shown for different values of the threshold ν in a semi-log plot. The slope of $\tilde{\xi}_\nu(r)$ for $r \gtrsim 50$ is $-1/r_c$, independent of ν i.e. the correlation length of the system does not change for the sets above the threshold (From [89])

$(k_s r_c)^{\gamma/2}$ one finds $r_0(\nu) \ll r_c$ and $r_0(\nu) \sim k_s^{-1}\nu$. On the other hand, if $\nu \gg \nu_c$ we have $r_0(\nu) \sim r_c \log(\nu)$ and in this case the statistics is dominated by shot noise (see below). For this reason we assume $r_0(\nu) < r_c(\nu)$ in the following.

We note that in the range of scales $r \leq r_0(\nu)$ the amplification of $\tilde{\xi}_\nu(r)$ is strongly non-linear in ν and scale dependent. Therefore, if the original correlation function $\tilde{\xi}(r)$ has a power-law behavior, $\tilde{\xi}_\nu(r)$ does not for $r \leq r_0(\nu)$: this is better shown directly in the case in which $r_c \to \infty$. In this case the correlation function of the original Gaussian field is

$$\tilde{\xi}(r) = \frac{\sigma^2}{(1+(rk_s^{-1})^\gamma)} \ .$$

Clearly on scales $k_s^{-1} < r < r_c$ the previous example does not differ from the present one. The amplification of $\tilde{\xi}_\nu$ for this example is plotted in Fig. 13.5. In order to investigate whether $\tilde{\xi}_\nu(r)$ is of the form $\tilde{\xi}_\nu(r) \sim (r/r_0(\nu))^{-\gamma_\nu}$, we plot $-d\log(\tilde{\xi}_\nu(r))/d\log(r) \sim \gamma_\nu$ in Fig. 13.6. Only in the regime where $\tilde{\xi}_\nu(r) \ll 1$ does γ_ν become constant and roughly independent of ν. This behavior is very different from the so-called galaxy-cluster mismatch where the linear amplification is linear already at scales where both correlation functions of galaxies and galaxy clusters are larger than 1 (see Chap. 12).

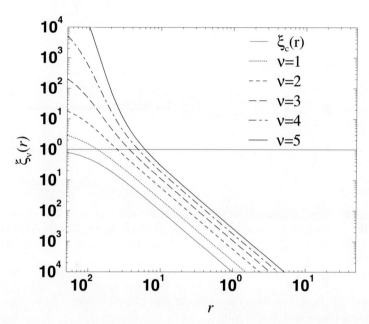

Fig. 13.5. The behaviors of $\tilde{\xi}(r) \sim \sigma^2/(1 + (k_s r)^\gamma)$ (with $\gamma = -2$, $k_s^{-1} = 0.01$) and $\tilde{\xi}_\nu(r)$ are shown for different values of the threshold ν in a log-log plot (From [89])

Let us now clarify how the amplification of $\tilde{\xi}_\nu(r)$ is related to the behavior of the peak sparseness as a function of the threshold ν. For a Gaussian random field, the peaks above a threshold ν have a mean size $D_p(\nu)$ and are separated from each other by a mean distance L_p, which are respectively given by [48, 237]:

$$D_p(\nu) \simeq \frac{D_0(k_s, r_c)}{\nu} \tag{13.17}$$

and

$$L_p(\nu) \simeq D_0(k_s, r_c) \exp(\nu^2/6) \nu^{-2/3} \tag{13.18}$$

so that

$$\frac{L_p}{D_p} \simeq \nu^{1/3} \exp(\nu^2/6) \text{ for } \nu \gg 1 \,. \tag{13.19}$$

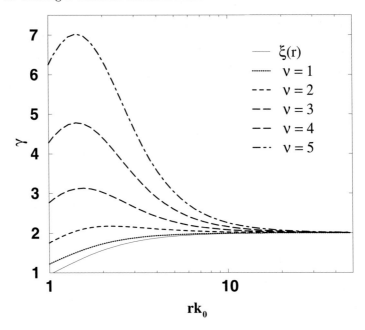

Fig. 13.6. The behavior of $\gamma_\nu(r)$ is shown for different values of the threshold ν for the correlation function shown in the previous figure. Clearly γ_ν is strongly scale dependent on all scales where $\tilde{\xi}_\nu \gtrsim 1$, corresponding to $r < 1$ in our units. (From [89])

The size $D_0(k_s, r_c)$ is the typical linear size of persistence of the sign of the fluctuation field and is given by

$$D_0^2 = \frac{\int_0^{+\infty} dk P_1(k)}{\int_0^{+\infty} dk k^2 Q_1(k)} \qquad (13.20)$$

where $P_1(k)$ is the Fourier transform of $\tilde{\xi}(r)$ along a straight line in space (in $d = 1$ it coincides with the PS $P(k)$). Equation (13.19) shows the strong enhancement of the sparseness of peaks (object) with increasing ν. It is this increase of sparseness which is at the origin of the amplification by biasing. In the light of (13.9)–(13.10)–(13.19), we see that increasing ν corresponds to a very particular sampling of fluctuations: the typical size of the surviving peaks D_p is slowly varying with ν while the average distance between peaks L_p is more than exponentially amplified, and finally the scale $r_c(\nu)$, over which the fluctuations are structured, is practically unchanged.

In summary, we have argued that bias does not influence significantly the correlation length $(r_c(\nu) \simeq r_c)$. It amplifies the correlation function basically because the mean density, $Q_1(\nu)$, is reduced more strongly than the conditional density, $P_2(\nu, r)/Q_1(\nu)$. According to (13.13), this amplification is strongly non-linear in $\tilde{\xi}(r)$ (exponential) at scales where $\nu^2 \tilde{\xi}_c(r) \geq 1$ and thus $\tilde{\xi}_\nu(r) > 1$.

13.4 Biasing and the Power Spectrum

We now discuss a different aspect of this model for bias. We are interested in understanding the effect of such biasing on the PS of the biased sub-field $\theta_\nu(\mathbf{x})$ [72]. In particular we are interested in the effect of biasing on Gaussian fields representing the matter distribution in standard cosmological models. As shown in Chap. 6, all these models share the same small k behavior of the PS $P(k) \sim k$. In Chap. 3 we have seen that this condition implies that the matter distribution typical of these models is super-homogeneous, that is with mass fluctuations increasing with scale more slowly than in a Poisson one. In particular $P(k) \sim k$ at small k implies $\sigma^2(R) \sim R^{-4}$ at large R (while $\sigma^2(R) \sim R^{-3}$ for a substantially Poisson matter distribution). After an explanation of the effect of biasing on a general Gaussian field, we focus on the case of a typical cosmological density field. In this context we show that the PS of the biased field $\theta_\nu(\mathbf{x})$ is significantly distorted at small k with respect to that of the original Gaussian field. In particular it no longer tends to zero as k does so. This implies that, even though the Gaussian field is super-homogeneous, the sub-field is substantially Poisson (using the classification introduced in Chap. 3). Furthermore, it is shown that, in this biasing scheme, for values of the threshold ν which are thought to represent the relation between the visible matter and the total matter in CDM models, the turn-over between the increasing part of the PS at small k and decreasing part at large k disappears. This shows the usefulness of determining from observations not just the PS of visible objects, but also their real space correlation properties, as we are going to discuss.

Let us now start by considering how biasing changes the value of the PS at $k = 0$ for a general Gaussian field. This modification changes the distribution in relation to the classification we have given in terms of $P(0)$ (see Chap. 3). We call $P(k)$ the PS of the original Gaussian field (i.e. the Fourier transform (FT) of $\tilde{\xi}_c(r)$) and $P_\nu(k)$ the PS of the biased sub-field $\theta_\nu(\mathbf{x})$ (i.e. the FT of $\tilde{\xi}_\nu(r)$). We start by showing that there is a universal value of the biasing threshold such that, for ν larger than this characteristic value, we have

$$P_\nu(0) > P(0) \tag{13.21}$$

independently of the shape of $\tilde{\xi}_c(r)$ (i.e. of $P(k)$). Let us start from (13.14), and consider the expansion of $\tilde{\xi}_\nu(r)$ up to second order in $\tilde{\xi}_c$. From (13.15) it is simple to show that $b_1(\nu)$ is a positive and increasing function of ν, with the asymptotic behavior $b_1(\nu) \simeq \nu^2$ for large ν. Analogously it is simple to show that also $b_2(\nu)$ is a positive function of ν. Let us call $\nu_o \simeq 0.303$ the value of the threshold such that $b_1(\nu_o) = 1$. We can say that if $\tilde{\xi}_c$ is sufficiently small at all r, then $\tilde{\xi}_\nu(r) > \tilde{\xi}_c(r)$ always. This then implies (13.21) by direct integration. Clearly this demonstration using the Taylor expansion (13.14) only up to the second order is limited to the cases in which $\tilde{\xi}_c(r) \ll 1$ at all scales (recall that $\tilde{\xi}_c(r) \leq 1$ by definition). To show that there is a value of ν above which (13.21) is indeed satisfied for all permitted values of $\tilde{\xi}_c$ (i.e. for all permitted functions $\tilde{\xi}_c(r)$), it suffices to find the threshold value $\nu_1 \geq \nu_o$ such that for all $\nu \geq \nu_1$ one has

$$\text{sign}\left[\frac{d\tilde{\xi}_\nu(r)}{d\tilde{\xi}_c(r)} - b_1(\nu)\right] = \text{sign}[\tilde{\xi}_c] \qquad (13.22)$$

independently of $\tilde{\xi}_c$. In fact, it is simple to show that this is a sufficient condition in order to have the exact curve $\tilde{\xi}_\nu(\tilde{\xi}_c)$, given by (13.9), all above the line $\tilde{\xi}_\nu = \tilde{\xi}_c$ for all $\tilde{\xi}_c$. The value ν_1 can be found numerically, using (13.9), and the result is $\nu_1 \simeq 0.38$.

Note that, if the condition given by (13.22) holds, this means simply that, relative to the asymptotic ($|\tilde{\xi}_c| \ll 1$ and $|\tilde{\xi}_\nu| \ll 1$) linearly biased regime in which $\tilde{\xi}_\nu \approx b_1(\nu)\tilde{\xi}_c$, the anti-correlated regions are less amplified ($|\tilde{\xi}_\nu| < b_1(\nu)|\tilde{\xi}_c|$) than the positively correlated regions ($|\tilde{\xi}_\nu| > b_1(\nu)|\tilde{\xi}_c|$). Thus the integral over the biased correlation function is always positive, and the bound given by (13.21) thus holds. Further it is easy to see that $P_\nu(0)$ is finite if $P(0)$ is: $\tilde{\xi}_\nu$ is bounded for any value of ν, and, as we have just seen, has the same convergence properties as $\tilde{\xi}_c$ at large distances. This implies that, if the integral of $\tilde{\xi}_c$ over all space converges, then also that of $\tilde{\xi}_\nu$ does.

In terms of the classification (see Chap. 3) of density fields through the relation between the behavior of the PS at small k and the mass variance at large r we thus draw the following conclusion: Both the critical (with $P(0) = \infty$) and Poisson-like (with $P(0) = \text{const.} > 0$) Gaussian fields give under biasing sub-fields $\theta_\nu(\mathbf{x})$ remaining in the same class; on the other hand super-homogeneous Gaussian fields (with $P(0) = 0$) become substantially Poisson ($P_\nu(0) = \text{const.} > 0$).

To further clarify the breakdown of the super-homogeneous condition, let us consider the following example. If we have a stochastic mass distribution, and we perform a random sampling of this distribution, biased or not, this process introduces a source of Poisson noise such that, if the original field is super-homogeneous, the sampled sub-field becomes essentially Poisson. Consider for example the case of a perfect lattice, which is a super-homogeneous distribution ($P(0) = 0$, see Chap. 3) in which the normalized variance in a sphere of radius R decays asymptotically as

$$\sigma^2(R) = \frac{\langle(\Delta M(R))^2\rangle}{\langle M(R)\rangle^2} \sim \frac{1}{R^4}. \qquad (13.23)$$

The distribution obtained by taking (or rejecting) each point with probability p (or $1-p$) is described by a simple binomial distribution, with a variance

13.4 Biasing and the Power Spectrum

$$\sigma^2(R) \propto \frac{p(1-p)}{N} \propto \frac{1}{R^3} \tag{13.24}$$

(N being the mean number of points inside a sphere).

It is important to note that the present biasing scheme is a *deterministic* process on a *disordered* system. In other words, once the stochastic realization of the Gaussian field is given, the process to decide which parts of the field survive on biasing (i.e. which are above the threshold) is not random, but deterministic. Despite this feature, the effect produced by this deterministic sampling is of the same kind of that given by a random sampling. This was not evident a priori. It suggests that any kind of biasing, deterministic or random, gives such a transformation of super-homogeneous fields into essentially Poisson ones.

Let us now turn to the implications of this result for cosmological models. Since such models have $P(0) = 0$ in the full matter PS, it is evident that we cannot have the behavior

$$P_\nu(k) \propto b_1(\nu) P(k)$$

for small k which one would naively infer from the fact that

$$\tilde{\xi}_\nu(r) \approx b_1(\nu) \tilde{\xi}_c(r)$$

for large separations. Inevitably a non-linear distortion of the biased PS at small k relative to the underlying one is induced. How important can the effect be qualitatively for a realistic cosmological model? To answer this question we consider the simple model PS

$$P(k) = Ak e^{-k/k_c}.$$

The differences with a CDM model – which has the same linear Harrison-Zeldovich (HZ – see Chap. 6) form at small k but a different (power-law) functional form for large k – are not fundamental here, and this PS allows us to calculate the correlation function $\tilde{\xi}_c$ analytically (see Chap. 6). This greatly simplifies the numerical calculation of the biased PS $P_\nu(k)$, which can be done by direct integration of (13.12). In Fig. 13.7 we show $P_\nu(k)$ for various values of the threshold $\nu = 1, 2, 3$. It shows a well defined flattening for $k \to 0$ even for a small value of the threshold ν. We see that the shape of the PS at small k is completely changed with respect to the underlying PS. Indeed the main feature of the latter in this range – a clear maximum and "turn-over" – is completely modified. Qualitatively it is not difficult to understand why this is so. The only characteristic scale in the PS (and also in the correlation function) is given by the turn-over (specified in our case by $k = k_c$). On the other hand, the value of $P_\nu(0)$ is just the integral over all space of $\tilde{\xi}_\nu$ which is proportional to the overall normalization A and (since it is strictly positive) must be given on dimensional grounds by Ak_c times

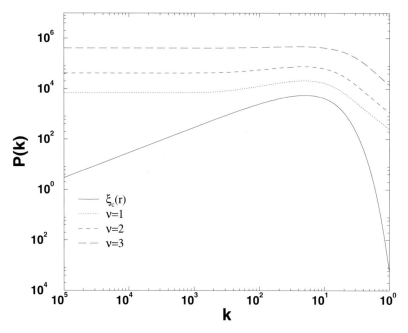

Fig. 13.7. The PS $P_\nu(k)$ derived from the biased correlation function $\tilde{\xi}_\nu(r)$ for values of the threshold $\nu = 1, 2, 3$ is shown. The underlying correlation function which gives $\tilde{\xi}_c(r)$ is that derived from the $P(k) = Ake^{-k/k_c}$ and the approximation given in (13.12) for $\tilde{\xi}_\nu$ is used. The clear distortion of the PS at small k is seen: the "turn-over" in the underlying PS essentially disappears already for $\nu = 1$. The constants k_c and A are fixed by $\tilde{\xi}_c(0) = 1$ and by the requirement that $\tilde{\xi}_c(r) = 0$ at $r = 38$ Mpc. The latter is taken as in a typical CDM model (see e.g. [170]). We could alternatively fix the wave-number at the maximum of the PS. (From [72])

some function which depends on ν. For $\nu \simeq 1$ this function is of order one, so that $P_\nu(0) \sim \max[P(k)]$.

This last point is better illustrated by considering the integral

$$J_3(r, \nu) = 4\pi \int_0^r x^2 \tilde{\xi}_\nu(x)dx \qquad (13.25)$$

which converges to $P_\nu(0) = \lim_{r\to\infty} J_3(r, \nu)$. In Fig. 13.8 the value obtained for it by numerical integration of the *exact* expression given by (13.9) for $\tilde{\xi}_\nu(r)$ is shown for $\nu = 1, 2, 3$. Also shown is the same integral for $\tilde{\xi}_c$ which converges to $P(0) = 0$. While the latter decreases at large r, converging very slowly to zero (as $1/r$ since $\tilde{\xi}_c(r) \propto -1/r^4$ at large scales), the former all converge toward a constant non-zero value. We see that the integral picks up its dominant contribution from scales around (and above for $\nu = 1$) $r \sim 10$ Mpc (see caption for explanation of the normalizations, which are irrelevant for the present considerations). From the inset in the figure, which shows both $\tilde{\xi}_c(r)$

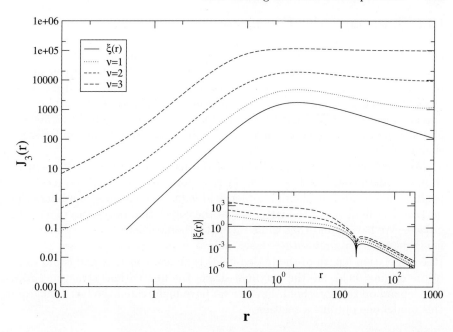

Fig. 13.8. The integral $J_3(r)$ for the same underlying correlation function as in Fig. 13.7 and for the same range of values of ν, calculated with the exact expression (13.13) for $\tilde{\xi}_\nu(r)$. Also shown is the analogous integral of $\tilde{\xi}_c$. While the latter converges slowly to its asymptotic value of $P(0) = 0$, the other integrals converge to constant non-zero $P_\nu(0)$. They are dominated by the range $r \sim 10\,\mathrm{Mpc}$ where, as can be seen from the inset which shows both $\tilde{\xi}_c$ and each of the $\tilde{\xi}_\nu(r)$, the correlation functions $\tilde{\xi}_\nu(r)$ are amplified non-linearly. The contribution from the highly amplified region at small r is small (because of the r^2 factor in the integral), which also makes the approximation in calculating the PS with (13.12) very accurate, as can be checked by comparing the asymptotic values with those of $P_\nu(0)$ in Fig. 13.7. (From [72])

and $\tilde{\xi}_\nu(r)$, we see that this is the range of scale at which the correlation function is non-linearly amplified. Moreover it is shown that the smaller scales at which $\tilde{\xi}_\nu(r)$ is most distorted relative to $\tilde{\xi}_c(r)$ do not contribute significantly to J_3 (because of the r^2 factor). This fact also explains the accuracy of the PS obtained using the approximation (13.12) for $\tilde{\xi}_\nu(r)$, which can be seen by comparing the asymptotic values of the integrals in Fig. 13.8 with $P_\nu(0)$ in Fig. 13.7. Note that for $\nu = 1$ the distortion away from linear is relatively weak in the part of the correlation function which dominates the integral in $J_3(r, \nu)$, and that there is even a non-negligible contribution from the larger scales at which the correlation function amplification is extremely close to linear.

13.5 Summary and Discussion

Let us first summarize what we have found in our analysis of the model of bias introduced in [123], and then consider what conclusions we can draw. In particular we will return to the issue which we discussed in Sect. 6.6.1 of Chap. 6: the normalization of the fluctuations at large scales probed by observations of the CMBR and those at smaller scales probed by observations of the distribution of visible matter, in particular of galaxies and galaxy clusters.

We have discussed the properties of the model in which one constructs from a random Gaussian field $\delta_\rho(\boldsymbol{x})$, by selection of fluctuations over a certain positive amplitude threshold, another field $\theta_\nu(\boldsymbol{x})$. The former represents the dark matter field, while the latter traces the correlation properties of a set of objects. We have found that while $r_0(\nu)$ (the homogeneity scale of the biased field) depends strongly on ν – essentially due to a sparseness effect – $r_c(\nu)$ (the correlation length of the biased field) is almost constant and equal to r_c of the original field, i.e. the maximal size of the fluctuations structures does not depend on the threshold. Further we have highlighted the fact that from this model one obtains a relation

$$\tilde{\xi}_{\nu'}(r) \approx b^2(\nu',\nu)\tilde{\xi}_\nu(r) \qquad (13.26)$$

only in the range of weak correlations (when $\tilde{\xi}_{\nu,\nu'}(r) \ll 1$). The observed galaxy-cluster mismatch and luminosity bias, discussed in Chap. 9 and Chap. 12, correspond to such a linear amplification in a range of scales $r_1 < r < r_2$ where $\tilde{\xi}_{\nu,\nu'}(r_1) > 1$ and $\tilde{\xi}_{\nu,\nu'}(r_2) > 1$. We have also considered the behavior of the PS of the biased field $\theta_\nu(\boldsymbol{x})$. Focusing on cosmological models for the dark matter field, with the super-homogeneous characteristics we have discussed in Chap. 6, we have shown that this characteristic is in fact modified. The biased field has a PS which does not tend to zero as k does so. One of the implications of this is that the linear relation

$$P_\nu(k) \approx b^2 P(k) \qquad (13.27)$$

cannot hold at small k. This is a relation which one might tend to infer to hold precisely at small k, from the fact that $\xi_\nu(r) \approx b^2 \xi(r)$ at large r (since $\xi_\nu(r) \ll 1$ in this limit). Applying the biasing model to a typical cosmological model we have shown in fact that the approximation (13.27) is not good in any range of k. In particular we found that the "turn-over" in the underlying PS to a HZ behavior ($P(k) \sim k$) is completely obliterated in the PS of the biased field.

In drawing conclusions from these findings various caveats need to be stressed. Firstly there are some with respect to the specific model we have analyzed. We note that the biased field $\theta_\nu(\boldsymbol{x})$ does not describe a set of discrete objects. It is a continuous field which is different qualitatively also from the underlying weighted density field $\delta_\rho(\boldsymbol{x})$, in that it simply defines a set of equal weight. One could envisage discretizing it in various ways. For example, the regions where $\theta_\nu(\boldsymbol{x}) = 1$ could be substituted by a set of equal mass points with number density proportional to the length of the connected region where $\theta_\nu(\boldsymbol{x}) = 1$. Or, alternatively, one could substitute each connected non-zero region by a single point at its center, possibly with a mass proportional to the mass in this region in the underlying dark matter field. While such a discretization would evidently change the correlation properties compared to those calculated, the important point is that we would expect these changes only to be at small scales. Thus the conclusions we have drawn on large scales (i.e. larger than the typical size of the selected regions) should be valid.

A further more general caveat is that, even with the possible elaborations for discretization we have just described, what we have examined is still a specific model of bias. Since its introduction many variants and modifications have been studied e.g. "stochastic bias" which introduced a further stochasticity in the selection, "non-local bias" which relaxes the strictly local relation of the selected and underlying field (see [179] for a review). More recently, in the context of a popular phenomenological model for dark matter clustering known as the "halo model" (for a review see [50]) biasing has been reformulated as a probability for selection of objects in regions of dark matter aggregation ("halos") of different mass. Fundamentally however all these models share the central idea of the original one i.e. that of a selection, which is essentially stochastic[4]. As we discussed in our consideration of the biased PS, this factor alone is essentially what leads to the qualitative change in the properties of the underlying and biased distribution. If, as in cosmological models, the underlying distribution is super-homogeneous, the biased one will not be. The reason for this is simplest to understand in real space, where super-homogeneity corresponds to the fact that the fluctuations in mass in a volume are proportional to the surface of the volume. The stochasticity of the selection introduces necessarily a term in the variance proportional to the volume, which then dominates. Thus the behavior of the PS corresponding to this super-homogeneity must disappear in the biased field, as indeed is observed in the specific model we have analyzed. The conclusion that the PS is not linearly amplified beyond the "turn-over" should therefore be robust in any bias model. A corollary of this is that the two-point correlation function $\xi(r)$ cannot be amplified linearly at all scales, and in particular we expect that the rupture of the super-homogeneity can be understood as being due generically to a scale-dependent modification of the correlation function (as the PS at $k = 0$ is the integral of the correlation function). On the basis of this analysis we cannot exclude, certainly, the possibility that, in a different model, there may be linear amplification in some more extended range of

---

[4] As discussed in Sect. 13.4 the Kaiser model we have discussed is actually a deterministic selection on a stochastic field, but the net effect is like that of a stochastic sampling.

scales. However the examination of this model suggests no reason to expect such a behavior other than in the region of weak correlation where an appropriate Taylor expansion of the biased correlation function in terms of the underlying one will probably be valid generically.

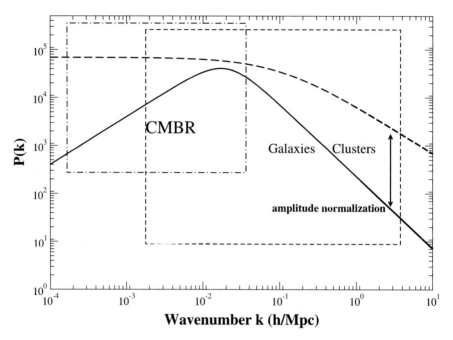

Fig. 13.9. In this figure we show a typical CDM type power spectrum (*solid line*) and the range associated to the primary observational constraints. The *left hand side*, the Harrison Zeldovich part of the PS ($P(k) \sim k$), is constrained by observations of the anisotropies of the CMBR (*dash-dotted box*). Current galaxy and galaxy cluster surveys gives constraints at smaller scales (*dashed box*). The normalization of the amplitude of the galaxy or cluster PS to the one observed in the CMBR is fundamentally important. It is usually determined by a *linear* re-scaling on the y-axis, ascribed to the effect of bias. This simple assumption is not consistent with the canonical model for the biasing of a Gaussian field, which introduces a non-linear distortion both at small and large wave-number. This is illustrated by the *dashed line*, which shows what is actually obtained for the PS of the biased field. On small scales (large k) there is a non-linear distortion and at large scales (small k) the behavior is typical of a substantially Poisson system with $P(k) \sim \text{const}$

We conclude thus that the findings we have made are for these essential points: neither the linear amplification of the correlation function $\xi(r)$ between galaxies and galaxy clusters, nor between galaxies of different luminosities, can be explained naturally with the framework of biasing. If a particular model manages to produce such an approximate relation over a range of scales (e.g. in a "halo" biasing model, where the selection probability is determined a posteriori to fit the data) this is a inevitably a result of a fine-tuning. Further the relation between the PS of the dark matter field and that of discrete objects such as galaxies related as envisaged by such a mechanism will generically not be linear for any k, and certainly will not be linear beyond the "turn-over". The importance of this latter point is considerable: such a linear renormalization of the PS of galaxies and galaxy clusters is in fact almost universally applied in cosmology (e.g. [141]). In the theory of structure formation this is essential since it is through such a normalization that one relates the fluctuations at large scales probed by observations of the CMBR and those at smaller scales probed by observations of the distribution of visible matter (discussed in Sect. 6.6.1 of Chap. 6). As the central aim of the theory of structure formation is in fact to relate these two kinds of data – the "initial conditions" at large scales probed through the CMBR and the observed properties of the distribution of visible matter today – the importance of our conclusions is evident (see Fig. 13.9).

We mention finally one other point which follows from our analysis. We have highlighted in Chap. 6 that the dark matter density field of standard cosmological models has very specific and interesting properties in the context of correlated systems i.e. it is what we call super-homogeneous. We have now seen that biasing destroys this characteristic in visible matter. Does this mean there is no direct trace of the super-homogeneity in galaxy distributions? The answer is in the negative: we have seen that biasing does lead to a simple linear amplification of the reduced two-point correlation function $\xi(r)$ in the region of weak correlations. At the same time we have seen in Chap. 6 that the super-homogeneous behavior of the PS $P(k) \sim k$ gives $\tilde{\xi}(r) \sim -r^{-4}$ at large r (beyond the inverse of the turn-over scale in k space). Thus we expect this characteristic behavior to survive biasing, and to describe the correlation function of galaxies at sufficiently large scales. The amplitude of the predicted correlation function in this region is very small, typically $\sim 10^{-2} \div 10^{-3}$ (see Fig. 6.2 in Chap. 6) and the detection of such a behavior in current galaxy surveys is impossible (since typically one expects at least an estimator variance of order $1/\sqrt{N}$ where N is the number of points – see Appendixes F–G). With the full SDSS survey – containing a million galaxies – it may however be possible to attain such a direct measurement, providing that galaxy distribution becomes homogeneous well inside the sample size. To determine whether it will be – after it is first established that the $\xi(r)$ analysis is the appropriate one – a detailed treatment is required of both estimators of $\xi(r)$ and the passage to the discrete distribution of galaxies from the underlying model.

14 The Gravitational Field in Stochastic Particle Distributions

14.1 Introduction

The knowledge of the statistical properties of the gravitational field in a given distribution of point-particles is very useful in many cosmological and astrophysical applications in which such particles are treated as elementary objects. In particular it is useful in the context of stellar dynamics and of cosmological N-body simulations to study the formation and the evolution of structures from some initial mass density perturbations [12, 13, 14]. Similar studies are useful in other domains of physics, such as the statistics of the *dislocation-dislocation* interaction in the analysis of crystal defects in condensed matter physics [244]. Until now a complete study of this problem has been accomplished only in the case of an uncorrelated Poisson particle distribution [43, 100]. Partial results have been found more recently in a few other cases: 1) fractal point distribution [88], 2) radial density profile of particles [62], and 3) in weakly correlated statistically homogeneous particle distributions [93].

As discussed by Chandrasekhar [43] one of the main problems of the dynamics of a self-gravitating particle distribution concerns the statistical analysis of the force acting on a single test particle of the system. In general, it is possible to show that this force is composed of two different contributions: the first is due to the system as a whole and the second is due to the influence of the immediate neighborhood of the particle. The former is a smoothly varying function of position and time while the latter is subject to relatively rapid fluctuations. These fluctuations are related to the underlying statistical properties of the particle distribution: their effect can be evaluated in a stochastic sense. We discuss here the problem of the computation of the stochastic force probability density for a set of points with given statistical properties.

The actual value of the gravitational force acting on a fixed "test" particle belonging to a given system, supposed arbitrarily to be at the origin of coordinates due to all the other N system particles is

$$\boldsymbol{F} = GM_0 \sum_{i=1}^{N} \frac{M_i}{|\boldsymbol{r_i}|^3} \boldsymbol{r_i} \qquad (14.1)$$

where M_i is the mass of the ith particle, $\boldsymbol{r}_i$ its position vector, and M_0 the mass of the test particle (G is the gravitational constant). The sum in (14.1) is extended to all the particles other than the test one, which are contained in the system volume V (with $V \to \infty$). The actual value of $\boldsymbol{F}$ clearly depends on the nature of the local particle distribution, and hence, it will be in general subjected to stochastic fluctuations. These fluctuations determine a more or less broad force probability density function (PDF) $P(\boldsymbol{F})$. Clearly if the point-particle stochastic system is statistically stationary (i.e., invariant under spatial translation) the statistical properties of $\boldsymbol{F}$ do not depend on the choice of the test particle. We will assume throughout this chapter this stationarity as well as the statistical isotropy of the distribution. Moreover we assume that all particles have equal mass ($M_i = M_j = M$ for any $i, j = 0, N$). Because of statistical isotropy the direction of $\boldsymbol{F}$ is completely random with equal probability in each direction, i.e. $P(\boldsymbol{F})$ depends only on $F = |\boldsymbol{F}|$. Therefore we can limit the analysis to the PDF $W(F)$ of F. The relation between $W(F)$ and $P(\boldsymbol{F})$ in three-dimensional space is simply

$$W(F) = 4\pi F^2 P(\boldsymbol{F}) . \tag{14.2}$$

14.2 Nearest Neighbor Force Distribution

The simplest approximation in which one can evaluate the force $\boldsymbol{F}$ acting on a fixed particle of the system is given by taking the contribution only due to its nearest neighbor (nn) particle. We denote the PDF of this contribution by $W_{nn}(F)$. Even though this approximation is in general very crude and inaccurate, the calculation of the nn contribution in some simple stochastic cases is an instructive exercise which can help to elucidate some basic properties of the total force distribution $W(F)$. For this simple exercise, we suppose that we have a statistically stationary and isotropic (i.e. with no preferred a priori position and direction in the statistical sense) fractal point-particle distribution with fractal dimension $D \leq 3$. Moreover we use the approximations given in Chap. 4 ((4.22)–(4.25)), which are quite strong for real fractals, but for $D = 3$ are the exact expressions for the isotropic Poisson case (see (2.64)–(2.66)). Therefore for $D = 3$ we will obtain the exact nn gravitational interaction for the Poisson case.

If r is the distance between the fixed test particle and its nn we have

$$F = |\boldsymbol{F}| = \frac{GM^2}{r^2} . \tag{14.3}$$

Denoting, as usual (see Chap. 2), by $\omega(r)$ the PDF for the random variable r, we can write

$$W_{nn}(F)|dF| = \omega(r)|dr| . \tag{14.4}$$

By using (2.71) for the conditional average density

14.2 Nearest Neighbor Force Distribution

$$\Gamma(r) \equiv \langle n(\mathbf{r}) \rangle_p = \frac{DB}{4\pi} r^{D-3},$$

and by replacing (14.3) in (14.4), we obtain

$$W_{nn}(F) = \frac{D}{2} F_{\langle \Lambda \rangle}^{\frac{D}{2}} F^{-\frac{D+2}{2}} \exp\left[-\left(\frac{F_{\langle \Lambda \rangle}}{F}\right)^{\frac{D}{2}}\right], \qquad (14.5)$$

where we have defined

$$F_{\langle \Lambda \rangle} = B^{\frac{2}{D}} G M^2 = \frac{GM^2}{\langle \Lambda \rangle^2} \left(\Gamma_e\left(1 + \frac{1}{D}\right)\right)^2, \qquad (14.6)$$

and $\langle \Lambda \rangle$ is the average distance between nn

$$\langle \Lambda \rangle = \int_0^\infty dr\, r\, \omega(r).$$

In the limit $F \to \infty$, we obtain the so called *strong field approximation* of (14.5):

$$W_{nn}(F) \simeq \frac{D}{2} F_{\langle \Lambda \rangle}^{\frac{D}{2}} F^{-\frac{D+2}{2}}. \qquad (14.7)$$

The strong field approximation, in practice, is quite good for $F \gg F_{\langle \Lambda \rangle}$. Indeed, in this limit, the exponential factor in (14.5) can be assumed to be equal to 1. The force $F_{\langle \Lambda \rangle}$ gives the order of magnitude of the interaction separating the strong field from the weak field limit.

We can now compute the first and the second moment of the force distribution by introducing a lower cut-off $l > 0$ in the nn distance (i.e. l is the minimal distance between nn particles). The effect of this cut-off can be taken into account by introducing an exponential decay for $F > l^{-2}$. Then we can also study the limit $l \to 0$. Hence we consider instead of (14.5)

$$W_{nn}(F) = \frac{D}{2} F_{\langle \Lambda \rangle}^{\frac{D}{2}} F^{-\frac{D+2}{2}} \exp\left[-\left(\frac{F_{\langle \Lambda \rangle}}{F}\right)^{\frac{D}{2}}\right] \exp\left(-l^2 F\right). \qquad (14.8)$$

By using (14.8) we obtain that the average value of the force is

$$\langle F \rangle_p = \int_0^\infty dF\, F W_{nn}(F) \sim \frac{1}{l^{2-D}}. \qquad (14.9)$$

Hence in the limit $l \to 0$ $\langle F \rangle_p$ is finite for $D > 2$, while it is infinite for $D < 2$. The mean square force can be computed easily in the same way and gives

$$\langle F^2 \rangle_p = \int_0^\infty dF\, F^2 W_{nn}(F) \sim \frac{1}{l^{4-D}}. \qquad (14.10)$$

In the limit $l \to 0$, $\langle F^2 \rangle_p$ is divergent for any possible value of the fractal dimension $D \leq 3$. Therefore the root mean square value (rms) of the force depends strongly on the lower cut-off l. Consequently, in this context, we expect to see large fluctuations of the observed values of the total gravitational field on different points of the system. This kind of divergence in the field moments is due to the fact that considering $l = 0$, no lower limit in the nn distance is imposed and the related gravitational force, being proportional to r^{-2}, is permitted to be arbitrarily large producing a power-law tail in $W_{nn}(F)$. In almost all physical problems it is quite reasonable to assume $l > 0$, in particular for the case of galaxy distribution (as galaxies are not point particles but extended objects). In gravitational N-body simulations, for example, a small scale smoothening of the gravitational force is always introduced. The physical results derived are, however, verified to be independent of this lower cut-off. The treatment of effects which depend on such a cut-off (e.g. tidal interactions leading to the formation of binary systems) are beyond the scope of the present treatment.

14.3 Gravitational Force PDF in a Poisson Particle Distribution

Chandrasekhar [43] has considered the behavior of the PDF of the Newtonian gravitational force $P(\boldsymbol{F})$ arising from a statistically isotropic Poisson distribution of sources. He showed that applying the Markov method, it is possible to compute exactly the PDF, known as the Holtzmark distribution, of the gravitational force acting on a test particle in the system. We briefly summarize the results of Chandrasekhar [43] for the computation of the Holtzmark distribution obtained by the Markov method. We then focus our attention on the strong and weak field limits (i.e. respectively $F \to \infty$ and $F \to 0$) deriving the approximate solutions in these cases. In particular, we show that in the large F limit, the total force distribution reduces to the nn approximation previously discussed.

In the case of a homogeneous and isotropic Poisson particle distribution, because of spherical symmetry, the average gravitational force acting on one particle due to the rest of the system vanishes. Any force acting on a particle is due to fluctuations away from exact (i.e. deterministic) spherical symmetry. Without lost of generality, let us suppose again that the particle from which we are calculating the gravitational force is at the origin of the system of coordinates. The force felt by this particle due to the other N particles contained in the system volume V (e.g. a spherical volume) can be written

$$\boldsymbol{F} = \sum_{i=1}^{N} \frac{GM^2}{|\boldsymbol{r}_i|^3} \boldsymbol{r}_i = \sum_{i} \boldsymbol{f}(\boldsymbol{r}_i) \qquad (14.11)$$

assuming, again, that all the N particles have equal mass M. We want to compute the probability $P(\boldsymbol{F})d^3F$ that the force in (14.11) takes a value in the element d^3F around $\boldsymbol{F}$.

14.3 Gravitational Force PDF in a Poisson Particle Distribution

The derivation of the PDF $P(\boldsymbol{F})$ for the Poisson case is simple and follows similar arguments to the derivation of the central limit theorem discussed in Chap. 2. If we call $p_c(\boldsymbol{r}_1, \boldsymbol{r}_2, \ldots, \boldsymbol{r}_N)$ the PDF for having the N particles in (14.11) occupying respectively the points $\boldsymbol{r}_1, \boldsymbol{r}_2, \ldots, \boldsymbol{r}_N$ conditioned on the fact that the origin of coordinates is occupied by another particle, we can write

$$P(\boldsymbol{F}) = \int_V \cdots \int_V \left(\prod_{i=1}^N d^3 r_i \right) p_c(\boldsymbol{r}_1, \boldsymbol{r}_2, \ldots, \boldsymbol{r}_N) \delta \left(\sum_i \boldsymbol{f}(\boldsymbol{r}_i) - \boldsymbol{F} \right), \tag{14.12}$$

where the integrals extends over the system volume V, and the Dirac delta-function picks up those particle configurations for which the sum of the $\boldsymbol{f}(\boldsymbol{r}_i)$ is equal to $\boldsymbol{F}$. In the particular case of a Poisson particle distribution the problem can be greatly simplified. In fact, since in a Poisson distribution there is no correlation between the position of different particles, $p_c(\boldsymbol{r}_1, \boldsymbol{r}_2, \ldots, \boldsymbol{r}_N)$ reduces simply to the product $\prod_{i=1}^N \tau(\boldsymbol{r}_i)$ where $\tau(\boldsymbol{r})$ is simply the unconditional PDF for finding a generic system particle at the point $\boldsymbol{r}$. Therefore, in the Poisson case, the problem consists in the evaluation of

$$P(\boldsymbol{F}) = \int_V \cdots \int_V \left(\prod_{i=1}^N d^3 r_i \, \tau(\boldsymbol{r}_i) \right) \delta \left(\sum_i \boldsymbol{f}(\boldsymbol{r}_i) - \boldsymbol{F} \right). \tag{14.13}$$

The PDF $\tau(\boldsymbol{r})$ in the homogeneous and isotropic Poisson case is just given by $1/V$ where V is the system volume. In order to evaluate (14.13) it is more convenient to study directly its Fourier transform (FT) $\tilde{P}(\boldsymbol{K})$ [170]:

$$\begin{aligned}
\tilde{P}(\boldsymbol{K}) &\equiv \int d^3 F \, P(\boldsymbol{F}) \exp\left(i \boldsymbol{K} \cdot \boldsymbol{F} \right) \\
&= \int_V \cdots \int_V \left(\prod_{i=1}^N d^3 r_i \, \tau(\boldsymbol{r}_i) \right) \exp\left[i \boldsymbol{K} \cdot \sum_i \boldsymbol{f}(\boldsymbol{r}_i) \right] \\
&= \prod_{i=1}^N \int_V d^3 r_i \, \tau(\boldsymbol{r}_i) \exp\left[i \boldsymbol{K} \cdot \boldsymbol{f}(\boldsymbol{r}_i) \right] \\
&= \left(\int_V d^3 r \, \tau(\boldsymbol{r}) \exp\left[i \boldsymbol{K} \cdot \boldsymbol{f}(\boldsymbol{r}) \right] \right)^N.
\end{aligned} \tag{14.14}$$

We now consider the large volume limit in which $V \to \infty$ and $N = nV$, with $n > 0$ being the (self-averaging) average number density characterizing the Poisson distribution[1] (see Sect. 2.4). In the statistically isotropic case (14.14) can be rewritten as

---

[1] Actually N can have statistical fluctuations from the average value nV, but it is simple to show that, for a Poisson distribution, and more generally for correlated uniform distributions (see Chap. 2), in the large V limit, they are vanishingly small relative to nV and give negligible contributions to $P(\boldsymbol{F})$.

$$\left(\int_V d^3r \tau(\mathbf{r})\exp\left[i\mathbf{K}\cdot\mathbf{f}(\mathbf{r})\right]\right)^N = \left(1 - \frac{1}{V}\int_V d^3r\left(1 - \exp\left[i\mathbf{K}\cdot\mathbf{f}(\mathbf{r})\right]\right)\right)^{nV}. \tag{14.15}$$

In the limit $V \to \infty$ one can use the definition of the exponential function, to obtain

$$\tilde{P}(\mathbf{K}) = \exp\left[-n\int d^3r\left(1 - \exp(i\mathbf{K}\cdot\mathbf{f}(\mathbf{r}))\right)\right] \equiv \exp\left[-nC(\mathbf{K})\right] \tag{14.16}$$

where the integral extends over all space, giving

$$C(K) = \frac{4}{15}(2\pi G M^2)^{3/2} K^{3/2} \equiv (F_o K)^{3/2}\frac{1}{n} \tag{14.17}$$

and

$$F_o = (4/15)^{2/3}(2\pi G M^2)n^{2/3} = \frac{GM^2}{\langle\Lambda\rangle^2}\Gamma_e^2(4/3)25^{-2/3}\pi^{\frac{1}{3}}. \tag{14.18}$$

The quantity F_o is called the *normalizing force*. Note that because of the statistical isotropy of the particle distribution $\tilde{P}(\mathbf{K})$ depends only on $K = |\mathbf{K}|$. Thus, by inverting the FT, we get

$$P(\mathbf{F}) = FT^{-1}\left[\tilde{P}(\mathbf{K})\right] = \frac{1}{(2\pi)^3}\int d^3K\,\exp\left(-i\mathbf{K}\cdot\mathbf{F} - (F_o|\mathbf{K}|)^{3/2}\right), \tag{14.19}$$

which can be simply shown to depend only on $F = |\mathbf{F}|$. Consequently, by using (14.2) we can write

$$W(F) = \frac{H(\beta)}{F_o} \tag{14.20}$$

where $\beta = F/F_o$ is the dimensionless force and

$$H(\beta) = \frac{2}{\pi\beta}\int_0^\infty \exp[-(x/\beta)^{\frac{3}{2}}]x\sin(x)dx. \tag{14.21}$$

This function is known as the Holtzmark distribution. The probability distribution $W(F)$ is characterized by the following asymptotic scaling behaviors

$$W(F) = \begin{cases} \frac{4F^2}{3\pi F_o^3} & \text{for } F \to 0 \\ \frac{15}{8}(2\pi)^{1/2}F_o^{3/2}F^{-5/2} & \text{for } F \to \infty. \end{cases} \tag{14.22}$$

In the Poisson case the nn approximation given by (14.5), and the exact Holtzmark distribution (14.21) agree very well in the large F region (see Fig. 14.1). The region where they differ most is when $F \to 0$. This is due to the fact that a weak force can arise only from a more or less symmetric configuration of particles around the test one in which fluctuations are

14.3 Gravitational Force PDF in a Poisson Particle Distribution

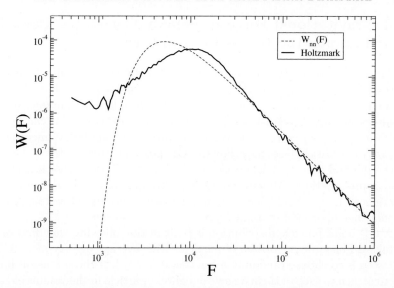

Fig. 14.1. Force distribution due to nearest neighbors and due to all particles for a statistically stationary and isotropic Poisson sample in $d = 3$. The *dotted line* represents the force distribution $W(F)$ computed by the nn approximation, while the *dashed line* is the Holtzmark distributions numerically computed in a realization of the Poisson process. The agreement is quite good at strong fields, while there is a clear deviation at weak fields $(F < F_{\langle\Lambda\rangle})$

determined by many particle effects, and hence the nn approximation fails. Instead in the strong field limit we may almost neglect the contribution to the force from far away points, because the main contribution is due to the limit $r \to 0$ in the elementary interaction (i.e. it comes from the nn). Note that, as in the nn case, because of the behavior of $W(F)$ for $F \to \infty$, $\langle F^2 \rangle$ diverges. This is due to the singularity of the particle-particle gravitational interaction at $r = 0$ together with the fact that in the Poisson distribution there is no explicit positive minimal distance between particles (i.e. no lower cut-off), as they are permitted to be at an arbitrarily small distance from one another. If such a positive minimal distance $l > 0$ is added to the generation algorithm of the particle distribution, as an additional ingredient, then $W(F)$ will be regularized at large F introducing effectively an upper cut-off in F proportional to $1/l^2$. This results in the finiteness of any moment $\langle F^m \rangle_p$ of the modulus of the force (in particular proportional to $1/l^{2m}$) with any $m \geq 0$.

14.4 Gravitational Force in Weakly Correlated Particle Distributions: the Gauss-Poisson Case

In this section we study the case of the so-called Gauss-Poisson (GP) point process, which generates particle distributions characterized fully by their two-point correlations, i.e. connected n-point correlation functions (or cumuants, see Chap. 2) vanish for $n \geq 3$ [128, 217]. In this sense it can be seen as the next step in correlated systems beyond the completely uncorrelated Poisson distribution (see [93]). For the GP point process we try to generalize the method used by Chandrasekhar [43] for the Poisson case introducing some approximations. Moreover we study the contribution to the total force experienced by a particle due to its nn, in order to evaluate the weight of the granular neighborhood of a fixed particle.

A GP point-particle distribution is built in the following way: first of all, we take a Poisson distribution of particles with average density $n_0 > 0$. The next step is to choose randomly a fraction $0 < q \leq 1$ of these Poisson points and to attach to each of them a new "daughter" particle in the volume element d^3r at a vectorial distance $\boldsymbol{r}$ from the "parent" particle with probability $p(\boldsymbol{r})d^3r$ for each "parent" particle independently of the others [128, 217]. Therefore the net effect of this algorithm is the substitution of a fraction q of the particles in the initial Poisson system with an equal number of correlated binary systems. This kind of point distribution is evidently very useful in physical applications characterized by the presence of binary systems.

The final average particle density of the GP distribution generated in this way is evidently $n = n_0(1 + q)$. It is also simple to show that the connected two-point correlation function is

$$\tilde{\xi}(\boldsymbol{r}) = \frac{\delta(\boldsymbol{r})}{n} + \frac{2q}{n(1+q)} p(\boldsymbol{r}) \tag{14.23}$$

and that all the other connected n-point correlation functions with $n \geq 3$ vanish [53]. All statistical information about a GP stochastic distribution is contained in the knowledge of n and of $\tilde{\xi}(\boldsymbol{r})$. For this reason the GP particle distribution can be seen as the discrete analog of continuous Gaussian stochastic fields (see Sects. 2.7 and 3.6). Moreover since $p(\boldsymbol{r})$ is a PDF, $\tilde{\xi}(\boldsymbol{r})$ is non-negative and integrable. In particular this implies that correlations are positive and short ranged (i.e. with finite correlation length as defined in Chap. 2). It is simple to show that (14.23) is finite: it is sufficient to use the definition of the average conditional density $\langle n(\boldsymbol{r}) \rangle_p$ of particles seen by a generic particle of the system at a vectorial distance $\boldsymbol{r}$ from it without counting the particle itself. As usual we write (see (2.52))

$$\langle n(\boldsymbol{r}) \rangle_p = n[1 + \xi(\boldsymbol{r})] \,, \tag{14.24}$$

where $\xi(\boldsymbol{r})$ is the non-diagonal part of $\tilde{\xi}(\boldsymbol{r})$. In the present case the conditional average density can be evaluated as follows: the number of particles seen in average by the chosen particle in the origin in the volume element d^3r around $\boldsymbol{r}$ is nd^3r if the chosen particle is neither a "parent" nor a "daughter" (i.e. with probability $(1-q)/(1+q)$) and $nd^3r + p(\boldsymbol{r})d^3r$ if it is either a "parent" or a "daughter" (i.e. with a complementary probability $2q/(1+q)$). This gives directly

$$\langle n(\boldsymbol{r})\rangle_p = n\left[1 + \frac{2q}{n(1+q)}p(\boldsymbol{r})\right],$$

which is equivalent to (14.23). Note that if $p(\boldsymbol{r})$ depends only on r (i.e. if it is spherically symmetric) then the particle distribution, in addition to being spatially stationary (i.e. translational invariant), is also statistically isotropic (i.e. rotationally invariant).

14.5 Generalization of the Holtzmark Distribution to the Gauss-Poisson Case

We can now try to generalize the Holtzmark distribution to this correlated case. Let us suppose we generate a GP distribution with fixed $n > 0$ and $0 < q \leq 1$ in a volume V. Let us choose the coordinate system so that the origin is occupied by a particle of the distribution, and assume that $G = M = 1$. As for the Poisson case in Sect. 14.3, we wish to calculate the PDF $P(\boldsymbol{F})$ of the total gravitational field $\boldsymbol{F}$ acting on this point due to all the other N particles in the system. As already seen in Sect. 14.3, once the conditioned N-particle PDF $p_c(\boldsymbol{r}_1, \boldsymbol{r}_2, \ldots, \boldsymbol{r}_N)$ is known, $P(\boldsymbol{F})$ is formally given by (14.12).

The problem is now more difficult since in the GP process, as the non-diagonal two-point correlation $\xi(r)$ is non-zero, $p_c(\boldsymbol{r}_1, \boldsymbol{r}_2, \ldots, \boldsymbol{r}_N)$ cannot be written as a product of N one particle PDF's as in the Poisson case. This means it is not possible to apply the Markov method, used for the Poisson case, to evaluate the $P(F)$ exactly. We therefore introduce the following approximation consisting in imposing the factorization

$$p_c(\boldsymbol{r}_1, \boldsymbol{r}_2, \ldots, \boldsymbol{r}_N) = \prod_{i=1}^{N} \tau(\boldsymbol{r}_i), \qquad (14.25)$$

taking into account that, as $\xi(\boldsymbol{r})$ is short ranged (being proportional to the PDF $p(\boldsymbol{r})$), and that the higher order connected correlation functions vanish, we can limit ourselves to using only the information about the conditional density of particles around the occupied origin. Doing so we take into account only the fact that on average the particle at the origin sees a density of particles at the point $\boldsymbol{r}$ given by (14.24) with $\xi(\boldsymbol{r})$ given by (14.23). This gives $\tau(\boldsymbol{r})$ proportional to $\langle n(\boldsymbol{r})\rangle_p$ with an appropriate normalization:

$$\tau(\boldsymbol{r}) = \frac{1 + \frac{2q}{n(1+q)}p(\boldsymbol{r})}{V + \frac{2q}{n(1+q)}}. \qquad (14.26)$$

Note that this is equivalent to approximating the given (statistically stationary) GP distribution seen by the particle in the origin with an inhomogeneous and radial Poisson distribution (see Sect. 4.3.1) around it generated by the following algorithm: once the space is partitioned in cells of volume d^3r, the cell around the point r is occupied by a particle with probability $\langle n(r) \rangle_p d^3r$ or stays unoccupied with the complementary probability $1 - \langle n(r) \rangle_p d^3r$ independently of the other cells.

This approximation makes it possible to use the Markov method to find $P(F)$ which, in the limit $V \to +\infty$, can be shown to be given by

$$P(F) = \frac{1}{(2\pi)^3} \int d^3K \exp\left(-i K \cdot F - n C_{GP}(K)\right), \qquad (14.27)$$

where

$$C_{GP}(K) = C(K) + \frac{2q}{n(1+q)} \int d^3r\, p(r) \left(1 - \exp\left[-i\frac{K \cdot r}{r^3}\right]\right). \qquad (14.28)$$

The function $C(K)$ is the analogue of $C_{GP}(K)$ for an isotropic Poisson distribution with the same average density n and is given by (14.17).

As shown below by a direct comparison with the results of numerical simulations, this approximation is quite accurate. Note that the function $\tilde{P}(K) = \exp\left(-n C_{GP}(K)\right)$ is the *characteristic* function (see Sect. 2.3.1) of the stochastic force F. As aforementioned, if the PDF $p(r)$ depends only on $r = |r|$, the particle distribution is statistically isotropic. Consequently, as for the Poisson case discussed in Sec. 14.3, $P(F)$ also depends on $F = |F|$ and $\tilde{P}(K)$ on $K = |K|$. This implies that the direction of F is completely random while the PDF of F is given by (denoting $p(r)$ by $p(r)$ to show the dependence only on r)

$$W(F) \equiv 4\pi F^2 P(F) = \frac{2F}{\pi} \int_0^\infty dK\, K \sin(KF) \times \qquad (14.29)$$

$$\times \exp\left\{-\frac{4(2\pi)^{\frac{3}{2}} n K^{\frac{3}{2}}}{15} - \frac{8\pi q}{1+q} \int_0^\infty dr\, r^2 p(r) \left[1 - \frac{r^2}{K} \sin\left(\frac{K}{r^2}\right)\right]\right\}.$$

14.5.1 Large F Expansion

We limit the rest of the discussion to the statistically isotropic case. As for the Poisson distribution, it is not possible to find an explicit form of $P(F)$ (or equivalently of $W(F)$). However we can connect its large F scaling behavior to that of the simple Poisson case (14.22) and to the small r behavior of $p(r)$. The fundamental point is to use the general properties of the Taylor expansion of the characteristic function $\tilde{P}(K)$ (see Appendix A) to the lowest order greater than zero. In particular in this isotropic case we use the fact that, if $P(F) \simeq C F^{-\alpha}$ at large F (note that $\alpha > 3$ in any case as $P(F)$ is a normalizable PDF) then [121]

14.5 Generalization of the Holtzmark Distribution to the Gauss-Poisson Case

$$\tilde{P}(\boldsymbol{K}) = \int d^3 F \exp\left(i\boldsymbol{K} \cdot \boldsymbol{F}\right) P(\boldsymbol{F}) =$$
$$= \begin{cases} 1 - \frac{1}{6}\mathcal{F}^2 K^2 & \text{if } \alpha > 5 \\ 1 - aK^{\alpha-3} & \text{if } 3 < \alpha \le 5 \,, \end{cases} \tag{14.30}$$

where

$$\mathcal{F}^2 = \int d^3 F \, F^2 P(\boldsymbol{F})$$

is the second moment of the force distribution, and $a > 0$ is a constant characterizing the singular part of the Taylor expansion. It is possible to show, using arguments similar to those explained in Appendix A, that the constant a is given by

$$a = 4\pi C \int_0^\infty dx\, x^{2-\alpha} \left(1 - \frac{\sin x}{x}\right). \tag{14.31}$$

Note that $\alpha > 5$ implies that $\mathcal{F}^2$ is finite, and that for the stationary and isotropic Poisson case $\alpha = 9/2$, so that

$$\tilde{P}(\boldsymbol{K}) \simeq 1 - \frac{4n}{15}(2\pi K)^{\frac{3}{2}}.$$

Therefore our strategy is to find α by connecting the expansion given in (14.30) to the form of $p(r)$ and in particular to its small r behavior. Let us suppose that $p(r) \simeq Br^\beta$ at small r ($B > 0$ and $\beta > -3$ as $p(r)$ is a PDF of a three-dimensional stochastic variable – see Chap. 2). It is quite simple to show that at small K the integral

$$I(K;\beta) = \int_0^\infty dr\, r^2 p(r) \left[1 - \frac{r^2}{K}\sin\left(\frac{K}{r^2}\right)\right]$$

behaves as follows:

$$I(K;\beta) \simeq \begin{cases} c_1 K^{\frac{3+\beta}{2}} & \text{if } \beta < 1 \\ c_2 K^2 & \text{if } \beta \ge 1 \,, \end{cases} \tag{14.32}$$

For $\beta = 1$ there will be logarithmic corrections to (14.32). The two constants c_1 and c_2 depend on $p(r)$ in the following way:

$$c_1 = \frac{B}{2}\int_0^\infty dx\, x^{-\frac{5+\beta}{2}}\left(1 - \frac{\sin x}{x}\right)$$
$$c_2 = \frac{1}{24\pi}\overline{\left(\frac{1}{r^4}\right)}, \tag{14.33}$$

where $\overline{(\ldots)} = \int d^3 r \, (\ldots) \, p(r)$ is the average over the PDF $p(r)$. Consequently, by inserting this result in (14.29), we can distinguish three possible asymptotic behaviors of $P(\boldsymbol{F})$:

- For $\beta > 0$ the dominating part in $\tilde{P}(\mathbf{K})$ at small K is exactly the same as in the stationary and isotropic Poisson case, i.e.

$$\tilde{P}(\mathbf{K}) \simeq 1 - \frac{4n}{15}(2\pi K)^{\frac{3}{2}}, \qquad (14.34)$$

which implies $P(\mathbf{F}) \simeq (n/2)F^{-9/2}$ (or equivalently $W(F) \simeq 2\pi n F^{-5/2}$) at large F. Note that not only the exponent, but also the amplitude, is found to be the same as in the stationary and isotropic Poisson case with the same unconditional average density.

- For $\beta = 0$ we have again substantially the same behavior as for the stationary and isotropic Poisson particle distribution, but the coefficient of the first non-zero order term picks up an additional contribution from the small scale two-point correlation $\xi(r)$:

$$\tilde{P}(\mathbf{K}) \simeq 1 - 8\pi n \left(\frac{(2\pi)^{\frac{1}{2}}}{15} + \frac{c_1 q}{n(1+q)} \right) K^{\frac{3}{2}}, \qquad (14.35)$$

which implies again $P(\mathbf{F}) \simeq CF^{-9/2}$ at large F, with

$$C = \frac{n}{2} + \frac{qB}{1+q}. \qquad (14.36)$$

which is larger than the coefficient found for an isotropic Poisson particle distribution with the same average density n. In practice, from (14.29) and (14.36), we have the same scaling behavior of $W(F)$ of the isotropic Poisson distribution but with a larger average density

$$n' = n + \frac{2qB}{1+q}.$$

- For $\beta < 0$ the small K behavior of $\tilde{P}(\mathbf{K})$ is completely changed from the isotropic Poisson case, being given by

$$\tilde{P}(\mathbf{K}) \simeq 1 - \frac{8\pi c_1 q}{1+q} K^{\frac{3+\beta}{2}}. \qquad (14.37)$$

This means (see (14.30)) that $2 - \alpha = -\frac{5+\beta}{2}$. Thus we can conclude that at sufficiently large F we have

$$P(\mathbf{F}) \simeq CF^{-\frac{9+\beta}{2}},$$

or equivalently

$$W(F) \simeq 4\pi CF^{-\frac{5+\beta}{2}},$$

with

$$C = \frac{q}{1+q}B. \qquad (14.38)$$

14.5 Generalization of the Holtzmark Distribution to the Gauss-Poisson Case

14.5.2 Small F Expansion

The small F behavior of $P(\boldsymbol{F})$ can be connected to the large K behavior of its FT. First note that

$$\lim_{K\to+\infty} 4\pi I(K;\beta) = 4\pi \int_0^{+\infty} dr\, r^2 p(r) = 1\ .$$

This simple observation implies (see (14.29)) that for any GP particle distribution, the asymptotically large K behavior of $\tilde{P}(\boldsymbol{K})$ is similar to that of the isotropic Poisson case with the same average density, but with an amplitude reduced by a factor $\exp(-2q/1+q)$. Consequently the small F behavior of $W(F)$ is the same as for the isotropic Poisson distribution but with an amplitude reduced by the same factor $\exp(-2q/1+q)$, i.e.

$$W(F) \simeq \exp\left(-\frac{2q}{1+q}\right)\frac{4}{3\pi}F_o^{-3}F^2\ , \qquad (14.39)$$

where F_o is given by (14.18).

14.5.3 Comparison with Simulations

The validity of these theoretical results is supported by the analysis of numerical simulations consisting in the generation of two kinds of GP particle

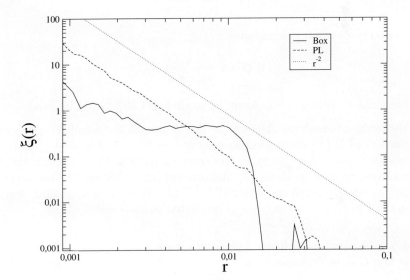

Fig. 14.2. Connected two-point correlation function measured in a single realization, with about 10^5 points in a box of volume $V=1$ for the Gauss-Poisson case where (Box) $p(r)$ is a *box − function* with cut-off at 0.01 and the label PL (power-law) corresponds to $p(r) = (1/4\pi r_0)(\exp(-r/r_0)/r^2)$ with $r_0 = 0.01$

distributions with two explicit choices of $p(r)$ (see Fig. 14.2), for which the PDF $W(F)$ of F is directly measured:

1. In the first case $p(r)$ is chosen to simply be a positive constant up to a fixed distance r_0 and zero beyond this distance:

$$p(r) = \begin{cases} \frac{3}{4\pi r_0^3} & \text{if } 0 < r \leq r_0 \\ 0 & \text{if } r > r_0 \,. \end{cases} \qquad (14.40)$$

or, in the notation of the previous section $B = \frac{3}{4\pi r_0^3}$ and $\beta = 0$, i.e., the probability of attaching a "daughter" particle at a distance between r and $r + dr$ from its "parent" is $3r^2 dr/r_0^3$ if $r \leq r_0$ and zero for $r > r_0$. As shown above this choice of $p(r)$ should give

$$W(F) \simeq \left(2\pi n + \frac{3q}{r_0^3(1+q)}\right) F^{-\frac{5}{2}}$$

at large F, i.e., with the same exponent but with a larger amplitude than the pure isotropic Poisson case. At small F, as shown above, the asymptotic behavior of $W(F)$ is given by (14.39).

2. In the second case $p(r)$ decays exponentially fast at large r but is singular as r^{-2} at small r, i.e.

$$p(r) = \frac{1}{4\pi r_0} \frac{\exp(-\frac{r}{r_0})}{r^2} \,. \qquad (14.41)$$

This choice of $p(r)$ should give

$$W(F) \simeq \frac{q}{r_0(1+q)} F^{-\frac{5+\beta}{2}}$$

at large F with $\beta = -2$. Again at small F (14.39) should be valid.

The results of these simulations for the large and the small F asymptotic behaviors of $W(F)$ show good agreement (see Figs. 14.3 and 14.4) with the theoretical predictions given in the previous section. Consequently the approximations used in these calculations appear to be valid. This suggests that the scaling relations obtained for $W(F)$ at large and small F can be extended to more general cases of correlated particle distributions.

14.5.4 Nearest-Neighbor Approximation for the Gauss-Poisson Case

As for the Poisson case, we now analyze, for GP distributions in the statistically isotropic case, the importance of the first nn contribution to the total force felt by a particle. To do this we use only the information that the average conditional density $\langle n(\mathbf{r})\rangle_p$ (which depends only on r) seen by

14.5 Generalization of the Holtzmark Distribution to the Gauss-Poisson Case

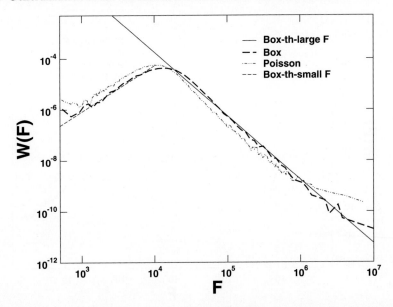

Fig. 14.3. Comparison between theoretical predictions (*solid lines*) and simulations of the tails of the PDF of the modulus of the gravitational force $W(F)$ for the case where $p(r)$ is a *box-function* given by (14.40). The theoretical behaviors at small and large fields (**th**) computed as explained in the text are shown as well. Finally for comparison it is also shown the behavior of the Holtzmark distribution for a Poisson distribution with the same number density

the particle in the origin is given by (14.24). Therefore we use, in this case, the approximations given by (4.22) and (4.25), and follow exactly the same steps as in the Poisson case. With $\langle n(\mathbf{r}) \rangle_p$ replacing the simple n of the isotropic Poisson case, we can write:

$$\omega(r) = \left(1 - \int_0^r \omega(x)dx\right) 4n\pi r^2 (1 + \xi(r)), \qquad (14.42)$$

where

$$\xi(r) = \frac{2q}{n(1+q)} p(r).$$

Equation (14.42) can be solved to give

$$\omega(r) = 4n\pi r^2 (1 + \xi(r)) \exp\left[-4n\pi \int_0^r dx\, x^2 (1 + \xi(x))\right]$$

By imposing $p(r) \simeq Br^\beta$ at small r and using again $F = 1/r^2$ in order to pass from $\omega(r)$ to $W_{nn}(F)$, it is simple to see that $W_{nn}(F)$ has the same aforementioned scaling behavior at large F of $W(F)$ for all the permitted values of β, with the same coefficient. Therefore also in the GP case we have

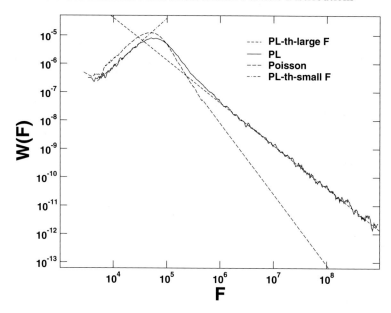

Fig. 14.4. The same as in the previous figure, but for the case where $p(r)$ is given by (14.41), i.e. it has an exponential upper cut-off r_0 at large r and is singular as r^{-2} at small r

that the main contribution to the force felt by a particle in the system is due to its first nn. This could have been expected as the main change introduced passing from Poisson to a GP distribution involves the introduction of additional density fluctuations in the neighborhood of any particle.

14.6 Gravitational Force in Fractal Point Distributions

Chandrasekhar's derivation of $P(\boldsymbol{F})$ in the Poisson case cannot be extended to the case of fractal distributions of particles with equal mass because such structures are characterized by strongly correlated fluctuations at all scales. In particular, in a fractal particle distribution, one can never use the basic approximation (exact only in the Poisson case) on which Chandrasekhar's method is based, consisting in assuming the validity of the factorization (14.25) of the joint N-particle (conditional and/or unconditional) PDF of the positions of the N particles.

It is important to note that assuming (14.25) with

$$\tau(\boldsymbol{r}) \sim r^{D-3}$$

and $0 < D < 3$ does not represent at all a fractal distribution[2], but rather an anisotropic and radial Poisson distribution (see Sect. 4.3.1) defined by the following algorithm: once the system volume is partitioned in elementary cells of volume d^3r, the cell around the point r is either occupied by a particle with probability $Ar^{D-3}d^3r$ or unoccupied with complementary probability $1 - Ar^{D-3}d^3r$, each cell being considered independently of all others. Such a radial and anisotropic Poisson distribution has a non-constant (unconditional) average density around the origin $\langle n(r) \rangle = Ar^{D-3}d^3r$. In this case, in analogy to (14.20), one obtains

$$W(F) = \frac{H\left(\frac{F}{F_o}, D\right)}{F_o}, \qquad (14.43)$$

where

$$H(\beta, D) = \frac{2}{\pi\beta} \int_0^\infty \exp[-(x\beta)^{\frac{D}{2}}]x \sin(x) dx, \qquad (14.44)$$

and

$$F_o = (4/15)^{2/D}(2\pi GM^2)A^{2/D}, \qquad (14.45)$$

The main change due to the radial anisotropic structure is that the scaling exponent in (14.44) is $D/2$ rather than $3/2$ characterizing the stationary and isotropic case. Hence in this case the tail of $W(F)$ has a slower decay (see Fig. 14.5). It is important to stress that such a result is a bad approximation for a real fractal particle distribution with $\langle n(r) \rangle_p = Ar^{D-3}$, because by neglecting correlations in a fractal, which is a *super-correlated* system, one erases their essential features.

An important limit is the strong field one ($F \to \infty$): In this case it is possible to show that the force distribution of (14.44) can be reduced to the one which can be derived under the nn approximation (14.5) with $A = \frac{DB}{4\pi}$. In Fig. 14.5 these theoretical results for the radial Poisson distribution are compared with estimates of $W(F)$ and $W_{nn}(F)$ from numerical simulations of a fractal particle distribution characterized by $\langle n(r) \rangle_p = Ar^{D-3}$.

14.7 An Upper Limit in the Fractal Case

In order to get a first insight about the statistical behavior of the gravitational force in a statistically stationary and isotropic fractal distribution with particles of equal mass M, we focus here on the determination of an upper limit to the total force acting on a generic particle of the system, that we can think to occupy the origin of coordinates, due to all other sources at a distance r between a lower limit $l > 0$ and an upper limit R_s. In the next

---

[2] Note that the proportionality constant between $\tau(r)$ and r^{D-3} depends appropriately on the system volume V to guarantee the normalization of probability.

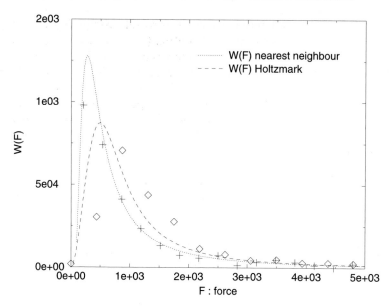

Fig. 14.5. Force distribution due to the nearest neighbors (*crosses*) and to all the field sources (*diamonds*) for a fractal distribution with $D = 2.41$ in $d - 3$ generated by the random β-model algorithm. The *dotted line* represents the force distribution $W_{nn}(F)$ given by (14.5) while the *dashed line* represents $W(F)$ given by (14.43)–(14.45) both valid for the radial Poisson case. Note that the theoretical $W_{nn}(F)$ is in quite good agreement with the analogous fractal quantity (in particular in the range $F > F_{\langle \Lambda \rangle}$). Instead the theoretical $W(F)$ is quite different from the fractal case. The reason is that, while (14.5) works quite well also for fractals, because it is based on the approximation from the neighborhood of the particle at the origin, in evaluating $W(F)$ we have omitted correlations at all scales, which is a bad approximation for the fractal case

section we study the behavior of the second moment of the total force itself. The force acting on the particle at the origin can be written

$$\boldsymbol{F}(l, R_s) = \sum_{l \leq r_i \leq R_s} \frac{GM^2}{r_i^3} \boldsymbol{r}_i , \qquad (14.46)$$

where the sum is extended to all particles in the system with a distance from the origin satisfying $l \leq r_i \leq R_s$. In a statistically stationary and isotropic stochastic fractal distribution the direction of $\boldsymbol{F}$ is completely random. Therefore we want to find the statistical properties of the modulus F. Clearly

$$F(l, R_s) \leq S(l, R_s) = \sum_{l < r_i \leq R_s} \frac{GM^2}{r_i^2} .$$

We want to study the average value of $S(l, R_s)$ with particular attention to its dependence on l and R_s. In this respect, we note again that in many

14.7 An Upper Limit in the Fractal Case

theoretical models and applications it is quite natural to consider the presence of a finite lower limit $l > 0$, and that instead the relevant physical information comes from the limit $R_s \to +\infty$.

Using the definition of the conditional density we have

$$\langle S(l, R_s) \rangle_p = GM^2 \left\langle \sum_{l < r_i \leq R_s} \frac{1}{r_i^2} \right\rangle_p = 4\pi GM^2 \int_l^{R_s} dr \frac{\Gamma(r)}{r^2} r^2 , \qquad (14.47)$$

where as usual for a fractal $\Gamma(r) = \langle n(r) \rangle_p = \frac{BD}{4\pi} r^{D-3}$ is the average conditional density and $D < 3$ is the fractal dimension. In the determination of (14.47) we have used that, for a general function $g(r)$, one can write

$$\sum_{l < r_i \leq R_s} g(r_i) = \int_{l < r \leq R_s} d^3r \, g(r) \sum_i \delta(r - r_i) = \int_{l < r \leq R_s} d^3r \, g(r) n(r) .$$
(14.48)

Therefore when we take the ensemble average (which is "conditioned" on the fact that the origin is occupied) with $g(r) = 1/r^2$, we obtain (14.47).

As a consequence we can write

$$\langle S(l, R_s) \rangle_p = C \int_l^{R_s} dr \, r^{D-3} , \qquad (14.49)$$

where $C = DBGM^2$. We can now distinguish three cases:

1. If $0 < D < 2$ (14.49) reduces to

$$\langle S(l, R_s) \rangle_p = \frac{C}{2 - D} \left(\frac{1}{l^{2-D}} - \frac{1}{R_s^{2-D}} \right) .$$

Therefore in this case, if we keep fixed l and take $R_s \to +\infty$, the quantity $\langle S(l, R_s) \rangle_p$ converges to the finite limit

$$\frac{C}{(2 - D) l^{2-D}} .$$

Since $\langle F(l, R_s) \rangle_p \leq \langle S(l, R_s) \rangle_p$, this means that for $0 < D < 2$, in the infinite volume limit, with a fixed lower cut-off for the nn distances, the average modulus of the force converges to a finite value. If moreover the system, as for many random fractals, has self-averaging properties, then we can say that in this limit, any particle of the system is subject to a finite gravitational force and, consequently, has a finite acceleration. This is somehow surprising, because, despite the "wild" mass fluctuations characterizing a fractal system at any scale, the gravitational force stays constant in the infinite volume (and mass) limit. The reason is that, if $0 < D < 2$, any particle sees on average that the system becomes empty sufficiently rapidly on large scales.

The limit $l \to 0$ is instead singular, giving an infinite value for $\langle S(l, R_s) \rangle_p$ for the simple reason that the elementary gravitational interaction GM^2/r^2 increases sufficiently rapidly when r goes to zero. For the same reasons similar behaviors have been found above in more uniform mass distributions such as Poisson and GP. As aforementioned, however, in many theoretical and real applications one indeed considers a small scale cut-off $l > 0$.

2. If $D = 2$ we can write simply

$$\langle S(l, R_s) \rangle_p = C \log\left(\frac{R_s}{l}\right),$$

i.e. $\langle S(l, R_s) \rangle_p$ is a logarithmically divergent function of the ratio R_s/l.

3. Finally, if $2 < D < 3$ we have

$$\langle S(l, R_s) \rangle_p = \frac{C}{D-2}\left(R_s^{D-2} - l^{D-2}\right).$$

This means that in this case $\langle S(l, R_s) \rangle_p$ diverges when the limit $R_s \to \infty$ is taken, with $l > 0$ fixed. This does not mean that also $\langle F(l, R_s) \rangle_p$ diverges in the same limit. Whether it does will depend on spatial correlations in the system (for instance we know that for $D = 3$ in the Poisson case $\langle F(l, R_s) \rangle_p$ is convergent in this limit, while in the GP case with $\xi(r) \sim r^{-2}$ at small r it is divergent). However in the following section we will show that, in the same limit, $\langle F^2(l, R_s) \rangle_p$ instead behaves qualitatively as $\langle S(l, R_s) \rangle_p$ in all real fractals with $D \in (0, 3]$.

In the limit $l \to 0$ the quantity $\langle S(l, R_s) \rangle_p$ is convergent because the singularity of the elementary gravitational interaction is regularized by the behavior of $\langle n(r) \rangle_p$ around $r = 0$.

In Fig. 14.6 are shown two examples of the behaviour of $S(l, R_s)$ for fractals with dimensions $D = 1.5$ and $D = 2.5$ and for the Poisson case.

14.8 Average Quadratic Force in a Fractal

Since, as aforementioned, it is (at least so far) impossible to generalize the Holtzmark distribution to the fractal case, we study here the rms force acting on a generic particle of a random fractal point-particle distribution in three dimensional Euclidean space. As in the previous section, we consider contributions to the force experienced by the particle at the origin given due to all the other particles between a minimal and a maximal distance, respectively l and R_s, from the origin of coordinates. Thus let, as above, $\boldsymbol{F}(l, R_s)$ be this total force given by (14.46). Under the assumption that the mass distribution is statistically isotropic, we have, as in the previous section, that

$$\langle \boldsymbol{F}(l, R_s) \rangle_p = 0, \tag{14.50}$$

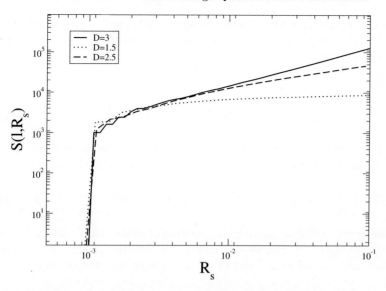

Fig. 14.6. Behavior of $S(l, R_s)$, as a function of R_s with ℓ fixed, for a fractal distribution with $D = 1.5, 2.5$ and for the purely Poisson case

because the direction of $\boldsymbol{F}(l, R_s)$ is completely random. Therefore the main statistical information about $\boldsymbol{F}$ comes from the average of its square modulus which can be written as:

$$\langle F^2(l, R_s) \rangle_p = G^2 M^4 \left\langle \sum_{l \leq |\boldsymbol{r}_i| \leq R_s} \sum_{l \leq |\boldsymbol{r}_j| \leq R_s} \frac{\cos(\theta_{ij})}{r_i^2 r_j^2} \right\rangle_p \qquad (14.51)$$

where θ_{ij} is the angle defined between by $\boldsymbol{r}_i$ and $\boldsymbol{r}_j$. This average can be performed by applying the following general reasoning. Let us take a generic two-point function $\phi(\boldsymbol{r}, \boldsymbol{r}')$ and consider the quantity

$$A(V) = \left\langle \sum_{i \in V} \sum_{j \in V} \phi(\boldsymbol{r}_i, \boldsymbol{r}_j) \right\rangle_p , \qquad (14.52)$$

where the sums are extended to all the particles inside a given volume V. The quantity $A(V)$ can be rewritten by considering that

$$\sum_{i \in V} \sum_{j \in V} \phi(\boldsymbol{r}_i, \boldsymbol{r}_j) = \int_V \int_V d^3r \, d^3r' \, \phi(\boldsymbol{r}, \boldsymbol{r}') n(\boldsymbol{r}) n(\boldsymbol{r}') , \qquad (14.53)$$

where as usual $n(\boldsymbol{r}) = \sum_i \delta(\boldsymbol{r} - \boldsymbol{r}_i)$. Consequently, taking the conditional average, we have

$$A(V) = \int_V \int_V d^3r \, d^3r' \, \phi(\boldsymbol{r}, \boldsymbol{r}') \langle n(\boldsymbol{r}) n(\boldsymbol{r}') \rangle_p , \qquad (14.54)$$

where $\langle n(r)n(r')\rangle_p$ is the two-point conditional density defined in Sect. 2.5.1 and 4.4. Now it is convenient to separate in $\langle n(r)n(r')\rangle_p$ the diagonal from the non-diagonal part in r and r': in fact, as shown in Sect. 2.5.1 (see (2.53)–(2.55)) for $r = r'$ we have a singular Dirac delta-like contribution proportional to $\langle n(r)\rangle_p$:

$$\langle n(r)n(r')\rangle_p = \delta(r-r')\langle n(r)\rangle_p + \Gamma^{(2)}(r,r'), \qquad (14.55)$$

where $\Gamma^{(2)}(r,r')$ is the regular non-diagonal part of the two-point conditional density for $r \neq r'$. This allows us to recast $A(V)$ in the following form:

$$A(V) = \int_V d^3r\, \phi(r,r)\langle n(r)\rangle_p + \int_V\int_V d^3r d^3r\, \phi(r,r')\Gamma^{(2)}(r,r'). \qquad (14.56)$$

In the case of a fractal particle distribution of dimension D we have as usual $\langle n(r)\rangle_p = \frac{DB}{4\pi}r^{D-3}$. For what concerns the function $\Gamma^{(2)}(r,r')$ we use the relation introduced in [34] and discussed in Sect. 4.4 (see (4.26)):

$$\Gamma^{(2)}(r,r') = \langle n(r)\rangle_p \langle n(r')\rangle_p \mathcal{L}\left(\frac{r}{r'},\theta\right), \qquad (14.57)$$

where θ is the angle between r and r', and $\mathcal{L}\left(\frac{r}{r'},\theta\right)$ is the so-called *lacunarity* function defined by (4.27)–(4.28).

By applying (14.55) and (14.57) to (14.51), we have

$$\langle F^2(l,R_s)\rangle_p = G^2 M^4 \left[\frac{DB}{4-D}\left(\frac{1}{l^{4-D}} - \frac{1}{R_s^{4-D}}\right) + \right.$$
$$\left. 8\pi^2(DB)^2 \int_l^{R_s}\int_l^{R_s} dr\, dr'\, r^{D-3}r'^{D-3}\int_0^\pi d\theta\, \sin\theta\cos\theta \mathcal{L}\left(\frac{r}{r'},\theta\right)\right]. \qquad (14.58)$$

Note that in the case of a statistically isotropic Poisson particle distribution ($D=3$) the function $\mathcal{L}\left(\frac{r}{r'},\theta\right) \equiv 1$ for any r/r' and θ; this implies that in this case the second term of (14.58) identically vanishes and $\langle F^2(l,R_s)\rangle_p$ is completely determined by the singular diagonal contribution to $\langle n(r)n(r')\rangle_p$, i.e. $\delta(r-r')\langle n(r)\rangle_p$. Instead for real fractals $\mathcal{L}\left(\frac{r}{r'},\theta\right)$ is never identically equal to unity but converges to it (see (4.28)) only for $r/r' \to 0,\infty$. This implies that the second contribution is important, and, in the limit of fixed small but positive l and large R_s, it can become dominant. Moreover it is important to note that, as already pointed out in Sect. 4.4, such behavior of $\langle n(r)n(r')\rangle_p$ for a fractal implies that conditional mass fluctuations in regions of size R are of the same order as the conditional average value of the mass itself in the same region for any R. In other words, this form $\langle n(r)n(r')\rangle_p$ takes into account the fact that for a fractal there is no length scale beyond which we can represent the system as a fluid with a well defined positive average mass density with superimposed small perturbations.

14.8 Average Quadratic Force in a Fractal

In order to simplify the discussion for fractal systems, we introduce a simple approximation for $\mathcal{L}\left(\frac{r}{r'},\theta\right)$ assuming that it depends only on θ. Let us rename, in this case $\mathcal{L}\left(\frac{r}{r'},\theta\right) \equiv \gamma(\theta)$. Note that $\gamma(\theta)$ is closely related to the angular correlation function of the fractal distribution. Under this assumption, we obtain

$$\langle F^2(l,R_s)\rangle_p = G^2 M^4 \left[\frac{DB}{4-D}\left(\frac{1}{l^{4-D}} - \frac{1}{R_s^{4-D}}\right) + \right.$$
$$\left. 8\pi^2 (DB)^2 C(D) \left(\int_l^{R_s} dr\, r^{D-3}\right)^2 \right], \quad (14.59)$$

where

$$C(D) = \int_0^\pi d\theta\, \gamma(\theta) \sin\theta \cos\theta \,. \quad (14.60)$$

Since in general $\gamma(\theta)$ is large at small θ and small for $\theta \simeq \pi$ then $C(D) > 0$. As for $S(l,R_s)$ we can distinguish three cases: (1) $0 < D < 2$, (2) $2 < D < 3$, and (3) $D = 2$.

1. For $0 < D < 2$ we have

$$\langle F^2(l,R_s)\rangle_p = \alpha(B,D)\left(\frac{1}{l^{4-D}} - \frac{1}{R_s^{4-D}}\right) +$$
$$\beta(B,D)\left(\frac{1}{l^{2-D}} - \frac{1}{R_s^{2-D}}\right)^2, \quad (14.61)$$

where

$$\alpha(B,D) = G^2 M^4 \frac{DB}{4-D}$$

and

$$\beta(B,D) = 8\pi^2 G^2 M^4 \left(\frac{DB}{2-D}\right)^2 C(D) \,.$$

It is interesting to note that both terms for fixed small $l > 0$ and $R_s \to \infty$ stay finite, similarly to $S(l,R_s)$ in the previous section in the same limit. In particular increasing R_s from l to $+\infty$, the quantity $\langle F^2(l,R_s)\rangle_p$ increases from 0 to the asymptotic value

$$\langle F^2(l,+\infty)\rangle_p = \frac{\alpha(B,D)}{l^{4-D}} + \frac{\beta(B,D)}{l^{4-2D}} \,.$$

For very small l this asymptotic value is dominated by the first "diagonal" term. The reason why $\langle F^2 \rangle$ is finite is that, when $D < 2$, although the system is characterized by large relative mass fluctuations at any scale, the mass seen from any particle of the system increases too slowly with distance, i.e., the system becomes empty too rapidly with distance, the average conditional density $\langle n(r)\rangle_p$ decreasing more rapidly than $1/r$.

Instead for $l \to 0$ the quantity $\langle F^2(l, R_s)\rangle_p$ diverges as l^{D-4} as the rapid decrease of the conditional average density from the origin is not able to regularize the singularity of the elementary gravitational interaction GM^2/r^2 at $r = 0$. This behavior for $l \to 0$ is shared by all three cases.

2. For $2 < D < 3$

$$\langle F^2(l, R_s)\rangle_p = \alpha(B, D) \left(\frac{1}{l^{4-D}} - \frac{1}{R_s^{4-D}}\right) + \beta(B, D) \left(R_s^{D-2} - l^{D-2}\right)^2 . \quad (14.62)$$

Therefore, fixing a small $l > 0$, we have that $\langle F^2(l, R_s)\rangle_p$ diverges for $R_s \to +\infty$ as R_s^{2D-4}. This divergence for large R_s is due to the fact that, even though l is positive and the system is statistically stationary and isotropic, the mass, and therefore the mass fluctuations, which in a fractal are proportional to the mass itself, seen by any particle grow rapidly enough to make $\langle F^2\rangle_p$ divergent. In other words at a certain scale R_s the particle at the origin will see a global mass fluctuation in a certain direction proportional to the average mass $\sim R_s^D$ generating a net force $\sim R_s^{D-2}$ in the same direction, and this is true for any sufficiently large value of R_s. Note that this effect is completely due to the second term in (14.62) which is the product of the typical strong three-point correlation and fluctuation properties of a fractal. In fact for an isotropic Poisson distribution ($D = 3$) this is no longer true because, as shown above, for the case $\beta = 0$ the random mass fluctuations on the scale R_s seen by an arbitrary particle increase only proportionally to $\sim R_s^{3/2}$ in a certain direction, giving rise to a small force fluctuation $\sim R_s^{-1/2}$ in the same direction.

More specifically, for a fractal, increasing R_s from l to $+\infty$ at a first step $\langle F^2(l, R_s)\rangle_p$ increases from zero following mainly the behavior of the first term (which tends to converge to $\frac{\alpha}{l^{4-D}}$) until around a distance r_0 defined by

$$r_0^{2D-4} \simeq \frac{\alpha(B, D)}{\beta(B, D)} \frac{1}{l^{4-D}} .$$

For $R_s > r_0$ the quantity $\langle F^2(l, R_s)\rangle_p$ starts to diverge as $\beta(B, D) R_s^{2D-4}$.

3. Finally for the marginal case $D = 2$ we have basically the same behavior as before except that the divergence in R_s proportional to R^{2D-4} is now logarithmic in R_s.

14.9 The General Importance of the Force-Force Correlation

Up to now we have discussed only the statistical one-point properties of the gravitational field in very different types of stochastic distributions of

14.9 The General Importance of the Force-Force Correlation

identical particles. In this context a characteristic shared by all the cases considered is that the contribution to the gravitational force experienced by a particle of the system due to its first nearest neighbor is dominant in determining large values of the force, while all the other particles of the distribution typically give a contribution at most of the same order. This observation can lead to two different conclusions:

- the gravitational field is a rapidly fluctuating quantity in space from particle to particle;
- the gravitational clustering dynamics of the system on scales of order of the nn average distance is governed by the nn interaction.

It would be incorrect to conclude that in general, on all scales, the gravitational dynamics of the particle system is determined only by interactions between each particle and its neighborhood: in fact we expect the large scale distribution of mass to play a role. Let us consider this point in some more details.

Let us start from the basic equation linking the gravitational field $\boldsymbol{E}(\boldsymbol{r})$ and the matter density field $\rho(\boldsymbol{r})$, the Poisson equation:

$$\nabla \cdot \boldsymbol{E}(\boldsymbol{r}) = -\rho(\boldsymbol{r}) \,, \tag{14.63}$$

where we have chosen the units, for simplicity, so that $4\pi G = 1$. By taking the average square modulus of the FT of (14.63), we can write

$$\left\langle |\boldsymbol{k} \cdot \tilde{\boldsymbol{E}}(\boldsymbol{k})|^2 \right\rangle = \left\langle |\tilde{\rho}(\boldsymbol{k})|^2 \right\rangle \,, \tag{14.64}$$

where $\tilde{\boldsymbol{E}}(\boldsymbol{k}) = FT[\boldsymbol{E}(\boldsymbol{r})]$ and, as usual, $\tilde{\rho}(\boldsymbol{k}) = FT[\rho(\boldsymbol{r})]$. On the right hand side of (14.64) we have, up to a normalization, the PS of the mass density field. Equation (14.64) implies that the PS of the gravitational field (considered as a scalar quantity, even though it is actually a 3×3 matrix as $\boldsymbol{E}(\boldsymbol{r})$ is a vector) is roughly k^{-2} times the PS of the mass density field. This implies that two-point correlations on large spatial scales between the values of the gravitational field at different points are much larger than those characterizing the density field. For instance by power counting it is simple to show that, if on large scales the mass density field is substantially Poisson (see Chap. 3), then the two-point correlation function (which is actually a matrix) of the gravitational field decays very slowly, as $1/r$. In general, by a similar power counting we can say that, if $P(k) \sim k^n$ for $k \to 0$ (recalling that we require $n > -3$ in order to have a mass density field which is a well defined three-dimensional stochastic process with a two-point correlation function going to zero at large distance), then in general the two-point correlation function of the gravitational field decays at large r as r^{-n-1}. Note that even for a super-homogeneous mass density field satisfying the Harrison-Zeldovich condition $P(k) \sim k$ at small k, presenting, as shown in Chap. 3, a sort of long range order, the correlation function of the gravitational field is mainly positive and decays slowly at large r as $1/r^2$. It is important also to note that, as the PS of the gravitational field scales for small k as k^{n-2}, there is a limitation to the permitted PS of the mass density fluctuations in order to have a gravitational field which is a well defined three-dimensional stochastic process (i.e. with two-point correlation going to zero at large distance): since to this aim we must have $n - 2 > -3$, we can deduce that the gravitational field is well defined only if $n > -1$.

We can conclude that, in general, while the gravitational field fluctuates greatly from particle to particle due to the large contribution from the nearest neighbor, it nevertheless shows long range two-point correlation. This means that the contribution to the force on a given particle from far away particles, which varies slowly in space, is not negligible. In approaching the problem of the gravitational dynamics of a stochastic particle distribution both aspects have in principle to be taken into account.

14.10 Summary and Discussion

We have discussed the statistical properties of the gravitational force $\boldsymbol{F}$ acting on a generic point in an uncorrelated Poisson particle distribution, a weakly correlated Gauss-Poisson (GP) distribution, and a fractal distribution.

For the Poisson case we have given [43] the exact PDF of $\boldsymbol{F}$ whose main features are the following: (1) for large values of F the statistical properties are mainly determined by the contribution of the first nn particle, while at small F the contribution of the system as a whole is important; (2) the momenta $\langle F^n \rangle_p$ diverge for $n \geq 3/2$ because there is no positive lower limit for the distance between two nn particles to cut-off the singularity at $r = 0$. Introducing artificially a lower cut-off for the nn distance, the PDF is regularized, making all the positive moments of F finite, but dependent on the cut-off for $n \geq 3/2$.

For the GP case only an approximate expression of the PDF of $\boldsymbol{F}$ is given, based on the fact that the main difference, introduced by correlations, with the Poisson case is in the neighborhood of any particle. This approximation gives good agreement with the numerical data for the exact PDF, in both regions of small and large F. The exponent characterizing the large F tail of the PDF depends strongly on the way in which "daughter" particles are attached to "parents". More precisely, it depends on the small r region of the connected two-point correlation function, i.e. on the PDF of the nn distance. Depending on the same small r behavior of $p(r)$, there is a power $n \leq 3/2$ such that all the moments $\langle F^m \rangle_p$ for $m > n$ diverge for analogous reasons to those in the Poisson case, and for similar reasons they can be made finite by introducing a lower cut-off in the nn distance. Instead for what concerns the small F behavior of $W(F)$ of a GP particle distribution we have found a universal relation with the small F behavior of the same function for a Poisson distribution with the same average density. This relation gives the same

14.10 Summary and Discussion

scaling exponent but with an amplitude smaller by a factor $A < 1$ which depends on the parameter q of the GP distribution. All these theoretical predictions have been tested by direct comparison with the results of numerical simulations of different types of GP distributions.

We have also studied the fractal case in the hypothesis of statistical stationarity and isotropy. Since a fractal is characterized by large mass fluctuations and strong correlations at any scale, it is impossible to find a good approximation for the PDF of $\boldsymbol{F}$ starting from the results for Poisson-like distributions. For this reason the study has been limited to the evaluation and analysis of $\langle F^2(l, R_s) \rangle_p$ with particular attention given to the dependence on the lower cut-off l and on the upper cut-off R_s of the distance between the "test" particle at the origin and the other particles in the system. We have seen that fixing a small $l > 0$, the dependence on R_s is different in the cases in which the fractal dimension is $D < 2$ or $2 < D < 3$ (with a marginal case when $D = 2$). In the first case that $\langle F^2(l, R_s) \rangle_p$ remains as R_s goes to infinity converging to an l dependent limit. This is due to fact that the conditional mass density decreases sufficiently rapidly to prevent the strong mass fluctuations (proportional at any scale to the entire average mass) making $\langle F^2(l, R_s) \rangle_p$ divergent. Instead, if $2 < D < 3$ this is no longer the case. In fact at large R_s, it is simple to see that

$$\langle F^2(l, R_s) \rangle_p \sim R_s^{2D-4} \ .$$

Moreover, in all fractal cases with $0 < D < 3$, the quantity $\langle F^2(l, R_s) \rangle_p$ diverges where $l \to 0$ for analogous reasons to the Poisson case with an exponent depending on D.

Finally we have analyzed briefly the general correlation properties of the gravitational field, and in particular the behavior of the field-field two-point correlation function in relation to the two-point statistical properties of the mass density field of the particle distribution. The main result is that, even though the values of the gravitational field (or force) acting on a system particle fluctuate strongly from particle to particle because of the large contribution from nearest neighbors, the long range nature of the elementary gravitational interaction and the absence of a screening effect give field-field correlations which are much more long-ranged than those of the mass density field.

Part III

Appendixes

A Scaling Behavior of the Characteristic Function for Asymptotically Small Values of k

We study here the small k behavior of the characteristic function $\hat{p}(\boldsymbol{k}) = FT[p(\boldsymbol{u})]$ of a general PDF $p(\boldsymbol{u})$ of a vectorial random variable $\boldsymbol{u}$ in d-dimensions. To do this we generalize the analysis given in Sect. 2.3.1 to the case in which $\hat{p}(\boldsymbol{k})$ cannot be expanded to all orders in Taylor series, i.e., when some higher order moment of $\boldsymbol{u}$ diverges. In order to make the treatment simple, we limit the discussion to the case in which $\boldsymbol{u}$ is a statistically isotropic variable, i.e., when $p(\boldsymbol{u})$ depends only on $u = |\boldsymbol{u}|$, and thus we write it as $p(u)$. Consequently also $\hat{p}(\boldsymbol{k})$ depends only on k and can be written as $\hat{p}(k)$. Let us assume that $p(u) \simeq Bu^{-\beta}$ at sufficiently large u with $B > 0$ ($\beta \to +\infty$ includes decay behaviors faster than any power). Note that in any case we require $\beta > d$ in order to have that $p(u)$ is a real PDF (i.e. an integrable function of $\boldsymbol{u}$).

First of all we rewrite the relation

$$\hat{p}(k) = \int d^d u\, p(u) e^{-i\boldsymbol{k}\cdot\boldsymbol{u}}, \qquad (A.1)$$

where the integral is over all space. Note that since $p(u)$ is a PDF its normalisation implies that

$$\hat{p}(0) = 1. \qquad (A.2)$$

We want to find the first correction to (A.2) at small k. In order to find it, let us partition the Fourier integral (A.1) into the sum of two terms: (1) the former is the integral over the sphere $\mathcal{C}(1/k)$ centered at the origin of radius $1/k$, and (2) the latter is the integral over the rest of space that we call $\overline{\mathcal{C}}(1/k)$:

$$\hat{p}(k) = \int_{\mathcal{C}(\frac{1}{k})} d^d u\, p(u) e^{-i\boldsymbol{k}\cdot\boldsymbol{u}} + \int_{\overline{\mathcal{C}}(\frac{1}{k})} d^d u\, p(u) e^{-i\boldsymbol{k}\cdot\boldsymbol{u}}. \qquad (A.3)$$

Let us now analyze the second term of (A.3) at small k. Since $p(u) > 0$, we can write

$$\left| \int_{\overline{\mathcal{C}}(\frac{1}{k})} d^d u\, p(u) e^{-i\boldsymbol{k}\cdot\boldsymbol{u}} \right| \leq \int_{\overline{\mathcal{C}}(\frac{1}{k})} d^d u\, p(u) \sim k^{\beta-d}, \qquad (A.4)$$

as $p(u) \simeq Bu^{-\beta}$ at large u.

A Scaling Behavior of the Characteristic Function

Let us now analyze the behavior of the first integral of (A.3). Since the integral is limited to a finite sphere, we can develop the complex exponential in Taylor series and take the sum outside the integral:

$$\int_{C(\frac{1}{k})} d^d u\, p(u) e^{-i\bm{k}\cdot\bm{u}} = \sum_{n=0}^{\infty} \frac{(-i)^n}{n!} \int_{C(\frac{1}{k})} d^d u\, p(u) (\bm{k}\cdot\bm{u})^n \; . \quad (A.5)$$

As $p(u)$ depends only on the modulus u, every term in the sum in (A.5) with odd n vanishes, and every other term with even n depends only on the modulus k, i.e.

$$\int_{C(\frac{1}{k})} d^d u\, p(u) e^{-i\bm{k}\cdot\bm{u}} = \sum_{n=0}^{\infty} \frac{(-1)^n A_{2n} k^{2n}}{(2n)!} \int_0^{\frac{1}{k}} du\, p(u)\, u^{2n+d-1} \; , \quad (A.6)$$

where

$$A_{2n} = \Omega_d d \int_{\Omega_d} d\Omega\, (\cos\theta)^{2n} > 0 \; ,$$

where Ω_d is the complete solid angle in d-dimensions, $d\Omega$ its infinitesimal, and θ the angle between $\bm{k}$ and $\bm{u}$.

Let us suppose that $d + m < \beta \le d + m + 1$ with m an arbitrary non-negative integer. It is simple to see that:

- All the integrals in the sum of (A.6) such that $2n \le m$ (i.e. $2n + d - \beta < 0$) converge to a finite positive constant in the limit $k \to 0$. In view of (A.4), the correction to this constant at small but finite k goes to zero as $k^{\beta-2n-d}$.
- Every integral with $2n > m$, (i.e. $2n + d - \beta \ge 0$) diverges in the same limit as $k^{-(2n+d-\beta)}$ (for the case in which $2n + d - \beta = 0$ exactly, the divergence in k is logarithmic).

These facts together with (A.4), allow us to conclude that, at small k to the first non-zero leading order,

$$\hat{p}(\bm{k}) \simeq 1 - Ak^\alpha \; , \quad (A.7)$$

where $\alpha = 2$ if $\beta > d + 2$ (i.e. if $\overline{u^2} < +\infty$). In this case, by direct analysis of the small k Taylor expansion of $\hat{p}(\bm{k})$, it is simple to show that $A = \overline{u^2}/2$. On the other hand, if $d < \beta \le d + 2$ we have $\alpha = \beta - d$, with $A > 0$ which depends on the singular part of the Taylor expansion. To find A, in this case, as a function of B and β, we note that is given by:

$$A = \lim_{k \to 0} k^{d-\beta} \left[1 - \int d^d u\, p(u) e^{-i\bm{k}\cdot\bm{u}} \right] \; .$$

Noting that $\int d^d u\, p(u) = 1$ and that $p(u)$ is a symmetric function, we finally can write:

$$A = B \int_\Omega d\Omega \int_0^{+\infty} dx\, x^{d-1-\beta} \left(1 - \cos(\hat{t}_d \cdot \bm{x})\right) \; ,$$

where Ω is the complete d-dimensional solid angle and $d\Omega$ its generic infinitesimal element, and $\hat{t}_d$ is the unit vector in the direction of the dth coordinate.

This kind of analysis can be generalized, in strict analogy with that given above, to find the first singular part of the Taylor expansion in the case $\beta > d + 2$.

A simple generalization is also possible for the case in which $p(\boldsymbol{u})$ depends on the modulus u of $\boldsymbol{u}$ only at sufficiently large values, which permits non-zero odd moments of the PDF.

B Fractal Algorithms

In this appendix we briefly describe some simple algorithms used to generate fractal point sets with a given dimension. We refer to Chap. 4 for a discussion of fractal geometry. We remark that, while changing the fractal dimension is straightforward, it is difficult to tune higher order correlations (or morphological properties) of the generated structures.

B.1 Cantor Set and Random Cantor Set

The construction of the *triadic* Cantor set is shown in Fig. B.1. We start with the closed interval $[0, 1]$. This starting form C_0 at the 0-th stage of the construction is called *the initiator*. The first stage of this construction consists in cutting the initiator in three equal pieces and deleting the middle third. This resulting C_1 is *the generator*. We then proceed by removing the middle thirds of the two remaining pieces, i.e. each line segment is replaced by a scaled down version of the generator. When we apply this cascade procedure ad infinitum we are finally left with a set of points $C = \lim_{k \to \infty} C_k$ which is the triadic Cantor set. Each piece of the set is, when we enlarge it appropriately, similar to the whole. This *scale-invariance*, which is rather obvious in this case as it is explicit in the construction, is called *self-similarity* under scale change. It is the basic feature of fractal structures.

The self-similarity in this example extends only to the length scale one. This *upper cut-off* can be easily removed by constructing an inverse cascade.

Fig. B.1. The first three iteration steps of the construction of the triadic Cantor set

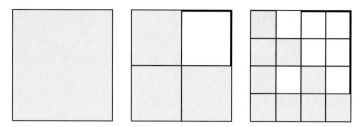

Fig. B.2. The first three iteration steps for the construction of the Sierpinski carpet

However in real cases (for example a computer generated cascade or fractals existing in nature) self-similarity is satisfied only between an upper and a lower cut-off.

One can create many different types of Cantor sets. Quite generally one can define a Cantor set that is completely determined by a generator which consists of b equal part of size $1/b$ and where n of these pieces are filled and $b - n$ are empty. For the triadic Cantor set one thus has $b = 3$ and $n = 2$.

In Fig. B.2 is shown the two-dimensional Cantor set also known as the *Sierpinski carpet*. The initiator here is a square. At the first stage of the construction one obtains the generator by dividing the square into four sub-squares of linear size $1/2$ of the original and deleting, for example, the one in the top right. At the next stage the three remaining sub-squares are replaced by three generators of linear size $1/2$ of the original and so forth. Of course one can build many different Sierpinski carpets through a generator that consists of a grid of b^2 sub-squares, n of them filled. For the triadic Cantor set we observe that if we choose $\epsilon = (1/3)^m$ ((with m an integer) then $N(\epsilon) = 2^m$ line segments of size ϵ are needed to cover the set. As a result we find for the box dimension of the triadic Cantor set

$$D_B = \lim_{m \to \infty} \frac{\log(2)^m}{\log(3)^m} = \frac{\log(2)}{\log(3)} = 0.6309 \ldots . \quad (B.1)$$

Similarly we find for the Sierpinski carpet that $N(\epsilon) = 3^m$ squares of linear size $\epsilon = (1/2)^m$ are needed to cover the set so that we find for the box dimension

$$D_B = \lim_{m \to \infty} \frac{\log(3)^m}{\log(2)^m} = \frac{\log(3)}{\log(2)} = 1.585 \ldots . \quad (B.2)$$

In general a d-dimensional Cantor set defined by a generator with n occupied boxes and $b^d - n$ empty boxes of size $1/b$ will have a box dimension

$$D_B = \frac{\log(n)}{\log(b)} . \quad (B.3)$$

We will now randomize the fractals introduced above and investigate their properties. A simple example of a random Cantor set is shown in Fig. B.3.

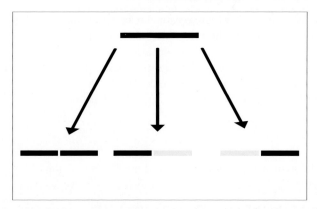

Fig. B.3. Schematic illustration of the construction of a random Cantor set. With probability p (*left*) both sub-boxes become occupied and with probability $1 - p$ (*center and right*) only one sub-box becomes occupied. These two last possibilities have probability $(1 - p)/2$, as the two different orientations of the empty box are possible

The construction of this set is defined by the three generators shown. At the first iteration the initiator is fragmented according to the first generator in $n_1 = 2$ segments of size $b = 1/2$ with a probability $p_1 = p$ or according to the other generators in $n_2 = 1$ segment of size $b = 1/2$ with probability $p_2 = 1-p$. At the next step the same procedure is repeated for the remaining segment(s) and so on. In general one can construct a random Cantor set using m generators. Each generator is defined on a one dimensional lattice with base b. The ith generator that consists of n_i full boxes and $b-n_i$ empty ones, has a probability of occurrence p_i. Similarly one can define a two dimensional (and three dimensional) random Cantor set with m generators, the ith generator consisting of a two dimensional grid with b^2 boxes with n_i filled and $b^2 - n_i$ empty (see Fig. B.4).

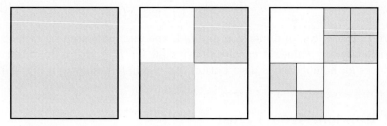

Fig. B.4. The first two iteration steps of a random Cantor set in $d = 2$. In this case $b = 1/2$ and the two generators have $n_1 = 2$ and $n_2 = 4$

The fractal dimension of these sets can be computed as follows. Consider an arbitrary realization at the kth iteration that consists of N_k boxes of linear size $\epsilon_k = b^{-k}$. If we take the fragmentation process one step further every box will fragment according to a generator i with probability p_i. If k is large enough so that $N_k(\epsilon_k) \gg 1$ one can make use of the law of large numbers so that we find for the number of boxes after $k+1$ iterations

$$N(\epsilon_k) = \sum_i n_i p_i N(\epsilon_k) \equiv \langle n \rangle N(\epsilon_k) . \tag{B.4}$$

If we take the logarithm of both sides and use the definition of box dimension, by considering that (B.4) is satisfied for $N(\epsilon) \sim \epsilon^D$, we see that

$$D_B = \frac{\log \langle n \rangle}{\log b} . \tag{B.5}$$

This result is easily proved equivalent to the self-averaging property. For the random Cantor set of Fig. B.3 we thus find

$$D_B = \frac{\log p_1 n_1 + p_2 n_2}{\log 2} = \frac{\log(1+p)}{\log(2)} . \tag{B.6}$$

B.2 Levy Flight

The Levy flight algorithm is a fractional random walk with a step size l drawn from a probability distribution [151]

$$p(l) = \begin{cases} 0 & \text{if } l < l_0 \\ D l_0^D l^{-(D+1)} & \text{if } l \geq l_0 \end{cases} . \tag{B.7}$$

This distribution is such that the probability that the step length λ is larger than l_0 is given by

$$p(\lambda > l) = \left(\frac{l}{l_0}\right)^{-D} \tag{B.8}$$

where D is the fractal dimension and l_0 is the minimum step size which can be scaled to any value. It can be shown that for $D \geq d - 1$ this algorithm produces an homogeneous distribution. An example is shown in Fig. 4.3

B.3 Random Trema Dust

The random trema dust (RTD) algorithm allows the construction of fractals with tunable dimensions and morphology [150, 151, 153]. In its simplest form it is given as follows in two dimensions. N points labelled $n = 1 \ldots N$ are distributed randomly in the unit square (or cube in d-dimensions). Using these points as centers one creates voids by removing circular (or spherical in d dimensions) regions of area (or volume in d dimensions) $C/2n$ where C is the co-dimension given by $C = d - D$, and D is the dimension of the fractal desired (and thus $0 < C < d$). Overlap is permitted of the voids with one another and with the region outside the unit square. For $C \leq 2$ (i.e. $D \geq d - 2$) it can be demonstrated that the probability of covering the unit square is zero as N tends to infinity, and the limit set is a fractal, up to an upper cut-off scale given by the size of the largest void. Numerically one generates a fractal with a finite lower cut-off after N iterations by filling the remaining volume randomly with a chosen density of points.

An important extension of the algorithm consists in replacing the circles forming the voids by regions, called *templates*, with a different geometry but the same area. Provided the area removed is conserved at each iteration any such modification is allowed. The morphological properties of the final set evidently depends on the choice of the template (while the dimension D remains unchanged). An example is the replacement of the circle by a chosen number of sub-circles (non-overlapping and with total area equal of that of the original circle). In numerical implementation of this algorithm (in two dimensions) one uses typically 10^5 to 10^6 iterations, which gives a fractal ranging over about three decades. In Fig. 4.2 is shown an example of a fractal generated in this way. It is clear that by increasing the number of sub-circles in which one fragments the area at each iteration step, more and more isotropic distributions are obtained. In this way the ratio between the size of the maximum void and the size of the sample can be reduced, while the scaling properties remain unchanged. As mentioned above scale-invariance is limited to the range of scales smaller than the size of the maximum void, above which one is biased by a finite size effect. This means that if we lower too much the size of the maximum void by fragmenting into too many sub-circles, then scaling is limited to a smaller range of scale. Therefore the "optimum" number of sub-circles must be a compromise between: (i) the total number of iteration steps, which determines the range of scale in which self-similarity is established, and (ii) the number of random points to be distributed in the regions not occupied by "voids". By varying these parameters one may obtain very isotropic structures which have uniform angular projections.

C Cosmological Models: Basic Relations

In Chap. 8 we have discussed the determination of statistical properties of the galaxy distribution in the approximation of a Euclidean space-time, which is a reasonable one at low redshifts. To extend this analysis to deeper galaxy samples one has to make assumptions about the geometry of space-time: the conversion of observational data, which give the angular coordinates, redshift and apparent magnitude of galaxies, into physical quantities such as the absolute magnitude and the distance depends on the cosmological model, which is canonically based on the Friedmann-Robertson-Walker (FRW) metric. Any given such model fixes the luminosity distance [206] as a function of redshift, in terms of the cosmological parameters characterizing it, i.e., standardly the Hubble constant H_0, the cosmological constant Λ and the matter density parameter Ω_M. Since the dependence of the redshift-distance relation on the Hubble constant is linear, an "incorrect" determination of H_0 corresponds to an overall rescaling of the distance scale which does not affect the statistical properties of the galaxy distribution. This is not true of Λ and Ω_0 since they induce a non-linear distortion of the distance scale. At small redshift (roughly $z < 0.1$) the corresponding corrections are not important, but they become so at increasing redshift.

This assumption of a particular set of cosmological parameters is important at different points in the data analysis: (i) in the projection from redshift space to real space (ii) in the computation of distances in real space (via e.g the curvature scale) and finally (iii) in the construction of a volume limited sample. In this Appendix we give the formulas needed for the application of such corrections. In Appendix D we consider the effect of such corrections on spatial galaxy counts.

In what follows we introduce some basic definitions and apply them to realistic cases. We do not enter here into the details of FRW models of the universe. We refer the interested reader to [17, 185, 206, 241] for a complete introduction to this subject.

C Cosmological Models: Basic Relations

C.1 Cosmological Parameters

The Hubble constant H_0[1] is the constant of proportionality between redshift and distance[2]:

$$z = \frac{H_0}{c}d \qquad (C.1)$$

where c is the speed of light. H_0 is usually written as $H_0 = 100h$ km sec^{-1} Mpc^{-1} where h is a dimensionless number which at present is measured to be in the range $0.55 < h < 0.75$ [226]. The inverse of the Hubble constant is the Hubble time t_H:

$$t_H = \frac{1}{H_0} = 9.8 \times 10^8 h^{-1} \text{yr} \qquad (C.2)$$

and the Hubble distance is defined as

$$D_H = \frac{c}{H_0} = 3000 h^{-1} \text{Mpc} . \qquad (C.3)$$

In FRW models there are two (dimensionless) parameters which determine the dynamics of the universe today, The mass density

$$\Omega_M \equiv \frac{8\pi G \rho_0}{3 H_0^2} \qquad (C.4)$$

where ρ_0 is the mass density of the universe today, and G Newton's constant of gravitation. The second is the cosmological constant

$$\Omega_\Lambda \equiv \frac{\Lambda c^2}{3 H_0^2} . \qquad (C.5)$$

The curvature of space is characterized by the parameter Ω_k which is related to the other two parameters by the constraint:

$$\Omega_k + \Omega_M + \Omega_\Lambda = 1 . \qquad (C.6)$$

Hence there are just two independent parameters which characterize the universe in these FRW models.

C.1.1 Comoving (Radial) Distance

The comoving (radial) distance D_C to an object observed at redshift z is given by

$$D_C = D_H \int_0^z \frac{dy}{E(y)} \qquad (C.7)$$

where

$$E(z) = \sqrt{\Omega_M (1+z)^3 + \Omega_k (1+z)^2 + \Omega_\Lambda} . \qquad (C.8)$$

All other distances can be simply expressed in terms of D_C.

---
[1] The subscript "0" refers to the present epoch.
[2] We use here the terminology and some of the results of [113], which gives a clear and compact review of distance measures in cosmology.

C.1.2 Comoving (Transverse) Distance

The comoving transverse distance D_M is simply related to the radial distance D_C:

$$D_M = \begin{cases} D_H \sinh\left(\sqrt{\Omega_k} D_C/D_H\right) & \Omega_k > 0 \\ D_C & \Omega_k = 0 \\ D_H \sin\left(\sqrt{|\Omega_k|} D_C/D_H\right) & \Omega_k < 0 \,. \end{cases} \quad (C.9)$$

The comoving distance between two objects at the same redshift but separated by an angle $\delta\theta$ is given by $D_M \delta\theta$. For $\Omega_\Lambda = 0$ there is an analytic solution for D_M:

$$D_M = D_H \frac{2\left[(2-\Omega_M)(1-z) - (2-\Omega_M)\sqrt{1+\Omega_M z}\right]}{\Omega_M^2 (1+z)} . \quad (C.10)$$

C.1.3 Luminosity Distance

The luminosity distance D_L is defined by

$$D_L = \sqrt{\frac{L}{4\pi S}} \quad (C.11)$$

where L the bolometric luminosity (i.e. integrated over all frequencies) and S is the apparent bolometric flux. There is a simple relation between D_L and D_M:

$$D_L = (1+z) D_M \,. \quad (C.12)$$

C.1.4 Magnitude

The apparent magnitude m of an object (in a photometric bandpass) is defined to be the ratio of the apparent flux of that source to the apparent flux of the bright star Vega. The distance modulus (DM) is defined as

$$DM = 5 \log\left(\frac{D_L}{10\mathrm{pc}}\right) . \quad (C.13)$$

It corresponds to the magnitude difference between an object's observed photometric flux and what it would be if it were at 10pc. Finally the absolute magnitude M is given by

$$m = M + DM + k(z) \quad (C.14)$$

where $k(z)$ is the k-correction (see Chap. 8).

C.2 Cosmological Corrections in the Analysis of Redshift Surveys

We consider three canonical cosmological models:

- FMD: "flat matter dominated" with $\Omega_{tot} = \Omega_M = 1$ ($\Omega_\Lambda = 0$)
- FLD: "flat Lambda dominated" with $\Omega_{tot} = 1$, $\Omega_M = 0.3$, $\Omega_\Lambda = 0.7$
- OBD: "open baryon dominated" with $\Omega_{tot} = \Omega_M = 0.05$ ($\Omega_\Lambda = 0$) ((i.e., a matter density equal to the baryon density inferred from nucleosynthesis).

Cosmological corrections enter in various places in the calculations described in Chaps. 8–12.

(1) The determination of absolute magnitudes, for the construction of VL samples. Here it is the luminosity distance (C.11) which enters in the relation between apparent and absolute magnitude (C.14):

$$M(z) = m(z) - 5\log D_L - 25 - k(z) \tag{C.15}$$

where D_L is in Mpc (C.12).

(2) The determination of the volume measure as a function of z. This enters when we distribute the "uniform" random sets for the measurement of correlation functions, and also when we examine the number counts to try to infer information about galaxy counts. The relevant quantity here is the comoving volume which is given by

$$dV_C = D_M^2 d\Omega dD_C \tag{C.16}$$

with respect to which the number densities of non-evolving objects should be constant.

(3) The determination of distances between points for the calculation of correlation functions (and in particular for the pair counts). Here we use again the comoving distance, but now measured between two points rather than with respect to the origin as for D_M.

C.2.1 Flat Cosmologies: FMD and FLD

For flat cosmologies we have $D_M = D_C$, and in terms of these the simple Euclidean volume element $dV_C = D_M^2 dD_M d\Omega$. Thus in these cases we need only calculate for each galaxy the comoving distance D_M given by

$$D_M = D_H \int_0^z \frac{dz}{\sqrt{\Omega_M(1+z)^3 + \Omega_\Lambda}} \tag{C.17}$$

which can be written as

$$D_M = D_H \int_0^z \frac{dz}{\sqrt{1 + \Omega_M(3z + 3z^2 + z^3)}}. \tag{C.18}$$

C.2 Cosmological Corrections in the Analysis of Redshift Surveys

Doing a Taylor expansion for $3\Omega_M z \ll 1$ we find

$$D_M = D_E \left[1 - \frac{3\Omega_M}{4}z - \frac{\Omega_M}{4}(1 - \frac{9\Omega_M}{4})z^2 + \ldots\right] . \tag{C.19}$$

where $D_E = D_H z$ is the Euclidean distance. We can rewrite this in terms of the "deceleration parameter" [206]

$$q_o = \frac{\Omega_M}{2} - \Omega_\Lambda = \frac{3}{2}\Omega_M - 1 \tag{C.20}$$

as

$$D_M = D_E \left[1 - \frac{1}{2}(1 + q_o)z + \frac{1}{6}(1 + q_o)(1 + 3q_o)z^2 + \ldots\right] . \tag{C.21}$$

For FMD we can use the Mattig relation which is the exact expression for (C.17) when $\Omega_\Lambda = 0$, and is valid also for the case $\Omega_{tot} \neq 1$:

$$D_M = D_H \frac{2\left[(2 - \Omega_M)(1 - z) - (2 - z)\sqrt{1 + \Omega_M z}\right]}{\Omega_M^2(1 + z)} \tag{C.22}$$

or equivalently, in terms of q_o it is

$$D_M = D_H \frac{[q_o z + (q_o - 1)(\sqrt{1 + 2q_o z} - 1)]}{q_o^2(1 + z)} \tag{C.23}$$

which Taylor expanded for $2q_o z \ll 1$ gives

$$D_M = D_E \left[1 - \frac{1}{2}(1 + q_o)z + \frac{1}{2}q_o(1 + q_o)z^2 + \ldots\right] . \tag{C.24}$$

In general the Mattig formula must agree with the exact expression we had above for the case $q_o = 1/2$, so one would expect (C.24) to agree with (C.21) above. The fact that they disagree at quadratic order is due to the fact that we have done a different expansion in the two cases (one before integration, the other after).

Hence we may proceed as follows. To calculate D_M use the Mattig relation in its full form for FMD (with $q_o = 0.5$) while for FLD use the expansion to linear order of the exact expression (C.21) i.e.,

$$D_M \approx D_E \left[1 - \frac{1}{2}(1 + q_o)z\right] \tag{C.25}$$

with $q_o = -0.55$. This agrees extremely well with the result obtained by numerical integration of the full expression even up to a redshift of almost $z \sim 1$ (plausibly given the smallness of the quadratic correction in (C.21)). Once we have D_M, the volume element in these coordinates is, as has been noted, the Euclidean one and (2) and (3) are dealt with completely by working everywhere with α, δ and $D_M(z)$ as spherical coordinates in flat space. It is in these coordinates that the random points of artificial catalogs, useful in many applications (e.g. when computing the correlation function by pair counting), should be generated, the number counts examined for their dimension, and the distances between points calculated (as the Euclidean distance in the corresponding rectilinear system x, y, z).

C.2.2 Open Model: OBD

First D_M is given by the Mattig relation, with $q_o = 0.025 \approx 0$. For (2) and (3) we need also D_C, which is related in an open cosmology to D_M by

$$D_M = \frac{D_H}{\sqrt{1-\Omega_M}} \sinh\left[\frac{\sqrt{1-\Omega_M} D_C}{D_H}\right]. \tag{C.26}$$

Now D_C is given by

$$D_C = D_H \int_0^z \frac{dz}{(1+z)\sqrt{1+\Omega_M z}} \tag{C.27}$$

which one can expand in $\Omega_M z$, and then integrate over z (without expanding also in z) to find, at linear order in $\Omega_M z$,

$$D_C = D_H \left[\left(1+\frac{1}{2}\Omega_M\right)\ln(1+z) - \frac{1}{2}\Omega_M z\right] \tag{C.28}$$

which expanded in z gives

$$D_C = D_H z \left[1 - \frac{1}{2}(1+q_o)z + \frac{1}{3}(1+q_o)z^2\right] \tag{C.29}$$

which expanded in the argument of the sinh $x = x - x^3/3! + ..$ keeping terms to order z^3 gives

$$D_M = D_H z \left[1 - \frac{1}{2}(1+q_o)z + \frac{1}{6}(1+4q_o)z^2 + ..\right] \tag{C.30}$$

which we note agrees with the expansion of the exact expression (C.21) for the flat case, up to terms of order $q_o^2 z^3$ which we have neglected here. This is so since they are both effectively expansions at small z and $q_o z$ of the Mattig relation.

D Cosmological and k-Corrections to Number Counts

In Appendix C we have given the formulas which are needed to compute the distance-redshift and magnitude-redshift relations for Friedmann-Robertson-Walker (FRW) models. In the analysis of real data which we have discussed in Chaps. 9–12 we have restricted ourselves to low redshifts at which the effect of these corrections is negligible. To extend our methods to higher redshift these corrections must of course be taken into account. As discussed briefly in Chap. 9 there are also other corrections which must be considered in this case – specifically the so-called k-corrections, as well as possible evolutionary corrections (which should take into account the effect of the intrinsic evolution of the objects being observed).

In principle the calculation of all these corrections is possible and the methods we have described can then be generalized relatively straightforwardly to much deeper samples. In this appendix we consider very briefly one question which is of importance when one undertakes such an analysis: to what extent are the results obtained sensitive to the corrections applied? Clearly the answer to this question needs to be quantified and taken into account when interpreting the results obtained.

To illustrate a little more we consider here the case of the exponent of clustering as inferred from number counts in real space, which we discussed in Chap. 10. Our conclusion is that, at larger redshift, the dimension inferred is in fact very sensitive to what is assumed about both cosmological and k-corrections. This is important in practice because the dimension observed in such counts (e.g. in the ESP survey [210]) have sometimes been used as evidence for homogeneity. We find that the uncertainty in the corrections can actually change the inferred dimension very significantly (see also [138] for more detail).

D.1 k-Corrections

In order to construct a volume limited (VL) sample we need to determine the absolute magnitude M of a galaxy at redshift z from its observed apparent magnitude m. This requires in general the assumption of a particular cosmological model and the application of an appropriate k-correction:

$$M = m - 5\log(D_L(z)) - 25 - k(z) \tag{D.1}$$

where now we explicitly use the luminosity distance (see Appendix C) $D_L(z) = D_M(1+z)$ in FRW models, where D_M is the comoving distance (see Appendix C). By the Euclidean case we mean the choice $D_L = (c/H_o)z$.

The k-correction is required to account for the fact that, because of redshift, the flux from a galaxy in the observed band of wave-lengths comes from a different (bluer) range of its emission spectrum. Depending on the spectral properties of the galaxy this means that it may appear to have an absolute magnitude brighter or fainter than it would if it were at lower redshift. The appropriate correction can in principle be determined from observations of the spectral properties of galaxies, and they have been calculated for various galaxy types as a function of redshift (e.g. see [87]). When applying them to a redshift survey, we usually need to make various assumptions as we do not have all the necessary information. In particular in most redshift surveys there has been no measure permitting the determination of galaxy types. This is currently changing radically with the advent of multi-band photometry. SDSS, in particular, observes in five photometric bands which allows an accurate direct determination of the k-correction.

D.2 k-Corrections and the Radial Number Counts

It is not difficult to understand how k-corrections – whether themselves correct or incorrect – can affect the radial number counts (Chap. 10) systematically. The number count from a single point in a VL sample corresponding to an absolute magnitude limit M_{lim} can be written schematically as

$$N^i(<R) = \int_0^R n^i(r) d^3r \int_{-\infty}^{M_{lim}} \phi(M,r) dM \tag{D.2}$$

where $n^i(r)$ is the galaxy density as seen from the origin (our galaxy), and $\phi(M,r)$ is the appropriately normalized luminosity function (LF) in the radial shell at r. If all corrections have been appropriately performed this function should be independent of r (neglecting evolution) i.e. the fraction of galaxies brighter than a given absolute magnitude is independent of r. In this case the second integral is just an overall normalization and the exponent of the number counts in the VL sample shows the behavior of the true (average) with $\langle N(<R)\rangle_p \sim R^D$ corresponding to the average behavior $\langle n(r)\rangle_p \sim r^{D-3}$ (for a fractal of dimension D, or homogeneous distribution with $D=3$). The effect of applying different k-corrections (or cosmological corrections, as we will see below) through the relation (D.1) is effectively to change the number count through the LF $\phi(M,r)$. Using an inappropriate correction will induce a spurious r dependence in this function, which can distort the relation between radial dependence of the number counts and the density.

When one applies a k-correction and observes a significant change in the number counts, there are thus two possible interpretations – (i) that one has applied the physical correction required to recover the underlying behavior for the galaxy number density, *or* (ii) that one has distorted the LF to produce a radial dependence unrelated to the underlying density. How can one check which interpretation is correct? Consider the effect of applying a k-correction which is *too large*. To a good approximation the net effect is a linear shift in the magnitude M with redshift, so that $M \to M - kz$. The second integral in (D.2) can in this case be written

$$\int_{-\infty}^{M_{lim}} \phi_p(M + kz)dM = \int_{-\infty}^{M_{lim}+kz} \phi_p(M)dM \qquad (D.3)$$

where ϕ_p denotes the physical, redshift independent LF: This is a function whose shape is well known to be fitted by a very flat power-law with an exponential cut-off at the bright end (see Chap. 11). If the upper cut-off of the integral is $M_{lim} + kz$, one can see from (D.2), taking $z \sim r$, that the number count picks up an additional contribution going as R^{D+1}, so that we expect the slope to increase by one power at some sufficiently large scale. As we go to the brighter end of the luminosity function, where it turns over, the fractional number of galaxies being added by the correction is even greater and we expect to see a growing effect on the slope with depth of sample. To illustrate the effect in a quantitative we have computed the integral (D.2) numerically for a distribution $\rho(r)$ which has an average $D = 2$ behavior in Euclidean coordinates, taking $\phi(M, r) = \phi_p(M + k(z))$, where $\phi_p(M)$ is the luminosity function and $k(z)$ the average k-correction of the type $k(z) = az$ where $a \simeq 3$ (see Fig. D.1)[138].

D.3 Dependence on the Cosmological Model

What about the dependence on cosmological model? The effect of changing cosmological model (and in particular the deceleration parameter q_o) is twofold: (i) it changes the relation between redshift and the co-moving distance in which we should see homogeneity, and (ii) it changes the absolute magnitude through relation (D.1). The effect of the former is relatively minor for the modest redshifts we consider here ($z < 0.3$), increasing the slope by at most about 0.2. The latter effect has the same form as that of the k-correction since, at linear order in redshift,

$$D_L = D_E \left[1 + \frac{1}{2}(1 - q_o)z + \ldots\right] \qquad (D.4)$$

where

$$D_E = (c/H_o)z \qquad (D.5)$$

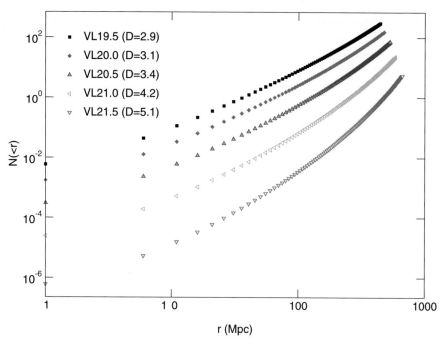

Fig. D.1. A plot of average number counts in a survey with the LF of the ESP survey ($\alpha = 1.2$ and $M^* = -19.6$), if the underlying distribution has fractal dimension $D = 2$ in Euclidean coordinates, with an unphysical shift to the LF corresponding to the linear k-correction discussed in the text. The different VL samples are labeled according to the cut in absolute magnitude (i.e. VL20.5 corresponds to a cut at $M = -20.5$). The distance r is that for the flat FRW cosmology ($q_o = 0.5$). At scales $r \gtrsim 300\,\mathrm{Mpc/h}$ we see that the effect is to produce an exponent $D \approx 3$, increasing systematically to larger values as we go to deeper VL samples with an absolute magnitude cut $M \gtrsim M^*$. (From [138])

is the Euclidean distance, and therefore from (D.1) we see that it is equivalent to an effective linear k-correction with $k_{eff} = 5(1 - q_o)/2 \ln 10$. So in the example shown in Fig. D.1, taking any FRW model with a sub-critical matter density essentially adds an even larger k-correction to the one already considered and leads to steeper slopes with an even more unstable behavior as a function of depth [138].

E Fractal Matter in an Open FRW Universe

E.1 Introduction

As discussed in Chap. 9 at small scales (up to $\sim 20\,\mathrm{Mpc/h}$) it is well established that the distribution of galaxies is fractal to a very good approximation. Beyond this scale, with currently available data, it is not possible to definitively conclude on whether this behavior continues – as suggested by the full analysis of number counts described in Chap. 10 – or whether there is indeed the clear convergence to homogeneity expected in standard cosmological models. It is evidently a question which must be settled empirically by future data, in particular by the full SDSS survey.

Despite the naturalness of envisaging the possibility that such a behavior might continue to larger scales, reflection in this direction is usually impeded by the conviction that it is incompatible with the very framework of standard theories (see e.g. [49]), and in particular with the high degree of isotropy of the cosmic microwave background radiation – CMBR – (i.e. [27, 58, 107]). With respect to the latter we note that in standard models the origin of radiation and baryonic matter is completely separate, with the latter being created in a dynamical process ("baryogenesis") completely distinct from the origin of the primordial radiation bath. The isotropy of the latter is therefore not fundamentally tied to the distribution of the matter, and the only real constraint is how much any such distribution actually perturbs the radiation. In this Appendix, which draws extensively on [139] (where further details can be found), we describe a model which illustrates that this supposed incompatibility of a fractal matter distribution, even extending to arbitrarily large scales, and the Friedmann-Robertson-Walker (FRW) framework is not well founded. The central point is that the self-gravity of a fractal, because it is asymptotically empty, becomes negligible at a finite scale compared to the effect of expansion associated to a homogeneous component. This means that it can actually be treated above a finite scale as a perturbation to a FRW cosmology. We discuss the case that the only homogeneous component is the radiation of the CMBR, and give an overview of the cosmological history of this model in which we suppose the baryonic matter is fractally distributed at all scales (in the initial conditions). While this specific simple model has since been ruled out by CMBR data at degree and smaller scales (which are not in accord with the observations of the the acoustic oscillations discussed in

Sect. 6.6.1) it is important conceptually and may be useful in further model building in this direction.

A fractal is a self-similar and intrinsically fluctuating distribution of points at all scales, which appears to preclude the description of its gravitational dynamics in the framework of the FRW solutions to general relativity [185]. The problem is often stated as being due to the incompatibility of a fractal with the cosmological principle, where this principle is identified with the requirement that the matter distribution be isotropic and homogeneous [49]. This identification is in fact very misleading for a non-analytic structure like a fractal, in which all points are equivalent statistically, satisfying what has been called the conditional cosmological principle [47, 150, 151, 153]. The obstacle to applying the FRW solutions has in fact solely to do with the lack of homogeneity. One of the properties of a fractal of dimension D, however, is that the average density of points in a radius r about any occupied point decreases as r^{D-3}, so that asymptotically the mass density goes to zero. An approximation which therefore *may* describe the large scale dynamics of the universe in the case that the matter has such a distribution continuing to all scales is given by neglecting the distribution of matter at leading order, relative to the small but homogeneous component coming from the CMBR. We will now show that is indeed a good perturbation scheme, and calculate the physical scale characterizing its validity.

E.2 Friedmann Solution in an Empty Universe

Consider first the standard FRW model with contributions from matter and radiation, for which the expansion rate is

$$H^2(t) = \left(\frac{\dot{a}}{a}\right)^2 = \frac{8\pi G}{3}(\rho_{\text{rad}} + \rho_{\text{mat}}) - k/a^2 \tag{E.1}$$

where $a(t)$ is the scale factor for the expansion, and $\rho_{\text{rad}} \propto 1/a^4$ is the radiation density, and $\rho_{\text{mat}} \propto 1/a^3$ the (homogeneous) matter density. The constant is

$$k = -H_o^2 a_o^2 (1 - \Omega_r - \Omega_m),$$

where H_o (a_o) is the expansion rate (scale factor) today and Ω_r (Ω_m) is the ratio of the radiation (matter) energy density today to the "critical" density $\rho_c = \frac{3}{8\pi G} H_o^2$. The sign of k determines whether the universe is closed ($k > 0$) or open ($k < 0$), with $k = 0$ corresponding to a "critical" spatially flat universe. Given the temperature of the CMBR [27], we have[1] $\Omega_r h^2 \approx 2.3 \times 10^{-5}$ (where h is the Hubble constant in units of $100\,\text{Mpc/km/s}$, with a typical measured value of $h \approx 0.65$ – see Chap. 8). If we make the simple

---

[1] We will neglect here, for simplicity, the minor modifications due to massless or low mass neutrinos, which can easily be incorporated in our analysis.

and natural assumption that galaxies trace the mass distribution the value of Ω_m depends directly on the determination of the scale of the cross-over to homogeneity. If the observed fractal distribution continues to a scale λ_0, above which it turns over to homogeneity, one has (see Chap. 9)

$$\Omega_m = \Omega_{10}\left(\frac{10}{\lambda_0}\right)^{3-D} \tag{E.2}$$

where Ω_{10} is the average density of matter (relative to critical) in a sphere of radius $10\,\mathrm{Mpc/h}$ about a galaxy, D is the fractal dimension, and λ_0 is measured in Mpc/h. For λ_0 sufficiently large that $\Omega_m < \sqrt{\Omega_r}$ the ρ_m term in (E.1) is always sub-dominant, and there is no matter dominated era. For simplicity we now consider the limit in which $\lambda_0 \to \infty$. Solving for the scale factor we then have

$$a(t) = a_o\sqrt{2H_o\Omega_r^{\frac{1}{2}}t}\sqrt{\left(1 + \frac{1-\Omega_r}{2\Omega_r^{\frac{1}{2}}}H_ot\right)} \tag{E.3}$$

which shows how the early time radiation dominated behavior $(a \propto t^{\frac{1}{2}})$ changes to the linear law $a \propto t$ at $t \approx 2H_o^{-1}\sqrt{\Omega_r}$ (redshift $z \sim 1/\sqrt{\Omega_r}$ where $1+z = a_o/a$). We now discuss how in each of these two phases (dominated respectively by the radiation and curvature) the fractal can be treated as a perturbation to this solution.

When we make numerical evaluations below we will use data from galaxy catalogs as discussed in Chap. 9, which give $D \approx 2$ and $\Omega_{10} = 0.007$. The latter assumes a mass to luminosity ratio of $10h$, in solar units, (that estimated for a typical spiral galaxy [81]). Note that since $\sqrt{\Omega_r} = 0.005h$ this value only requires $\lambda_0 > 10\,\mathrm{Mpc/h}$ for a direct transition from radiation to curvature domination. If instead we take the global mass to luminosity ratio to be that estimated in clusters ($\approx 300h$) we require $\lambda_0 > 250\,\mathrm{Mpc/h}$.

E.3 Curvature Dominated Phase

First consider the curvature dominated phase. The radiation is negligible and, at scales well within the horizon, we can use Newtonian gravity to describe the solution and its perturbations when the self-gravity of the matter is included. The leading solution is simply the free expansion of the fractal, with every point moving radially away from its neighbor at a constant velocity proportional to its distance i.e. $\dot{r} = H_o r(t_o) = H(t)r(t)$. To estimate the deviation from this flow due to the self-gravity of the fractal, we take a point in the flow and integrate the work done against gravity along its trajectory (in the leading order unperturbed flow). If the particle moves from an initial position $\boldsymbol{R_o}$, where it feels a total gravitational acceleration $\boldsymbol{F(R_o)}$, to a final position $\boldsymbol{R} = x\boldsymbol{R_o}$, this work done (per unit mass) is simply

$$W(\boldsymbol{R_o}, x) = \boldsymbol{F}(\boldsymbol{R_o}) \cdot \boldsymbol{R_o} \left(1 - \frac{1}{x}\right).$$

The integral is performed along the *unperturbed* trajectory using the fact that the force on the chosen point simply scales as $1/x^2$. A sufficient condition for the Hubble flow to apply to a good approximation at all subsequent times is simply that $W(\boldsymbol{R_o}, x = \infty)$ be much less than the kinetic energy of the particle i.e.

$$\boldsymbol{F}(\boldsymbol{R_o}) \cdot \boldsymbol{R_o} \ll \frac{1}{2} H_o^2 R_o^2. \tag{E.4}$$

Noting that the force at the origin of the flow was implicitly taken to be zero, we see that the validity of the criterion (E.4) will be determined in a fractal by *the difference between the gravitational force on two occupied points as a function of the distance between them*. The gravitational force on a point in a fractal has been discussed in Chap. 14. Its behavior can be understood as the sum of two parts, a local or "nearest neighbors" piece due to the smallest cluster (characterized by the lower cut-off ℓ in the fractal) and a component coming from the mass in other clusters. The latter is bounded above by the scalar sum of the forces

$$\langle |\boldsymbol{F}| \rangle \leq \lim_{L \to \infty} \int_\ell^L \frac{G\rho_m(r)}{r^2} 4\pi r^2 dr \sim L^{D-2} \tag{E.5}$$

so that for $D < 2$ it is convergent, while for $D > 2$ it may diverge. If there is a divergence, it is due to the presence of angular fluctuations at large scales, described by the three-point correlation properties of the fractal. For the difference in the force between two points the local contribution will be irrelevant well beyond the scale ℓ, while it is easy to see that the "far-away" contribution will now converge as L^{D-3}, and its being non-zero is a result of the absence of perfect spherical symmetry. Noting that as a function of distance R between points this component is bounded above by the same behavior as the force, we write

$$\langle |\boldsymbol{F}(\boldsymbol{R})|_{R \gg \ell} \rangle = A_3 \frac{GM(R)}{R^2} \propto R^{D-2} \tag{E.6}$$

where the pre-factor A_3 contains non-trivial information about the three-point correlation function of the fractal. These convergence properties of the relative force on two points are enough to draw a simple conclusion from the criterion (E.4): For a fractal in Hubble flow there is always a scale above which its evolution will be well described by continued Hubble flow for all subsequent times.

We now apply this to the Universe, and estimate the physical scale today R_o up to which the unperturbed "no matter" Hubble flow can be maintained right through the curvature dominated era. Given that this era begins at a redshift $z \approx 1/\sqrt{\Omega_r} \approx 200h$, we require

$$F(R)R|_{z=200h} \approx (200h)F(R_o)R_o < \frac{1}{2}H_o^2 R_o^2 \quad \rightarrow \quad R_o > 20\Omega_{10}(200h)A_3 \,. \tag{E.7}$$

What is observed is the Hubble flow with deviations (peculiar velocities) only at "cluster" scales ~Mpc. Taking our estimate for Ω_{10} we thus require $A_3 \lesssim 1/20h$. A fractal with a very weak three point correlation is one which has a very isotropic angular projection, so if the Universe is indeed a fractal a small value of A_3 would be expected. It is simple also to derive an expression for the peculiar velocity (small compared to the Hubble flow velocity v_H) which implies a simple linear relation, just as in standard perturbed homogeneous cosmology [218], between the local force and the velocity perturbation

$$(\Delta \boldsymbol{v}/v_H)(\boldsymbol{R}) \propto (\boldsymbol{F}(\boldsymbol{R})/R) \,.$$

This relation, for which there is apparently observational support, is usually used to determine an unknown constant (the "bias" factor [218] – see discussion in Chap. 13). In the present framework it can in principle be used to extract information about the total mass density and the constant A_3, which in turn can be related to angular data (and ultimately measured directly in forth-coming redshift surveys).

Here we have assumed that the fractal extends to arbitrarily large scales. For a finite λ_0 the analysis can be easily modified, by breaking the integrals at the appropriate scale. The part from scales greater than λ_0 will give a contribution which can be re-absorbed within the Hubble flow, while the perturbation will maintain the same scaling at smaller scales. The central point which we emphasize is that the scale R_o is only indirectly related in this case to the homogeneity scale, *and remains finite as* $\lambda_0 \to \infty$.

We have thus seen that an open FRW universe is always a good approximation beyond some finite scale if matter is distributed as a simple fractal up to an arbitrarily large scale. In particular such an open model – because it is dominated by the kinetic energy of the Hubble flow – can explain naturally how large structures can co-exist with an almost perfect Hubble flow. We further note a few other of its striking features: *(i)* Since the Universe is to a good approximation in completely free expansion at large scales with $a(t) \propto t$, we have a deceleration parameter $q_o \approx 0$. This is a good fit to recent supernovae observations [187]. Rather than being due to the effect of an unknown "anti-gravitational" component which mysteriously cancels the decelerating effect of the matter on the expansion, the effect is due to the decay toward zero of the matter density on such scales. *(ii)* The expansion age of the Universe is $t_o = H_o^{-1} \approx 10h^{-1} \approx 15$ billion years, larger by 50% than in the standard matter dominated case. This value is comfortably consistent with the estimated age of globular clusters (the oldest known astrophysical objects) 11.5 ± 1.3 billion years [42]. *(iii)* The size of the horizon today is $R_H(t_o) \approx -\frac{1}{2}cH_o^{-1}\ln\Omega_r \approx 20,000$ Mpc/h, a factor of about three larger than in the standard case.

E.4 Radiation Dominated Era

So far we have considered the model only in the curvature dominated era i.e. back to redshift $z \approx 200h$. For this last point above however we have extrapolated the model back to the radiation dominated era, assuming that the effect of the matter distribution can also be consistently treated as a correction in this epoch to the FRW solution without matter. We now justify this assumption, and then discuss some of its consequences for a specific cosmology of this type. For the former we simply treat the fractal as a set of perturbations to the energy density in a manner analogous to the way such perturbations may be treated in the standard framework. There the criterion one would use to apply the uniform Hubble flow to describe the growth of the horizon is simply that such perturbations be small at the horizon scale i.e. $\frac{\delta\rho}{\rho}|_{hor}$ be small (where ρ is the homogeneous energy density i.e. that in the radiation). In a fractal perturbations are non-analytic and $\delta\rho$ has no meaning as defined in the standard case. We can however write down the mean mass (or energy) at the scale of horizon about an occupied point. Taking this as the appropriate $\delta\rho_{hor}$ is clearly the right adapted criterion, as the fractal is simply made of voids and structures, and voids clearly will not perturb the flow. We thus require that the fractal obey

$$\delta_H(z) \equiv \frac{\rho_{mat}}{\rho_{rad}}|_{hor}(z) = \frac{1}{1+z} \frac{\Omega_{10}}{\Omega_r} \left(\frac{10}{R_H(z)}\right)^{3-D} < 1 \quad (E.8)$$

where for ρ_{mat} we have taken the mass inside the (comoving) horizon $R_H(z)$ at redshift z. In the radiation dominated era (for $z > 1/\sqrt{\Omega_r}$) we have $R_H(z) \approx \frac{cH_0^{-1}}{z\sqrt{\Omega_r}}$. At $z = 10^3$ this gives $R_H \approx 600\,\mathrm{Mpc}$ so that, for a fractal with $D = 2$, we have at this redshift $\delta_H \approx \Omega_{10}$. In order that the fractal matter be indeed a small correction at this redshift we require, approximately, $\Omega_{10} < 1$, which holds comfortably even if there is much more dark matter than we assumed in obtaining our estimate $\Omega_{10} \approx 0.01$. We can thus continue to use the FRW solution back to an arbitrary redshift for a fractal distribution of matter extending to the corresponding scales, provided that the condition (E.8) holds.

To make a link with central observations in cosmology such as the CMBR and nucleosynthesis, we need to specify a precise model. In the spirit of this approach we now consider here a radical (but very simple) possibility for a cosmology which makes use of the results we have presented: We consider a universe which at very early times (deep in the radiation dominated era) is a radiation bath at a given temperature with superimposed *fractal perturbations in baryon number* up to the arbitrarily large scale λ_0, and down to some scale Λ. Note that positing a very different distribution for matter and radiation does not represent a loss of simplicity in comparison to standard models, which generically envisage the (almost) homogeneously distributed matter as coming from a dynamical process ("baryogenesis") completely distinct from the origin of the primordial radiation bath. Instead of fixing the initial condition on a homogeneous baryon to photon ratio ($n_B/n_\gamma \sim 10^{-9}$), with some independent superimposed spectrum of analytical fluctuations, we specify our fractal in baryon number from the properties observed in the distribution of visible matter at large scales today. This is the appropriate normalization given that these perturbations are simply frozen at all but very small scales in the curvature dominated era. In particular we take $D = 2$ and the normalization of the mass given by Ω_{10}. Interestingly, stated in terms of the parameter δ_H above, these values correspond to the special case that δ_H is constant, and of order one. Below the lower cut-off scale ℓ we take the distribution to be smoothed, with the corresponding density $\Omega_{10}(10/\ell)$ (where ℓ is the comoving scale given in units of Mpc/h). A natural lower bound for this scale is that characterizing the baryon diffusion (up to the corresponding time), which will smooth any inhomogeneity at smaller scales.

E.5 Fluctuations in the CMBR

The fluctuations in the temperature of the CMBR depend on the "intrinsic" fluctuations imprinted at time of decoupling of the photons, plus the fluctuations induced in their propagation from their last scattering. We consider here only the former as they are typically the dominant effect for perturbations which are essentially frozen on the scales we are interested in. Relative to the standard critical or near critical mass universe there are two main features to note here. First, photon decoupling will be modified greatly: While in the standard case there is a global baryon density which determines the time/temperature of decoupling, here the relevant baryon density varies enormously – for a photon in a void it is zero, while for one in a structure it is the local density of baryons associated with the lower cut-off scale ℓ, a density which can be several orders of magnitude greater than in the standard case (if we take ℓ as small as the baryon diffusion distance). However, since the decoupling temperature is only logarithmically sensitive to this parameter, the decoupling of photons in structures will still occur around redshift $z \sim 10^3$. On the other hand, if the scale λ_0 is so large that there are voids of the order of the horizon scale at this time (at $z = 10^3$ we found $R_H \approx 600\,\text{Mpc}$), most photons will decouple at the much earlier time of electron-positron annihilation since after this time they find themselves in a neutral environment. Thus the "thickness" of the surface of last scattering will be very much greater than in the standard case, essentially consisting of two stages of "void decoupling" (at a redshift of $\sim 10^9$) and "structure decoupling" at redshifts comparable to the standard one.

The other main difference relative to the standard case is that, because of the extremely low background density in the model, the effect of the hyperbolic geometry is much greater. In particular, at high redshift ($z\sqrt{\Omega_r} \gg 1$), the angle θ subtended by a scale of given physical size ℓ_θ is

$$\theta \approx \ell_\theta / H_o^{-1} \sqrt{\Omega_r} \,,$$

which means that the physical scale corresponding to $1°$ on the sky is 10^4 Mpc i.e. of the order of the horizon scale today, and a factor of about 100 larger than in the standard case. Considering the possibility that the fractal extends to such enormous scales, we can make a naive estimate of the amplitude of fluctuations on the microwave sky: Adapting the Sachs-Wolfe formula as derived for the standard case of analytical fluctuations, the *maximum* amplitude of temperature variation between two photons sub-tending some angle should be estimated by the energy density fluctuation represented by a structure at the corresponding physical scale. At $10°$ on the sky we then have

$$\frac{\delta T}{T}\Big|_{10°} \sim \frac{1}{3}\frac{\delta\rho}{\rho}(\ell_{10}, z_d) = \frac{1}{3}\frac{\rho_B(\ell_{10})}{\rho_{radn}(z_d)} = \frac{1}{1+z_d}\frac{\Omega_B(\ell_o)}{3\Omega_r}\Big|_{t_o} \approx \frac{\Omega_{10}}{z_d} \quad (E.9)$$

where z_d is the decoupling redshift. Thus, the largest effect will come from the photons which decouple last, for which this maximum amplitude will be $\sim 10^{-5}$, which is comparable to the average amplitude observed at this scale by COBE [27]. On the other hand, for a modest value of the crossover scale to homogeneity (e.g. $\lambda_0 \sim 100$ Mpc/h), the effect of the fractal distribution of matter on the photons at decoupling will only be visible at scales much smaller even than those which will be probed future missions. A detailed study of the perturbations induced by propagation of photons through an expanding structure of this kind – requiring techniques quite different to the standard ones treating analytical perturbations – will be required to see if it is possible to produce perturbations at the levels observed by experiment at the angles which have been probed to date.

E.6 Other Remarks

Finally a few brief comments on nucleosynthesis and structure formation. The cooling rate of the plasma is the same as in the standard case, and the results of nucleosynthesis will depend on the local baryon to entropy ratio, which as discussed above is related to the scale Λ. If this scale is larger or comparable to the horizon scale at nucleosynthesis the amount of helium produced will not differ much from the standard case – it is essentially independent of the baryon to entropy ratio – while the residual densities of deuterium etc. will be lower. For a smaller (i.e. sub-horizon) Λ the effect of inhomogeneities will be important, but a reliable calculation for the effect becomes very difficult to perform. It is unlikely however to modify the tendency for lower values of the "trace" elements, since this arises due to the fact that the elements are synthesized in denser regions compared to the standard case. If, on the other hand, the fractal (up to a finite sub-horizon λ_0) is formed after nucleosynthesis, the appropriate value for the local density would be that at λ_0. We note that for modest values of this scale ($\lambda_0 \sim 50 \div 100\,\mathrm{Mpc/h}$) this would correspond to standard nucleosynthesis if there is ratio of dark to visible baryons comparable to the mass to light ratio inferred in clusters. Clearly some physics quite different to that at work in standard models would be required to make it possible to generate such a structure between nucleosynthesis and the curvature dominated era when (as we have noted) the structure gets frozen in at all but small scales.

F Errors in Full Shell Estimators

We calculate here the typical errors for the full shell (or full-sphere) estimators of the conditional density and reduced two-point correlation function, and for the average density and unconditional mass variance. We derive explicit formulas for the Poisson distribution, which will apply also in the case of weakly correlated distributions. In Appendix F we discuss some of the common (other than full-shell) estimators used in the literature.

We suppose that we have a finite sample of N particles in a volume V, extracted from a point distribution which is statistically stationary and isotropic (Chap. 2). Given a certain function F of the number density $n(\mathbf{r})$, let us call as usual $\langle F \rangle$ the ensemble average and $\overline{F}$ the volume average in the finite sample. Analogously we call $\langle F \rangle_p$ the *conditional* ensemble average and $\overline{(F)}_p$ the *conditional* volume average in the sample. The quantities $\overline{F}$ and $\overline{(F)}_p$ are the statistical estimators of $\langle F \rangle$ and $\langle F \rangle_p$ respectively. Moreover we assume that the quantities $\langle n \rangle$, $\xi(r)$ and the other correlation functions are self-averaging quantities of the ensemble (i.e. volume averages in the infinite volume limit are equal to the ensemble averages). We call $\bar{n}$, $\Gamma_E^*(r)$, $\Gamma_E(r)$ and $\xi_E(r)$ the estimators in the sample of $\langle n \rangle$ $\Gamma^*(r)$, $\Gamma(r)$ and $\xi(r)$ respectively. We analyze now the errors between these estimators and their real (ensemble) averages.

F.1 Bias and Variance of Estimators

In general a statistical estimator X_V of an average quantity X (defined in a single realization by the infinite volume limit) in a volume V is simply a quantity calculable from the finite sample. The only condition which it must satisfy, in order to be a valid estimator, is that it must approach the true ensemble value of X as the sample boundaries go to infinity i.e.

$$\lim_{V \to \infty} X_V = X \ . \tag{F.1}$$

Another, stronger, condition one can impose on an estimator is that its ensemble average *in the finite volume* be equal to the ensemble average i.e.

$$\langle X_V \rangle = X \ . \tag{F.2}$$

F Errors in Full Shell Estimators

An estimator is referred to as *unbiased* if this condition is satisfied. If it is not satisfied it means that there is some systematic offset in the finite volume relative to the real ensemble value (*bias*). Typically we are estimating quantities which depend on a length scale (e.g. a two-point correlation function as a function of separation), and such bias will then depend on the ratio of this scale to the characteristic scale (or scales) of the sample. Whether there is bias in an estimator, and what it is, depends on the statistical properties of the ensemble, and must be calculated for any given case. For measurement purposes it is obviously important to know the possible biases of the estimator in consideration. It is usually only possible to calculate such biases analytically in very particular cases (e.g. Poisson process). The variance of an estimator is simply $\langle X_V^2 \rangle - \langle X_V \rangle^2$, i.e. it measures fluctuations with respect to $\langle X_V \rangle$ (whether biased or not). The bias of an estimator can be supposed irrelevant if it is small compared to this variance.

F.2 Unconditional Average Density

The most common estimator of the unconditional average density is evidently

$$\bar{n} = \frac{N}{V}.$$

Note that, even applied to an essentially uniform distribution with a well defined mean density $\langle n \rangle$, this is generically a biased estimator: since there is always a point (our galaxy) at the apex of the sample, the average density measured in this way is actually the conditional density. It is simple to show that

$$\langle \bar{n} \rangle = \langle n \rangle \left[1 + \frac{1}{V} \int_V d^3r \; \tilde{\xi}(r) \right]. \tag{F.3}$$

This is related to the integral constraint on the estimator of the reduced two-point correlation function $\tilde{\xi}(r)$ discussed in Sect. 6.5 of Chap. 6. Only for the case of a Poisson distribution it is therefore unbiased. Evidently the effect becomes relatively less important as the scale of the sample increases. On scales sufficiently larger than the homogeneity scale λ_0 the effect becomes negligible on the amplitude of the correlation function, but it may still affect the estimation of the correlation length r_c (see Chap. 9).

Let us evaluate the error between $\bar{n}$ and $\langle n \rangle$, neglecting this bias i.e. neglecting the effect of the implicit condition that the apex of the sample is an occupied point. The ensemble variance of the number of particles in a volume V is (see Chap. 2):

$$\langle N^2(V) \rangle - \langle N(V) \rangle^2 = \langle n \rangle^2 \int_V d^3r \int_V d^3r' \; \tilde{\xi}(|\boldsymbol{r} - \boldsymbol{r}'|). \tag{F.4}$$

Our estimate of the error $\delta \bar{n} = E(\bar{n}, \langle n \rangle)$ is then given by

$$E^2(\overline{n}, \langle n \rangle) \simeq \frac{\overline{n}^2}{V^2} \int_V d^3r \int_V d^3r' \, \tilde{\xi}_E(|\boldsymbol{r} - \boldsymbol{r}'|) \, . \tag{F.5}$$

In the Poisson case we simply have $\tilde{\xi}(r) = \tilde{\delta}(\boldsymbol{r})/\langle n \rangle$ and then $\delta \overline{n} \simeq \frac{\sqrt{N}}{V}$ or

$$\frac{\delta \overline{n}}{\overline{n}} \simeq \frac{1}{\sqrt{N}} \, .$$

An alternative estimator is given by the mass density in the largest sphere which can be inscribed in the sample (which coincides only for a spherical sample). It is biased for the same reason in the previous case (but with a bias which is more difficult to quantify). Its variance for the Poisson case is given by the same expression, except that N is now the average number of points in the sphere. Thus it has a larger variance.

F.3 Conditional Number of Points in a Sphere

The average number of points in a sphere $C(R)$ of radius R around an occupied point can, in general, be written as (see Chap. 2)

$$\langle N(R) \rangle_p = \langle n \rangle \, \|C(R)\| + \langle n \rangle \int_{C(R)} d^3r \xi(r) \tag{F.6}$$

where $\|C(R)\| = (4\pi/3)R^3$ is the volume of the sphere. The sample estimate of $\langle N(R) \rangle_p$ is

$$\overline{(N(R))}_p = \frac{1}{N_c(R)} \sum_{i=1}^{N_c(R)} N_i(R) \, , \tag{F.7}$$

where $N_i(R)$ is the number of points in a sphere of radius R around the point i, and $N_c(R)$ is the number of "centers" over which one takes the sample average, i.e for which the sphere of radius R is completely included in the sample volume. For example, in a cubic volume of size L, for an homogeneous point distribution with homogeneity scale much smaller than R, one has approximately

$$N_c(R) = \overline{n}(L - 2R)^3 \simeq N \left(1 - \overline{n} \frac{6R}{L} \right) \, . \tag{F.8}$$

In other geometrically regular samples there is a similar behavior with a sharp decrease of $N_c(R)$ when R approaches the sample size. The quantity $\sqrt{\langle N^2(R) \rangle_p - \langle N(R) \rangle_p^2}$ gives the order of the difference between a single $N_i(R)$ and $\langle N(R) \rangle$. Using (2.59) one obtains

$$A^2(R) \equiv \langle N^2(R) \rangle_p - \langle N(R) \rangle_p^2 = \tag{F.9}$$

$$\langle n \rangle^2 \int_{C(R)} d^3r \int_{C(R)} d^3r' \left[\tilde{\xi}(|\boldsymbol{r} - \boldsymbol{r}'|) + \tilde{\zeta}(r, r', |\boldsymbol{r} - \boldsymbol{r}'|) - \xi(r)\xi(r') \right] \, ,$$

where $\tilde{\zeta}(r, r', |\boldsymbol{r} - \boldsymbol{r}'|)$ is the reduced three point correlation function.

In order to estimate the characteristic deviation $\overline{\delta(N(R))}_p$ of $\overline{(N(R))}_p$ with respect to $\langle N(R)\rangle_p$, we have to distinguish two regimes:

1. If R is sufficiently small so that the spheres around the $N_c(R)$ centers do not overlap significantly (in practice $R < \overline{n}^{-1/3}$, with the condition $\xi(r) \ll 1$ on these scales), then we can write:

$$E^2\left(\overline{(N(R))}_p, \langle N(R)\rangle_p\right) \simeq \frac{A^2(R)}{N_c(R)}. \tag{F.10}$$

2. If the spheres overlap significantly and $N_c(R)\|C(R)\| > V$ (in practice $R > \overline{n}^{-1/3}$ again with the condition $\xi(r) \ll 1$ on these scales) then spheres with centers at a distance $\Delta < R$ see almost the same set of particles. For this reason, in this case the *effective* number of centers will be $\approx V/\|C(R)\|$. Therefore

$$E^2\left(\overline{(N(R))}_p, \langle N(R)\rangle_p\right) \simeq \frac{\|C(R)\|}{V}A^2(R). \tag{F.11}$$

In a Poisson distribution $\tilde{\zeta}(r, r', |\mathbf{r}-\mathbf{r}'|) = \tilde{\xi}(r)\tilde{\xi}(r')$ (see Chap. 2), and therefore in this case one can write:

$$\overline{\delta(N(R))}_p \simeq \begin{cases} \sqrt{\frac{\overline{n}}{N_c(R)}}\|C(R)\| \sim R^{3/2} & \text{if } R < \overline{n}^{-1/3} \\ \frac{\|C(R)\|}{V}\sqrt{N} \sim R^3 & \text{if } R > \overline{n}^{-1/3} \end{cases}. \tag{F.12}$$

F.4 Integrated Conditional Density

By definition (see Chap. 4) we have that the integrated conditional density (i.e. the conditional density measured in balls rather than in shells) can be written as (see Fig. F.1)

$$\Gamma^*(R) = \frac{\langle N(R)\rangle_p}{\|C(R)\|}.$$

Consequently its estimator takes the form

$$\Gamma_E^*(R) = \frac{\overline{(N(R))}_p}{\|C(R)\|}.$$

Therefore, recalling the discussion in the previous section, it is very simple to see that:

1. If the spheres do not overlap ($R < \overline{n}^{-1/3}$ when $\xi(r) \ll 1$) then they can be considered as independent and one has

$$\delta\Gamma_E^*(R) = \sqrt{\frac{1}{N_c(R)\|C(R)\|^2}A^2(R)}. \tag{F.13}$$

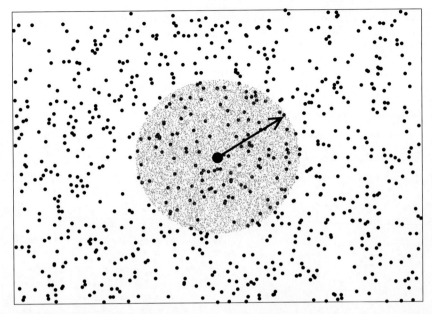

Fig. F.1. Estimation of the integrated conditional density: given a point of the distribution one counts the number of points contained in a ball of radius r and divides it by the volume $V(r)$ of the ball. Repeating the procedure for all points in the sample for which the ball of radius r is inside the sample volume (full-shell estimator), and making the average of all these determinations, one obtains the estimation of $\Gamma^*(r)$

2. In the case in which the spheres overlap and $N_c(R)\|C(R)\| > V$ ($R > \bar{n}^{-1/3}$ when $\xi(r) \ll 1$) then

$$\delta \Gamma_E^*(R) = \sqrt{\frac{\|C(R)\|}{V} A^2(R)} \ . \tag{F.14}$$

Repeating the arguments leading to (F.12) we have for a Poisson distribution

$$\delta \Gamma_E^*(R) \simeq \begin{cases} \sqrt{\frac{\bar{n}}{N_c(R)\|C(R)\|}} \sim R^{-3/2} & \text{if } R < \bar{n}^{-1/3} \\ \frac{1}{V}\sqrt{N} = \delta\bar{n} \sim \text{const} & \text{if } R > \bar{n}^{-1/3} \end{cases} \tag{F.15}$$

F.5 Conditional Average Density in Shells

The function $\Gamma(r) = \langle n(\mathbf{r}) \rangle_p$ can be written as

$$\Gamma(r) = \frac{\langle N(r+\Delta r) \rangle_p - \langle N(r) \rangle_p}{\|C(r, \Delta r)\|} \ , \tag{F.16}$$

where $C(r, \Delta r)$ is the spherical shell of radius r and thickness Δr, and $\|C(r, \Delta r)\| = \left[\frac{4\pi}{3}(r + \Delta r)^3 - \frac{4\pi}{3}r^3\right] \simeq 4\pi r^2 \Delta r$ is its volume. Its estimator is (see Fig. F.2)

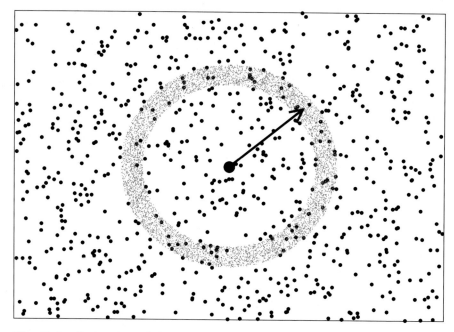

Fig. F.2. Estimation of the conditional density in shells: given a point of the distribution one counts the number of points contained in a spherical shell of radius r and thickness Δr and divides it by the volume $\Delta V(r)$ of the shell. Repeating that procedure for all points in the sample for which the shell is inside the sample volume (full-shell estimator), and making the average of all these determinations, one obtains the estimation of $\Gamma(r)$

$$\Gamma_E(r) = \frac{\overline{(N(r + \Delta r))}_p - \overline{(N(r))}_p}{\|C(r, \Delta r)\|} = \frac{1}{N_c(r + \Delta r)} \sum_{i=1}^{N_c(r+\Delta r)} \Gamma_i(r), \quad (F.17)$$

where $\Gamma_i(r) = (N_i(r + \Delta r) - N_i(r))/\|C(r, \Delta r)\|$ is the density of points in the shell $C(r, \Delta r)$ around the center i[1].

Very often logarithmic shells $\Delta r = \alpha r$ with $\alpha \ll 1$ are used. Analogously to the case of $\Gamma_E^*(r)$, the typical fluctuation of $\Gamma_i(r)$ from a single center and the "real" $\Gamma(r)$ is given by:

---

[1] More precisely this is the estimator of the conditional density at a distance $r + \Delta r/2$. However in what follows we neglect this correction, as it is irrelevant for the results on the noise estimation.

$$\delta\Gamma_i(r) \simeq \frac{\sqrt{B(r)}}{\|C(r,\Delta r)\|}, \qquad (F.18)$$

where

$$B(r) \equiv \langle\Delta N^2(r)\rangle_p - \langle\Delta N(r)\rangle_p^2 = \qquad (F.19)$$

$$\langle n\rangle^2 \int_{C(r+\Delta r)} d^3 r_1 \int_{C(r+\Delta r)} d^3 r_2 \left[\tilde\xi(|\boldsymbol{r}-\boldsymbol{r}'|) + \tilde\zeta(r,r',|\boldsymbol{r}-\boldsymbol{r}'|) - \tilde\xi(r)\tilde\xi(r')\right].$$

We can repeat the arguments developed for $\Gamma_E^*(r)$ and thus we find:

1. If the shells overlap only slightly [2]

$$N_c(r+\Delta r)\|C(r,\Delta r)\| < V,$$

and we find from (F.17)–(F.18)

$$\delta\Gamma_E(r) = \sqrt{\frac{\delta\Gamma_i(r)}{N_c(r+\Delta r)}}. \qquad (F.20)$$

2. If the shells overlap significantly and $N_c(r+\Delta r)\|C(r,\Delta r)\| > V$ then the "effective" number of centers is $V/\|C(r,\Delta r)\|$. Hence

$$\delta\Gamma_E(r) = \sqrt{\frac{\delta\Gamma_i(r)\|C(r,\Delta r)\|}{V}}. \qquad (F.21)$$

Note that now the crossover between the two regimes occurs at a distance r_{co} larger than in the case of $\Gamma^*(r)$ and that it is a function also of Δr. Again if one has good estimators of ξ and ζ one can deduce $\delta\Gamma_E(r)$ for any kind of stationary and uniform distribution.

In analogy with the case of $\Gamma_E^*(r)$ we find in Poisson-like distributions:

$$\delta\Gamma_E(r) \simeq \begin{cases} \sqrt{\frac{\overline{n}}{N_c(r+\Delta r)\|C(r,\Delta r)\|}} \sim r^{-3/2} & \text{if } r < r_{co} \\ \frac{1}{V}\sqrt{N} = \delta\overline{n} \sim \text{const.} & \text{if } r > r_{co} \end{cases} \qquad (F.22)$$

As in the case of $\Gamma_E^*(r)$, in the first regime, if $\Delta r \ll r$, $N_c(r+\Delta r) \simeq N(1 - 6r/V^{1/3}) \simeq N$. Therefore r_{co} is given approximately by the equation

$$r_{co}^2 = \frac{1}{4\pi\overline{n}\Delta r}. \qquad (F.23)$$

This is valid also in the case of non-zero but weak correlations. Note that the Poisson case can be considered a lower limit for any correlated stationary and isotropic distribution far from super-homogeneity.

---

[2] In a weakly correlated homogeneous distribution this means roughly

$$N\left(1 - \frac{6r}{L}\right)4\pi r^2 \Delta r < V.$$

F.6 Reduced Two-Point Correlation Function

One estimator of $\xi(r)$ in full shells is given simply by

$$\xi_E(r) = \frac{\Gamma_E(r)}{\bar{n}} - 1 . \tag{F.24}$$

The error (or absolute difference) between $\xi_E(r)$ and $\xi(r)$, can be found by the propagation of errors:

$$\delta\xi_E(r) \simeq \frac{\delta\Gamma_E(r)}{\bar{n}} + \frac{\Gamma_E(r)}{\bar{n}^2}\delta\bar{n} . \tag{F.25}$$

Therefore we can write

$$\xi_E(r) \simeq \xi(r) + \frac{\delta\Gamma_E(r)}{\bar{n}} + (\xi_E(r) + 1)\frac{\delta\bar{n}}{\bar{n}} . \tag{F.26}$$

The first perturbative term is an oscillating (noise-like) term and the second is a "shift plus a stretch". Note that both terms have the same amplitude over most of the range, with the first typically dominating at small separations (e.g. for the Poisson case (F.22)). Therefore when the signal $\xi_E(r)$ is smaller than this noise-like term we get a bad estimation of the real $\xi(r)$.

To illustrate these results more explicitly we generate a Poisson distribution of 10^6 points in a cubic box of unitary side. We then compute the estimators of the conditional density and of the reduced two-point correlation function $\xi_E(r)$ by using (F.17) and (F.24). In Fig. F.3 we show the result, together with the estimation of the errors as explained above. There is very good agreement both in the region where $r < \bar{n}^{-1/3}$ and $r > \bar{n}^{-1/3}$ where the plateau shown by (F.22) is evident.

Let us now discuss the effect of the integral constraint, which is a boundary condition imposed on the estimator $\xi_E(r)$ due to our ignorance of the ensemble average density. In a spherical sample of radius R_s, we have that we may estimate the average density as $\bar{n} = \Gamma_E^*(R_s)$. Therefore

$$\xi_E(r) = \frac{\Gamma_E(r)}{\Gamma_E^*(R_s)} - 1 . \tag{F.27}$$

This implies that

$$\int_{C(R_s)} d^3r\, \xi_E(r) = 0 . \tag{F.28}$$

As has been mentioned, this is an artificial constraint due to the fact that the average density is given by the sample estimator (see Chap. 6). Equation (F.27) then becomes

$$\xi_E(r) = \xi(r) + \frac{\delta\Gamma_E(r)}{\Gamma_E^*(R_s)} + \Gamma_E(r)\frac{\delta\Gamma_E^*(R_s)}{(\Gamma_E^*(R_s))^2} . \tag{F.29}$$

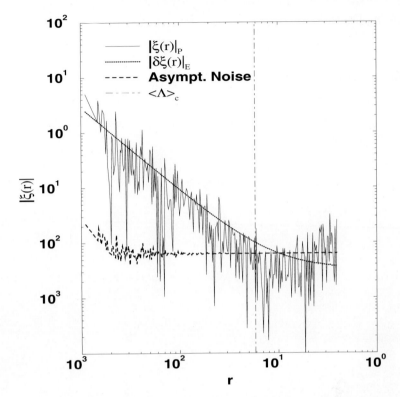

Fig. F.3. Behavior of the absolute value of the reduced two-point correlation function $|\xi_E(r)|$ for a Poisson distribution. The conditional density has been estimated by averaging over 2×10^4 points. The length scale $\langle \Lambda \rangle_c$ marks the average distance between nearest neighbors. The curves $|\xi_P(r)|$ and the "asymptotic noise" show the theoretical behavior of the noise

Therefore we can re-write (F.28) as

$$0 = \int_{C(R_s)} d^3r \, \xi_E(r) = \int_{C(R_s)} d^3r \, \xi(r) + \int_{C(R_s)} d^3r \, \frac{\delta \Gamma_E(r)}{\Gamma_E^*(R_s)} + \frac{\delta \Gamma_E^*(R_s)}{(\Gamma_E^*(R_s))^2} \int_{C(R_s)} d^3r \, \Gamma_E(r) \,. \tag{F.30}$$

The second and the third terms have the same amplitude. However the second term is noise-like (oscillating) and we can neglect it. Thus we can say that the distortion introduced by the integral constraint becomes significant at the distance R_{ic} such that

$$\left| \int_{C(R_{ic})} d^3r \, \xi_E(r) \right| \leq \left| \frac{\delta \Gamma_E^*(R_s)}{(\Gamma_E^*(R_s))^2} \right| \int_{C(R_{ic})} d^3r \, \Gamma_E(r) \tag{F.31}$$

F Errors in Full Shell Estimators

This implies that

$$\frac{\delta \Gamma_E^*(R_{ic})}{(\Gamma_E^*(R_{ic}))^2} \int_{C(R_{ic})} d^3r\, \Gamma_E(r) \qquad (F.32)$$

is of the same order as the intrinsic term $\int_{C(R_{ic})} d^3r\, \xi(r)$.

G Non Full-Shell Estimation of Two Point Correlation Properties

In Chap. 9 (Sect. 9.5) we have discussed the problem of the estimation in a finite sample of the conditional average density $\Gamma(r)$. We explained why, for an analysis in which one wants to minimize the a priori assumptions made about the nature of the underlying distribution, an appropriate estimator is the full-shell (FS) one. This choice bounds the scale at which one can extract constraints on this quantity to below the radius of the largest sphere which can be inscribed in the sample volume. Typically, because of the non-spherical geometry of galaxy samples, this is a scale which is much smaller than the largest separation between pairs of galaxies. The estimators of the reduced two-point correlation function $\xi(r)$ used standardly in the cosmological literature extend to this latter scale. Any such estimator $\xi_E(r)$ can be converted to an estimator

$$\Gamma_E(r) = \bar{n}_E \left(1 + \xi_E(r)\right) \tag{G.1}$$

for $\Gamma(r)$, where $\bar{n}_E$ is an estimator for the mean density. Thus one can use these same estimators straightforwardly to extend the analysis with the FS estimator of the conditional density to larger scales. In this appendix we describe briefly[1] the most common recipes used for $\xi_E(r)$, and discuss their relative merits when converted to estimators of $\Gamma(r)$. As they are pair counting algorithms they are actually much easier to implement practically in a sample of arbitrary geometry than the FS estimator (which requires a determination of the distances of all points to the boundaries). Thus, for practical purposes, using them may be a short-cut for the analysis we have described. We emphasize again, however, that great care must be taken in interpreting the results obtained with them from around the scale at which partial shells begin playing an important role in these estimators, i.e. beyond the scale up to which one can calculate the FS estimator. From that scale the variance (and, as we discuss below, also the bias in some cases) of such estimators is calculable only in very simple cases (e.g. for the uncorrelated Poisson case), and completely uncontrolled for a broader class of distributions (in particular irregular ones).

---

[1] We draw here considerably on the much more detailed treatment of estimators given in [128].

G.1 Estimators with Simple Weightings

Let us consider first estimators which can be cast in the form

$$\xi_E(r) = \frac{1}{\bar{n}_E} \sum_i n_i(r) w_i(r) - 1 \quad \text{with} \quad \sum_i w_i(r) = 1 \quad \text{(G.2)}$$

where $n_i(r) = \Delta N_i(r)/\Delta V_i(r)$, with $\Delta N_i(r)$ the number of points in the volume ΔV_i defined by the intersection between the sample and a shell of radius r and thickness Δ centered on the occupied point i. The $w_i(r)$ are (normalized) weights in the sum, which runs over all points in the sample, and $\bar{n}_E$ is an estimator of the mean density in the sample (e.g. $\bar{n}_E = \bar{n} = N/V$). The FS estimator for $\xi(r)$ is given by the weightings: $w_i(r) = \frac{1}{N_c(r)}$ if the spherical shell about the point i is fully enclosed in the sample, and $N_c(r)$ is the number of such points at radius r, $w_i = 0$ otherwise.

Using (G.2) in (G.1) we obtain the second class of estimators for $\Gamma(r)$ considered in Sect. 9.5 of Chap. 9. As discussed there these are unbiased estimators for $\Gamma(r)$ (up to an implicit assumption about the weakness of three-point correlations). The same is *not* true for the corresponding estimators (G.2) of $\xi(r)$. This is because of the introduction of the estimator $\bar{n}_E$ of the mean density. As discussed in Appendix F even the simplest estimator $\bar{n}_E = \bar{n} = N/V$ is not an unbiased quantity. Further in general we cannot write the equality

$$\lim_{\Delta \to 0} \left\langle \frac{n_i(r)}{\bar{n}_E} \right\rangle = \frac{\langle n(r) \rangle_p}{\langle \bar{n}_E \rangle} \quad \text{(G.3)}$$

which would be required to reduce all bias to that of the estimator of $\bar{n}_E$. This is because the fact that the mean density is estimated from the sample itself leads to built-in constraints linking the measured correlations deterministically to the mean density. In particular we have the constraints

$$\sum_j \Delta N_i(r_j) = N - 1 \quad \forall i \; . \quad \text{(G.4)}$$

Generically this kind of constraint is referred to as the integral constraint (see also Sect. 6.5 of Chap. 6). It is manifest in a very simple way if one estimates the mean density as

$$\bar{n}_E = \Gamma_E^*(R_c) = (3/R_c^3) \int_0^{R_c} \Gamma_E(r) r^2 dr \quad \text{(G.5)}$$

where $\Gamma_E(r)$ is the estimator of $\Gamma(r)$ (which does not require a determination of the mean density) and R_c is a chosen scale, which (i) depends on the sample size, and (ii) is much larger than the homogeneity scale λ_0 (at which $\Gamma(r)$ is observed to flatten). This is a good estimate of the mean density as it converges to the true mean (unconditional) density as the sample goes to infinity. (For the FS estimator one would take naturally $R_c = R_s$). One then has the constraint

$$\int_0^{R_c} \xi_E(r) r^2 dr = 0 \,. \tag{G.6}$$

In this case (and generically) therefore the estimated reduced correlation function (for this class of estimators) in any sample is therefore distorted systematically (i.e. biased) relative to the real correlation function at scales comparable to the sample size.

Going beyond FS estimators a simple estimator is given by an equal weighting of all centers i.e. $w_i = \frac{1}{N}$ in (G.2), where N is the total number of points in the sample, and taking $\bar{n}_E = (N-1)/V$ (excluding the point at the origin). This is known as the Rivolo estimator. Given that it gives equal weight to partial and full shells, it clearly will generically have a variance with respect to the ensemble value which increases strongly at scales comparable to R_s. To estimate it numerically the simplest method is to distribute randomly and without correlation an additional set of points (i.e. a Poisson distribution of points) over the sample volume. It can then be estimated as

$$\xi_E^R(r) = \frac{N_R}{N_D(N_D - 1)} \sum_i \frac{D_i(r)}{R_i(r)} - 1 \tag{G.7}$$

where $D_i(r)$ ($R_i(r)$) is the number of data (random) point in the distance range $(r - \Delta/2, r + \Delta/2)$ from the point i, and N_D (N_R) is the number of data (random) points in the volume.

A very widely used estimator for $\xi(r)$ used in the cosmological literature is the one introduced by Davis & Peebles [55]. It is very simple to calculate numerically as it is given by

$$\xi_E^{DP}(r) = \frac{2N_R}{N_D - 1} \frac{DD(r)}{DR(r)} - 1 \tag{G.8}$$

where DD ($DR(r)$) is the number of data-data (data-random) pairs with separation in the range $(r - \Delta/2, r + \Delta/2)$. It is easy to verify that this corresponds to (G.2) with the choice of weighting $w_i(r) = \Delta V_i(r)/\sum_i \Delta V_i(r)$, and $\bar{n}_E = (N-1)/V$. This is thus a weighting scheme which weights the partial shells in proportion to their volume. This may compensate in certain distributions for the additional variance associated to the partial shells.

The (non-zero) bias and variance of this estimator has been calculated [143] for an uncorrelated Poisson point distribution. The variance at large scales is proportional to $1/\sqrt{N}$, as for the the FS estimator considered in Appendix F.

G.2 Other Pair Counting Estimators

We mention two other estimators which are often considered in the cosmological literature, but which cannot be cast in the form of (G.2). Both of them have been shown [143] to be unbiased estimators for the Poisson process, and the second of them to minimize the variance for this same case.

The first is the so-called "natural" estimator which can be estimated in a finite sample as

$$\xi_E^N(r) = \frac{N_R(N_R-1)}{N_D(N_D-1)} \frac{DD(r)}{RR(r)} - 1 \,. \tag{G.9}$$

It can formally be cast in the form of (G.2) by dropping the normalization condition on the weights, as it can be shown to correspond to the weighting

$$w_i = \frac{1}{N} \frac{\Delta V_i}{4\pi r^2 \Delta} \frac{V}{\bar{V}(r)} \tag{G.10}$$

where $\bar{V}(r)$ is the mean volume of the overlap of the sample volume with itself displaced by a vector of length r (the mean is with respect to randomization over this vector's direction). This geometrical factor appears whenever one does a random-random pair counting since one can show that

$$RR(r) = n_R^2 \bar{V}(r) 4\pi r^2 \Delta \tag{G.11}$$

where n_R is the mean density of random points. This estimator thus gives a weight to each partial shell which is proportional to its volume but then inversely weighted in proportion to the displaced sample overlap volume $\bar{V}(r)$. Note that the normalization condition on the "weights" is recovered in the limit that the sample volume goes to infinity, so that the estimator is indeed a good one for $\xi(r)$.

Note that this estimator can equivalently be cast in the more general form than (G.2) as

$$\bar{\xi}_E(r) = \sum_i \frac{n_i(r) - \bar{n}w(r)}{\bar{n}w(r)} w_i(r) \qquad \sum_i w_i(r) = 1 \tag{G.12}$$

with

$$w_i(r) = \frac{\Delta V_i(r)}{\sum_i \Delta V_i(r)} \qquad w(r)^{-1} = \frac{V}{\bar{V}(r)} \frac{\sum_i \Delta V_i}{(4\pi r^2 \Delta) N} \,. \tag{G.13}$$

The function $w(r)$ effectively modifies the estimation of the mean density as a function of scale ($\bar{n} = N/V$). With respect to the Davis & Peebles estimator, the modification thus changes only how the mean density is estimated as a function of scale. Effectively the estimator changes the effective volume ($V \to V/w(r)$) over which the N points would be distributed to estimate the mean.

The second estimator is the one introduced by Landy & Szalay [143] which is given by

$$\xi_E^{LS}(r) = \frac{N_R(N_R-1)}{N_D(N_D-1)} \frac{DD(r)}{RR(r)} - 2\frac{N_R-1}{N_D} \frac{DR(r)}{RR(r)} + 1 \,. \tag{G.14}$$

It cannot be written in the simple forms used above but requires a further generalization of the form (G.12) to

$$\bar{\xi}_E(r) = \frac{1}{\bar{n}w(r)} \sum_i [n_i(r) - \bar{n}v(r)]w_i(r) \qquad \sum_i w_i(r) = 1 \qquad \text{(G.15)}$$

in terms of which it is given by the same expressions as for (G.13), and $v(r) = 2 - w(r)$. Since $w(r) \to 1$ in the infinite volume limit, we recover in both cases the correct result $(\xi(r))$ in this limit. Here therefore one changes not only the overall density normalization as a function of the scale, but also uses a different estimate of the mean density with respect to which the fluctuations are estimated.

The Landy-Szalay (LS) estimator is popular in the cosmological literature because it is the minimal variance estimator for a Poisson distribution of points, with a variance which is proportional to $1/N$ rather than $1/\sqrt{N}$ for the other estimators considered.

G.3 Estimation of the Conditional Density Beyond R_s

We have noted that any of these estimators can be used to estimate the conditional density $\Gamma(r)$ through (G.1). Those in the first section are biased as estimators of $\xi(r)$, but unbiased as estimators of $\Gamma(r)$. Given the facility with which it can be calculated the Davis & Peebles one i.e.

$$\Gamma_E^{DP}(r) = 2n_R \frac{DD(r)}{DR(r)}, \qquad \text{(G.16)}$$

where n_R is the density of the random points, is a convenient choice if one wishes to extend the conditional density analysis to scales beyond R_s (where the FS estimator terminates). We have underlined that the subtlety in its interpretation beyond this scale lies in the unknown character of its variance in a wider class of distributions.

The estimators we have considered in Sect. G.2 have the property of being unbiased estimators of $\xi(r)$ for a Poisson process, with the LS estimator having the further property of a reduced variance in this case. Converted to an estimator of $\Gamma(r)$ through (G.1) with $\bar{n}_E = \bar{n} = N/V$, the LS estimator can be written, from (G.15), as

$$\Gamma_E(r) = \frac{1}{w(r)} \sum_i [n_i(r) + 2\bar{n}(1 - w(r))] w_i(r) \qquad \text{(G.17)}$$

with $w_i(r)$ and $w(r)$ given as in (G.13). This would appear inevitably therefore to be a biased estimator for $\Gamma(r)$. Further there is no reason to expect the fact that it is corresponds to a minimum variance estimator of $\xi(r)$

for the Poisson case to mean that its variance will be any more controllable for a wider class of distributions (in particular irregular distributions). To extend the conditional density estimation to scales larger than R_s using pair-counting algorithms, our recommendation is, therefore, to use the DP estimator (G.16).

H Estimation of the Power Spectrum

Let us recall the basics of the power spectrum (PS) analysis. We suppose that the sample is periodic in a volume V_u, with V_u much larger than the homogeneity scale. As discussed at length in Chap. 3 only in this case is it meaningful to study correlations between density fluctuations with respect to the average density. The survey volume $V \in V_u$ contains N galaxies at positions r_i, and the galaxy density contrast is

$$\delta_n(r) = \frac{n(r)}{\bar{n}} - 1 \tag{H.1}$$

where $n(r) = \sum_{i=1}^{N} \delta(r - r_i)$ and where $\bar{n}$ is the estimator of the average number denisty which, for simplicity, we take to be equal to the ensemble average number density. Expanding the density contrast in its Fourier components we have

$$\delta_k = \frac{1}{V} \int_V \delta_n(r) \exp^{-ikr} d^3r = \frac{1}{N} \sum_{j \in V} e^{ikr_j} - W(k), \tag{H.2}$$

where

$$W(k) = \frac{1}{V} \int_V W(r) e^{-ikr} d^3r \tag{H.3}$$

is the Fourier transform of the survey window function $W(r)$, defined to be unity inside the survey region, and zero outside. The variance of δ_k is

$$\langle |\delta_k|^2 \rangle = \frac{1}{N} + \frac{1}{V} P_c(k). \tag{H.4}$$

The first term is the usual shot noise term which comes from the diagonal term of $\tilde{\xi}(r)$ (see Chap. 2). The second term is the Fourier transform $\hat{\xi}(k)$ of the non-diagonal part of the reduced two-point correlation function of the infinite system, convoluted with the square modulus of the sample window function (e.g. [173]).

$$P_c(k) = \frac{1}{(2\pi)^3} \int \hat{\xi}(k') |W(k - k')|^2 d^3k', \tag{H.5}$$

which can be written as

$$P_c(\mathbf{k}) = \int \hat{\xi}(\mathbf{k}')F(\mathbf{k}-\mathbf{k}')d^3k' , \tag{H.6}$$

where

$$F(\mathbf{k}-\mathbf{k}') = \frac{1}{(2\pi)^3}|W(\mathbf{k}-\mathbf{k}')|^2 . \tag{H.7}$$

In practice, one averages the spectrum over k-shells of thickness $\Delta \le 2\pi/R_s$:

$$P_a(k) = \frac{1}{k^2\Delta}\int_k^{k+\Delta} \tilde{P}_c(k')k'^2 dk' \tag{H.8}$$

As in the case of the estimation of the reduced two-point correlation function $\tilde{\xi}(r)$, the fact that the average density comes from the sample itself results in a artificially built-in constraint on the PS. One would anticipate that the effect of this in the PS would be to suppress power at small k. Indeed this is what is imposed if the constraint on the estimator of $P(k)$ takes the same form as the integral constraint discussed in Appendix G. As discussed in Chap. 6 (Sect. 6.5) such a suppression of power at small k in a finite sample should not be confused with true super-homogeneous behavior of the distribution. To establish this through a PS analysis one needs to be sure that such behavior is truly independent of the sample size.

In the estimation of the PS we have an additional complication when the sample shape is not spherical [2]. We are able to estimate only the convolution of the PS with the sample window function as given by (H.5)–(H.6). In general the window function has sharp edges in real space which may introduce spurious oscillations in the PS. To disentangle the effect of the window function is very delicate, in particular in the case that the sample is a small angular slice of a sphere (i.e. as for many galaxy redshift surveys). Even for a distribution with weak correlations it is to be recommended that the PS analysis be complemented by a real space analysis of correlations.

Finally it is interesting to note what happens when the PS analysis is applied in the case where the distribution is fractal, i.e. without having tested that the average density in the sample is a well defined positive quantity. In such a situation the PS shows the same finite size effects which we have discussed for the reduced two-point correlation function (see Chap. 9) [222]. More specifically, in the case we consider a fractal of dimension D in a spherical sample of radius R_s it is simple to show that [222]

$$P_E(r) \approx a_k(R_s, D)\frac{R_s^{3-D}}{k^D} - \frac{b_k(R_s, D)}{k^3}$$

where $a_k(R_s, D)$ and $b_k(R_s, D)$ are oscillating functions. This implies that (i) the amplitude depends explicitly on the sample size and (ii) the PS does not present a single power-law behavior but has a cut-off a scales of order R_s. This latter feature is the counterpart of the "integral constraint" discussed in Chap. 9 for the reduced two-point correlation function.

References

1. Abdalla, E., & Chirenti, C.B.M.H., Physica A **336**, 433 (2004)
2. Amendola, L., IX Brazilian School of Cosmology and Gravitation, Ed. Novello, M. (Atlantica, Paris, 2004)
3. Amit, D., *"Field Theory, the Renormalization Group and Critical Phenomena"*, (Mc Graw-Hill, New York, 1978)
4. Anderson, P.W., Science, **177**, 393, (1972)
5. Antal, T., Droz, M., Györgyi, G., Rácz, Z., Phys. Rev. Lett., **87**, 240601, (2001)
6. Aracangelis, L. de, Redner, S. & Coniglio, A., Phys. Rev. B, **31**, 4725, (1985)
7. Arnouts S., et al., Astron. Astrophys. Suppl., **124**, 163, (1997)
8. Ashcroft, N.W. & Mermin, N.D. *"Solid state physics"*, (Saunders College, 1976)
9. Bahcall, N.A., & Soneira, R.M., Astron. Astrophys., **270**, 20, (1983)
10. Bahcall, N.A., Astrophys. J., **585**, 182, (2003)
11. Badii, R., & Politi, A., Phys. Lett. A, **104**, 303, (1984)
12. Baertschiger, T. & Sylos Labini, F. Europhys. Lett., **57**, 322 (2002)
13. Baertschiger, T., Joyce, M. & Sylos Labini, F., Astrophys. J. Lett, **581**, L63 (2002)
14. Baertschiger T.& Sylos Labini, F., Phys. Rev. D, **69**, 123001-1 123001-14 (2004)
15. Bardeen, J., Bond, J.R., Kaiser, N. & Szalay, A., Astrophys. J., **304**, 15, (1986)
16. Bartlett, M.S., J. Roy. Statist. Soc. Ser. B, **25**, 264 (1963)
17. Baryshev, Y., Sylos Labini, F., Montuori, M., Pietronero, L., Vistas in Astron., **38**, 419, (1994)
18. Baryshev, Yu.V. & Teerikorpi, P., *"The Discovery of Cosmic Fractals"*, (World Scientific, Singapore, 2002)
19. Baryshev, Yu.V. & Buhkmastova, Yu.L., Astronomy Letters, **30**, 444, (2004)
20. Bashinsky, S., & Bertschinger, E., Phys. Rev. Lett., **87**, 081301, (2001)
21. Batusky, D.J. & Burns, J.O., Astrophys. J, **299**, 5, (1985)
22. Baus, M. & Hansen, J.-P., Phys. Rep., **59**, 1, (1980)
23. Bothun, G.D., & Cornell, M.E., Astron. J., **99**, 1004, (1990)
24. Beck, J., Acta Mathematica **159**, 1-878282, (1987)
25. Bendat, J.S. & Piersol, A.G., *"Random Data"*, (Wiley & Sons, New York, 1971)
26. Bennett, C.L., et al., Astrophys. J. Suppl., **148**, 1, (2003)
27. Bennett, C.L., et al., Astrophys. J., **436**, 423, (1994)
28. Benoist, C., et al., Astrophys. J., **472**, 452, (1996)
29. Benzi, R., Paladin, G., Parisi, G., Vulpiani, A., J. Phys. A., **17**, 3251, (1984)

30. Bernasconi, A., & Schnider, W., Proc. of the Conference *"Fractals in Physics"*, Pietronero, L. & Tosatti E., Eds. (1986)
31. Bertin, E. & Dennefeld, M., Astron. Astrophys., **317**, 43, (1997)
32. Binggeli, B., Sandage, A., Tammann, G. A., Astron. Astrophys. Ann. Rev., **26**, 509, (1988)
33. Blanton, M.R., et al., Astrophys. J., **592**, 819, (2003)
34. Blumenfeld, R. & Ball, R.C., Phys. Rev E., **47**, 2298, (1993)
35. Bottaccio, M., Capuzzo-Dolcetta, R., Miocchi, P., Montuori, M., Pietronero, L., Europhys. Lett., **7**, 315, (2002)
36. Bottaccio, M., Montuori, M., Pietronero, L., Europhys. Lett., in the press (2003)
37. Bothum, G.D., et al., Astrophys. J., **308**, 510, (1988)
38. Buchert, T., Mon. Not. R. Astron. Soc., **254**, 729, (1992)
39. Budavari, T. et al., Astrophys. J., **595**, 59, (2003)
40. Cappi, A., et al., Astron. Astrophys, **335**, 779, (1998)
41. Castellan, C. & Peliti, L., J. Phys. A, **19**, L429, (1986)
42. Chaboyer, B., Phys. Rep., **307**, 23, (1998)
43. Chandrasekhar, S., Rev. Mod. Phys., **15**, 1, (1943)
44. Charlier, C.V.L., Arkiv for Mat. Astron. Physik, **4**, 1 (1908)
45. Charlier, C.V.L., Arkiv for Mat. Astron. Physik, **16**, 1 (1922)
46. Chown, M., New Scientist, **2200**, 22, (1999)
47. Coleman, P.H. & Pietronero, L., Phys. Rep., **231**, 311, (1992)
48. Coles, P., Mon. Not. R. Acad. Soc., **222**, 9p, (1986)
49. Coles, P., Nature, **391**, 120, (1998)
50. Cooray, A. & Sheth, R., Phys. Rep., **372**, 1, (2002)
51. Cowie, L., Gardner, J. P., Lilly, S. J., McLean, I. Astrophys. J., **360**, L1, (1990)
52. Da Costa, L.N., Vogeley, M., Geller, M.J., Huchran J., Park, C., Astrophys. J., **437**, L1, (1994)
53. Daley, D. J. & Vere-Jones, D., *"An introduction to the theory of point processes"*, (Springer Verlag, Berlin, 1988)
54. Davis, M. & Huchra, J., Astrophys. J., **254**, 437, (1982)
55. Davis, M. & Peebles, P.J.E., Astrophys. J., **267**, 46, (1983)
56. Davis, M. et al., Astrophys. J. Lett., **333**, L9, (1988)
57. Davis, M., in the Proc. of the Conference *"Critical Dialogues in Cosmology"*, Ed. Turok, N., p.13, (World Scientific, Singapore, 1997)
58. De Bernardis, P., et al., Nature, **404**, 955, (2000)
59. De Lapparent, V., Geller, M. & Huchra, J., Astrophys. J., **302**, L1, (1986)
60. De Lapparent, V., Geller, M. & Huchra, J., Astrophys. J., **332**, 44, (1988)
61. Dekel, A., et al., Astrophys. J. Lett., **388**, L5, (1989)
62. Del Popolo, A., Astron. Astrophys., **311**, 715, (1996)
63. De Vaucouleurs, G., Science, **167**, 1203, (1970)
64. de Vega, H. J. & Sanchez, N. in Procs of the 7th course of the Chalonge School, Sanchez, N. Ed., NATO 562 C, 433, Kluwer (2001)
65. de Vega, H.G., & Sanchez, N., Nucl. Phys. B, **625**, 409, (2002)
66. de Vega, H.G., & Sanchez, N., Nucl. Phys. B **625**, 460, (2002)
67. Di Nella, H., Montuori, M., Paturel, G., Pietronero, L., Sylos Labini, F., Astron. Astrophys. Lett., **308**, L33, (1996)
68. Doobs, J.L., *"Stochastic Processes"*, (John Wiley & Sons, New York, 1953)

69. Driver, S.P., Phillips, S., Davies, J.O., Morgan, I., Disney, M.J., Mon. Not. R. Astron. Soc., 266, 155, (1994)
70. Durrer, R., Eckmann, J.-P., Sylos Labini, F., Montuori, M., Pietronero, L., Europhys. Lett., **40**, 491, (1997)
71. Durrer, R. & Sylos Labini, F., Astron. Astrophys. Lett., **339**, L85, (1998)
72. Durrer, R., Gabrielli, A., Joyce, M., Sylos Labini, F., Astrophys. J. Lett, **585**, L1, (2003)
73. Eckmann, J.-P. & Ruelle, D., Rev. Mod. Phys., **57**, 617, (1985)
74. Efstathiou, G., Ellis, R.S. & Peterson, B., Mon. Not. R. Astr. Soc., **232**, 431, (1988)
75. Einasto, J., et al., Mon. Not. R. Astr. Soc., **269**, 301, (1994)
76. Eiseinstein, D.J. & Hu, W., Astrophys. J., **496**, 605, (1998)
77. El-Ad, H., & Piran, T., Mon. Not. R. Astr. Soc., **313**, 553, (2000)
78. Ellis, R.S., Ann. Rev. Astron. Astrophys., **35**, 389, (1997)
79. Erzan, A., Pietronero, L., Vespignani, A., Rev. Mod. Phys., **67**, 554, (1995)
80. Evertsz, C.J.G. Ed., *"Fractal geometry and Analysis"*, (World Scientific, Singapore, 1996)
81. Faber, S.M. & Gallagher, J.S., Astron. Astrophys. Ann. Rev., **17**, 135, (1979)
82. Falco, E.F., Kurtz, M.J., Geller, M.J., Huchra, J.P., Peters, J., Berlind, P., Mink, D.J., Tokarz, S.P., Elwell, B., Pub. Astron. Soc. Pac., **111**, 438, (1999)
83. Falconer, K., *"Fractal geometry"*, (John Wiley & Sons, New York, 1990)
84. Feller, W., *"An Introduction to Probability Theory and its Applications"*, (John Wiley& Sons, New York, 1965)
85. Felten, J., Astron. J., **82**, 861, (1977)
86. Frisch, U. & Parisi, G., in the Proc. of the Conference *"Turbolence and predictaibility of geophysical flows and climiatc dysnamics"*, Eds. Ghil, N., Benzi, R. & Parisi, G., (North-Holland, Amstrerdam, 1985)
87. Fukugita, M., Shimasaku, K. & Ichikawa, T., Pub. Astr. Soc. Pac., **107**, 945, (1995)
88. Gabrielli, A., Sylos Labini, F. & Pellegrini, S., Europhys. Lett., **46**, 127, (1999)
89. Gabrielli, A., Sylos Labini, F. & Durrer, R., Astrophys. J. Lett., **531**, L1, (2000)
90. Gabrielli, A. & Sylos Labini, F., Europhys. Lett., **54**, 1, (2001)
91. Gabrielli, A., Joyce, M., Sylos Labini, F., Phys. Rev. D, **D65**, 083523, (2002)
92. Gabrielli A., Jancovici B., Joyce M., Lebowitz J., Pietronero L., Sylos Labini F., Phys. Rev., **D67**, 043406,(2003)
93. Gabrielli A., Masucci, P., Sylos Labini F., Phys. Rev. **E69**, 031110, (2004)
94. Gabrielli A., in preparation (2004)
95. Gabrielli A., et al., in preparation (2004)
96. Gács, P. & Szász, D., Annals of Probability, **3**, 597, (1975)
97. Gaite, J., Dominguez, A., & Perez-Mercader, J., Astrophys. J. Lett., **522**, L5, (1999)
98. Gardiner, C. W., *"Handbook of stochastic methods"*, (Springer Verlag, Berlin, Second Edition, 1997)
99. Gardner, J.P., Publ. Astron. Soc. Pac., **110**, 291, (1998)
100. Gnedenko, B., *"The theory of probability"*, (Mir Publishers, Moscow, 1975)
101. Gott J.R. III, et al., `astro-ph/0310571`
102. Grassberger, P. & Procaccia, I., Physica D, **13**, 3, (1984)
103. Grujic, P.V., Serb. Astron. J., **165**, 45 (2002)

104. Haggerty, M.J., Astrohys. J., **166**, 257 (1971)
105. Haggerty, M.J., Wertz, J.R., Mon. Not. R. Astron. Soc., **155**, 495 (1971)
106. Halsey, T.C., Jensen, M.H., Kadanoff, L.P., Procaccia, I., Shraiman B.I., Phys. Rev. A., **33**, 1141, (1986)
107. Hanany, S., et al., Astrophys. J., **545**, L1, (2000)
108. Hansen, J.P. & McDonald, I.R., *"Theory of simple fluids"*, (Academic Press, London, 1976)
109. Harrison, E.R., Phys. Rev. D, **1**, 2726 (1970)
110. Hentchel, H.G.E. & Procaccia, I., Physica D, **8**, 435, (1983)
111. Hertz, P., Math. Ann., **67**, 387, (1909)
112. Hockney, R.W. & Eastwood, J.W., *"Computer simulation using particles"* (McGraw-Hill, New York, 1981)
113. Hogg D., astro-ph/9905116
114. Hradecky, V., Jones, C., Donnelly, R.H., Djorgovski, S.G., Gal, R. R., Odewahn, S.C., Astrophys. J., **543**, 521, (2000)
115. Hu, W. & Dodelson, S., Ann. Rev. Astron. Astrophys., **40**, 171, (2002)
116. Huang, K., *"Statistical Mechanics"*, Second Edition, (John Wiley and Sons, New York, , 1987)
117. Hubble, E., Astrophys. J., **79**, 8, (1934)
118. Hubble, E., Astrophys. J., **84**, 517, (1936)
119. Huchra, J., Davis, M., Latham, D., Tonry, J., Astrophys. J. Suppl., **52**, 89, (1983)
120. Huchra, J.P., Vogeley, M.S. & Geller, M.J., Astrophys. J. Suppl., **121**, 287, (1999)
121. Hughes, B.D., *"Random Walks and Random Environments"* Vol. 1, (Oxford Science Publications, England, 1995)
122. Kagan, Y.Y., Pure App. Geophys., **155**, 233, (1999)
123. Kaiser, N., Astrophys. J. Lett., **284**, L9, (1984)
124. Kauffmann, G. &, Fairall, A. P., Mon. Not. R. Astr. Soc., **248** , 313, (1991)
125. Kendall, D. G. & Rankin, R. A., Quart. J. Math. Oxford (2), **4**, 178, (1953)
126. Kerscher, M., Astron. Astrophys., **343**, 333, (1999)
127. Kerscher, M., Szapudi, I., Szalay, A., Astrophys. J., **535**, L13, (2000)
128. Kerscher, M., Phys. Rev. E, **64**, 056109, (1999)
129. Kerscher, M., Europhys. Lett., **61**, 856, (2003)
130. Kolb, E.W. & Turner, M.S., *"The Early Universe"*, (Addison-Wesley Publishing Company, 1992)
131. Koo, D.C., & Kron, R.G., Ann. Rev. Astron. Astrophys., **30**, 613, (1992)
132. Kuhlman, B., Melott, A., & Shandarin, S.F., Astrophys. J., **470**, L41, (1996)
133. Jenkins, A., et al., Astrophys. J., **499**, 20, (1998)
134. Jensen, L. & Szalay, A., Astrophys. J., **305**, L5, (1986)
135. Jensen, M.H., Paladin, G., Vulpiani, A., Phys. Rev. Lett., **67**, 208, (1991)
136. Jorgensen, H.E., Kotok, E., Naselsky, P., Novikov, I., Mon. Not. R. Astron. Soc., **265**, 261, (1993)
137. Joyce, M., Montuori, M., Sylos Labini, F., Astrophys. J., **514**, L5, (1999)
138. Joyce, M., Montuori, M., Sylos Labini, F., Pietronero, L., Astron. Astrophys., **344**, 387, (1999)
139. Joyce, M., Anderson, P. W., Montuori, M., Pietronero, L. & Sylos Labini, F., Europhys. Lett., **50**, 416, (2000)
140. Joyce, M. & Sylos Labini, F., Astrophys. J. Lett., **554**, L1, (2001)

141. Lahav, O., et al., Mon. Not. R. Acad. Soc., **333**, 961, (2002)
142. Landau, L.D. and Lifsits, E.M., *"Statistical Physics"*, (Mir, Moscow, 1978)
143. Landy, A. & Szalay, A., Astrophys. J., **412**, 64, (1993)
144. Liddle, A.R. & Lyth, D.H., *"Cosmological Inflation and Large Scale Structures"*, (Cambridge University Press, Cambridge, 2000)
145. Lindner, U., Einasto, J., Einasto, M., Wolfram, F., Klaus, F., Tago, E., Astron. Astrophys., **301**, 329, (1995)
146. Liske, J., Lemon, D.J., Driver, S., Cross, N.J.G. & Couch, W.J. Mon. Not. R. Astron. Soc., **344**, 307, (2003)
147. Lucchin, F. & Coles, P., *"Cosmology"*, (John Wiley & Sons, New York, 1997)
148. Ma, S.K., *"The Modern Theory of Critical Phenomena"*, (Benjamin Reading, 1976)
149. Maddox, S.J., Sutherland, W.J., Efstathiou, G., Loveday, J., Peterson, B.A., Mon. Not. R. Astr. Soc., **242**, 43, (1990)
150. Mandelbrot, B.B., *"Fractals: Form, Chance and Dimension"*, (Freeman, New York, 1977)
151. Mandelbrot, B.B., *"The Fractal Geometry of Nature"*, (Freeman, New York, 1983)
152. Mandelbrot, B.B., in *"Fluctuations and Pattern Formation"*, Eds. Stanley, H.E., Ostrowski, N., Nijhoff, M. (Dordrecht, 1988)
153. Mandelbrot, B.B., in the Proc. of the *Erice Chalonge School, "Current Topics in Astrofundamental Physics: Primordial Cosmology"*, NATO ASI series C511, Sanchez, N. & Zichichi, A., Eds. (Kluwer, 1998)
154. Mann, R. G., Heavens, A. F. & Peacock, J. A., Mon. Not. R. Astron. Soc., **263**, 798, (1993)
155. Martin, Ph. A., Rev. Mod. Phys., **60**, 1075, (1988)
156. Martin, Ph.A. & Yalcin, T., J. Stat. Phys., **22**, 435, (1980)
157. Martinez, V.J., Science, **284**, 445, (1999)
158. Marzke, R.O., da Costa, L.N., Pellegrini, P.S., Willmer, C.N.A., Geller, M.J., Astrophys. J., **503**, 617, (1998)
159. Mattila, P., *"Geometry of Sets and Measures in Euclidean Spaces"*, (Cambridge University Press, Cambridge, 1995)
160. Melott, A., Comments Astrophys., **15**, 1 (1990)
161. Meneveau, C. & Sreenivasan, K.R., Phys. Rev. Lett., **59**, 1424, (1988)
162. Metcalfe, N., Fong R., Shanks, T., Mon. Not. R. Astr. Soc., **274**, 769, (1995)
163. Metcalfe, N., Shanks, T., Campos, A., McCracken H., Fong, E. Mon. Not. R. Astr. Soc., **323**, 795, (2001)
164. Montuori, M. & Sylos Labini, F., Astrophys. J., **487**, L21, (1997)
165. Naselsky, P. & Novikov, I., Astrophys. J. Lett., **413**, 14, (1993)
166. Neyman, J. & Scott., E.L., Astrophys. J., **116**, 144, (1952)
167. Neyman, J., Scott., E.L., Shane, C.D., Astrophys. J. Suppl., **1**, 269, (1954)
168. Norberg, P., et al., Mon. Not. R. Acad. Soc., **328**, 64, (2001)
169. Olive, K., Steigman, G. & Walker, T., Phys. Rep. **389**, 333, (2000)
170. Padmanabhan, T., *"Structure formation in the universe"*, (Cambridge University Press, Cambridge, 1993)
171. Paladin, G., Vulpiani, A., Phys. Rep., **156**, 147, (1987)
172. Paladin, G., Vulpiani, A., Phys. Rev. B, **35**, 2015, (1987)
173. Park, C., Vogeley, M.S., Geller, M., Huchra, J., Astrophys. J., **431**, 569, (1994)
174. Parisi, G., *"Statistical Field theory"* (Wiley & Sons, New York, 1989)

175. Paturel, G., Bottinelli, L., Di Nella, H., Fouqué, P., Gouguenheim, L. & Teerikorpi, P., Astron. Astrophys., **289**, 711, (1994)
176. Paturel, G., Bottinelli, L., Gouguenheim, L., Astron. Astrophys., **286**, 768, (1994)
177. Paturel, G., et al., Astron. Astrophys. Suppl., 124, 109, (1997)
178. Paul W., & Baschnagel, J., "Stochastic processes from physics to finance" (Springer-Verlag, Berlin, 1999)
179. Peacock, J.A., *"Cosmological physics"*, (Cambridge University Press, Cambridge, 1999)
180. Peacock, J.A., et al., Nature, **410**, 169, (2001)
181. Peacock, J.A. & Dodds, S.J., Mon. Not. R. Acad. Soc., **267**, 1020, (1994)
182. Pécseli, H.L., *"Fluctuations in Physical Systems"*, (Cambridge University Press, Cambridge, 2000)
183. Peebles, P.J.E. & Yu, J.T., Astrophys. J., **162**, 185, (1970)
184. Peebles, P.J.E., *"Large Scale Structure of the Universe"*, (Princeton Univeristy Press, Princeton, New Jersey, 1980)
185. Peebles, P.J.E., *"Principles Of Physical Cosmology"*, (Princeton University Press, Princeton, New Jersey, 1993)
186. Peliti, L. & Pietronero, L., Rivista del Nuovo Cimento, **10**, 1, (1987)
187. Perlmutter, S., et al., Astrophys. J., **517**, 565, (1999)
188. Phuam, N. & Bhatt, Ravin N., Eds., Proc. of the Conference *"More is Different"*, (Princeton Series in Physics, Princeton, New Jersey, 2001)
189. Picard, A., Astronomical J. **102**, 445, (1991)
190. Pietronero, L., Physica A, **144**, 257, (1987)
191. Pietronero, L. & Siebesma, A.P., Phys. Rev. Lett., **57**, 1098, (1986)
192. Pietronero, L. & Siebesma, A.P., Phys. Rev. Lett., **66**, 1038, (1988)
193. Pietronero, L., Montuori, M. & Sylos Labini, F., in the Proc. of the Conference *"Critical Dialogues in Cosmology"*, Ed. Turok, N., p.24, (World Scientific, Singapore, 1997)
194. Pietronero, L., in the Proc. of the Conference *"Order and chaos in non-linear physical systems"*, Eds. Lundquist, K., March, N.H., & Tosi, M.P., (Plenum Publishing Corporation, 1988)
195. Politzer, H.D. & Wise, M.B., Astrophys. J., **285**, L1, (1994)
196. Postman, M., Huchra, J.P., & Geller, M.J., Astrophys. J., **384**, 404, (1992)
197. Rácz, Z., & Plischke, M., Phys. Rev. E, **50**, 3530 (1994)
198. Radin, C., Notices Amer. Math. Soc., **42**, 26, (1995)
199. Radin, C., chapter in *"Geometry at Work"*, MAA Notes, **53** (Math. Assoc. Amar., Washington, DC, 2000)
200. Rammal, R., Tannous, C. & Tremblay, A.M.S., Phys. Rev. Lett., **54**, 1718, (1985)
201. Renyi, A., *"Probability Theory"*, (North-Holland, Amsterdam, 1970)
202. Ribeiro, M.B., Astrophys J **415**, 469, (1993)
203. Rice, S.O., in *"Noise and Stochastic Processes"*, Ed. Wax, N., (Dover Publications, New York, 1954)
204. Rousseau, J., Di Nella, H., Paturel, G., & Petit, C., Mon. Not. R. Astron. Soc., **282**, 144, (1996)
205. Ruelle, D. *"Statistical Mechanics"* (Benjamin, New York, 1969)
206. Sandage, A., in *"The deep universe"* Edited by Binggeli, B. & Buser, R., Saas-Fee Advanced Course 23, Lecture Notes, (Swiss Society for Astrophysics and Astronomy, Springer-Verlag, Berlin, 1995)

207. Sandage, A., Tammann, G.A. & Hardy, E., Astrophys. J., **172**, 253, (1972)
208. Saslaw, W.C., *"The Distribution of the Galaxies"*, (Cambridge University Press, Cambridge, 2000)
209. Saunders W., et al., Mon. Not. R. Acad. Soc., **317**, 55, (2000)
210. Scaramella, R., et al., Astron. Astrophys., **334**, 404, (1998)
211. Schecther, P., Astrophys. J., **203**, 297, (1976)
212. Schneider, P. & Bartelmann, M., Mon. Not. R. Astron. Soc., **273**, 475, (1995)
213. Siebesma, A.P., Ph.D. Thesis, (University of Groningen, Groningen, 1986)
214. Siebesma, A.P., Tremblay, R.R., Erzan, A. & Pietronero, L., Physica A, **156**, 613, (1989)
215. Sornette, D., Phys. Rep., **297**, 239, (1998)
216. Splinter, R.J., Melott, A.L., Shandarin, S.F.& Suto, Y., Astrophys. J., **497**, 38, (1998)
217. Stoyan, D., Kendall, W.S & Mecke, J., *"Stochastic geometry and its applications"*, (Wiley, New York, 1995)
218. Strauss, M.A., & Willick, J.A., Phys. Rep., **261**, 271, (1995)
219. Sylos Labini, F., Astrophys. J., **433**, 464, (1994)
220. Sylos Labini, F. & Pietronero, L., Astrophys. J., **469**, 28, (1996)
221. Sylos Labini, F., Montuori, M. & Pietronero, L., Physica A, **230**, 336, (1996)
222. Sylos Labini, F. & Amendola, L., Astrophys. J. Lett., **468**, 96, (1996)
223. Sylos Labini, F., Montuori, M. & Pietronero, L., Phys. Rep., **293**, 66, (1998)
224. Sylos Labini, F., Baertschiger, T. & Joyce, M., Europhys. Lett. **66**, 171, (2004)
225. Szapudi I. & Szalay, A., Astrophys. J., **494**, L41, (1998)
226. Tammann, G.A., in the Proc. of the Conference *"Critical Dialogues in Cosmology"* Ed. Turok, N. (World Scientific, Singapore, 1997)
227. Taylor, A.N. & Hamilton, J.S., Mon. Not. R. Astron. Soc., **282**, 767, (1996)
228. Teerikorpi, P., et al., Astron. Astrophys., **334**, 395, (1998)
229. Teerikorpi, P., Astron. Astrophys., in the press (2004)
230. Torquato, S., *"Random Heterogeneous Materials"* (Springer-Verlag, Berlin, 2002)
231. Torquato, S. & Stillinger, F. H., Phys. Rev., E, **68**, 041113, (2003)
232. Totsuji, H. & Kihara, T., Publ. Astron. Soc. Jpn, **21**, 221, (1969)
233. Tully, B.R., Astrophys. J., **303**, 25, (1986)
234. Tully, B.R., Astrophys. J, **303**, 25, (1986)
235. Tully, B.R., Scaramella, R., Vettolani G. & Zamorani G., Astrophys. J., **388**, 9, (1992)
236. Tyson, J.A., Astron. J., **96**, 1, (1988)
237. Vanmarcke, E., *"Random Fields"*, (MIT press, Cambridge, MA USA, 1983)
238. Vlad, M.O., Astrophys. and Space. Sci. **218**, 159, (1994)
239. Weaver, W. & Shannon, C. E., *"The Mathematical Theory of Communication"* (University of Illinois Press, Urbana, 1949)
240. Web page of pil group: *http://pil.phys.uniroma1.it*
241. Weinberg, S.E., *"Gravitation and Cosmology"*, (Wiley & Sons, New York, 1972)
242. Weir, N., Djorgovski, G., Fayyad, U.M., Astron. J., **110**, 1, (1995)
243. Wertz, J.R., Univ. Texas Publ. Astron., **N3**, 3 (1970)
244. Wilkens, M., Acta Metall., **17**, 1155, (1969)
245. Wilson, K.G., Phys. Rep., **12**, 75, (1974)
246. White, S.D.M., *"Lectures given at Les Houches"* astro-ph/9410093 (1993)

247. Wu, K.K., Lahav, O. & Rees, M., Nature, **225**, 230, (1999)
248. York, D., et al., Astron. J., **120**, 1579 (2000)
249. Zaldarriaga, M. & Seljak, U., Astrophys. J. Suppl., **129**, 431, (2000)
250. Zehavi, I., et al., Astrophys. J., **571**, 172, (2002)
251. Zeldovich, Ya.B., Astron. Astrophys., **5**, 84, (1969)
252. Zeldovich, Ya.B., Mon. Not. R. Acad. Soc., **160**, 1, (1972)
253. Ziman, J.M., *"Principles of the theory of solids"*, (Cambridge University Press, Cambridge, 1972, second edition)
254. Zwicky, F., Herzog, E., Wild, P., Karpowitz, M., Kowal, C.T., *"Catalogue of Galaxies and of Clusters of Galaxies"* (6. Vols.) (Pasadena, California Institute of Technology, 1961-68)

Index

Abell clusters catalog   304–307
acoustic oscillations   167, 175, 180, 183–185, 187
Anderson, P.W.   2
angular correlation function   134–136
anti-correlations   127, 169, 178, 244
auto-correlation function   35, 40

bias (in estimators)   247, 248, 396
bias (model of)   313, 321, 325, 330
biasing   231, 314, 316, 318, 321, 324–327, 330–333
binomial measure   145–148, 150–152, 156
Birkhoff theorem   34
Bochner theorem   41, 56, 93
bootstrap errors   253, 254
box dimension   102, 103, 106, 107, 109, 133
Braggs peaks   210
Bravais lattice   85, 87
Brillouin zone   210, 211

Cantor set
– deterministic   369, 370
– random   369–372
capacity   102
casual horizon   168, 169
center for astrophysics redshift catalog   228, 230, 251–254, 256, 258, 260, 263, 279–281, 285, 307
central limit theorem   42, 55, 64–66, 68, 69, 72, 130, 131
Chandrasekhar   173, 335, 338, 342, 350
characteristic function   36–38
clustering   127, 128, 295
– parameter   128, 129
clusters of galaxies   299

– conditional average density   303, 306
– fractal dimension   306
– reduced two-point correlation function   230, 304–307
co-dimension   108, 132, 136
coarse graining   59, 60, 62, 300
cold dark matter   168, 173, 175, 177, 178, 185, 194, 195, 198, 314, 315, 320, 325, 327, 328, 332
comoving distance   376–378, 382
complexity   2
conditional
– average   48
– average density   91, 107–109, 111–113, 116, 117, 119–121, 128, 135, 136, 142, 236, 237, 239, 246, 247, 249, 250, 253, 398, 399
– correlation function   48
– cosmological principle   31, 107, 386
– probability   48
– properties   48, 50
– two-point density   116–118, 134
– variance   50, 109, 118, 120, 123, 124
continuous stochastic process   31
correlation coefficients   37
correlation function   35
– connected   36, 38, 57
– Davis & Peebles estimator   407, 408
– definition   35
– full shell estimator   249
– Landy & Szalay estimator   408
– natural estimator   408
– reduced   36
– Rivolo estimator   407
correlation kernel   41, 56, 91

Index

correlation length 8, 39, 40, 42, 43, 51, 52, 58–62, 236, 243, 256, 318, 319, 321, 322, 330
correlation matrix 57, 91
cosmic microwave background radiation 27, 55, 167–169, 314, 332, 333, 386
- anisotropies 174, 175, 179, 180, 182, 183, 191
- multi-pole moments 181
- quadrupole moment 181, 182
cosmological constant 175
cosmological principle 31, 386
covariance function 36
critical
- behavior 46, 236, 243
- phenomena 4–6, 169
- systems 170, 189, 243, 244
cumulants expansion 36, 38

de Vaucouleurs 1
deceleration parameter 379
density contrast 76
dielectric breakdown model 6, 7
diffusion limited aggregation 6, 105
displacement field 199–204, 217
- correlated 212, 214, 216
- uncorrelated 206, 207, 209

effective depth 250–253, 263
ergodic theorem 7
ergodicity 34, 40, 71, 74, 107, 119
estimators 236, 239, 246, 247, 249, 250, 253, 395, 396, 401, 402
- errors 395
Euler equation 183

Feynman graphs 36, 38
fluctuation-dissipation theorem 8, 39
fractal 2, 5, 29, 30, 101–103, 105–109, 112–115, 117–121, 124–137, 139–142, 235–238, 369, 370, 372, 373
- angular projection 134, 136
- dimension 102, 103, 106–108, 112, 116, 120, 122–125, 128–138, 140, 141, 241, 369, 370, 372, 373
- dimension imaginary part of 123
- distribution 10, 240
- geometry 5, 6, 101
- intersection of 132, 133

- orthogonal projection 134, 136
Friedmann-Robertson-Walker
- cosmology 172, 386
- models 168, 171, 222, 223, 228, 375, 376, 381, 386, 389, 390
full shell estimator 235, 247, 249, 250, 263

galaxies
- angular catalogs 219, 222, 227, 228, 232, 285, 287
- conditional average density 221, 232, 235, 254–258, 282, 283, 291, 294
- field to field fluctuations 284, 285, 287
- fractal dimension 255, 257, 279, 285
- homogeneity scale 228
- large scale structures 227
- luminosity density 259–261, 263
- luminosity function 221, 255, 258, 259, 270, 283–285, 291–293, 296
- magnitude counts 265, 266, 268–273, 275–278, 283–289
- normalization problem in the magnitude counts 283–285
- power-law correlations 228, 254, 255, 257, 279, 283
- radial counts 265–267, 276–282, 284, 285, 288, 289
- redshift surveys 50, 174, 219, 220, 222, 224, 226–231, 233, 252
- reduced two-point correlation function 228, 256, 257
Gauss-Poisson 342–344, 347–350, 354, 360
Gaussian
- density field 31, 55–57, 72, 76, 91, 93–95, 97–99, 313–316, 319, 321–323, 325–327, 330, 332
- functional 56
- spheres 82, 172
- window function 194
generalized fractal dimensions 144, 147–149, 161
glass 89, 169, 193, 195
granularity 3

halo model 113, 331
Harrison Zeldovich

Index   423

– condition   84, 91, 173
– criterion   171, 172
– model   169
– power spectrum   168, 169, 171–174, 176, 190, 197, 198, 327, 330
– turnover of the power spectrum   169, 173, 174, 176, 198
Hausdorff dimension   103, 106, 137
Holtzmark distribution   338, 340, 341, 343, 354
homogeneity   34
– condition   33
– scale   9, 28, 34, 35, 39, 43–45, 50, 51, 53, 108, 128, 131, 132, 135, 142, 236–243, 310, 318, 321, 330
horizon scale   171, 172
hot dark matter   168, 173, 175
Hubble
– constant   222
– law   222, 223, 388–390
– radius   228
hyper-uniform
– distribution   80, 83, 169

inflation   168
information theory   55, 69, 72
integral constraint   170, 177, 238, 242–245, 402
Ising model   59–62, 101

k-correction   266, 281, 284, 288, 289, 377, 381–384
Kardar-Parisi-Zhang model   6
Khinchin theorem   34, 41, 56, 75, 93, 173, 175

lacunarity   129, 134, 135, 138, 139
lattice   79, 80, 83, 85–89, 169, 193, 195, 196
Legendre transformation   149, 150, 158
Levy flight   66, 104, 211, 372
log-periodic oscillations   121, 123–126
long-range order   169
luminosity bias   230, 257, 259, 292, 299, 305, 308, 309, 313, 330
luminosity distance   375, 377, 378, 382
luminosity segregation   292
Lyon-Meudon extragalactic database   281, 282, 285, 287

magnitude
– absolute   224
– apparent   222, 224
Mandelbrot, B.B.   6, 7, 102, 103, 106, 132
mass density   260–263
mass function   61, 63
mass to luminosity ratio   260, 261
mass variance   41, 78
– normalized   42, 43, 47, 78, 79, 81–83, 85, 97
– Poisson   45
mass-length
– dimension   107
– relation   107, 119, 123–126
Mattig relation   379, 380
maximum entropy principle   70
metric dimension   102, 103, 106
mismatch galaxy-cluster   299, 304, 306, 307, 311, 313, 321, 323, 330
mixed dark matter   168
morphological segregation   296
multifractal   2–4, 102, 143, 145–148, 151, 154, 156–158, 163, 291, 293–297, 301
– analysis   143, 158, 159, 294
– deterministic   143, 145, 147
– dimensions   147–149, 294
– formalism   143, 159, 294, 301
– generator   147, 148, 151, 152, 161, 163
– measure   143–152, 160, 163
– measure distribution   159–161
– random   143, 144, 151, 152, 296
– spatial correlations   158
– spectrum   143, 149, 150, 154, 158, 159, 162, 291, 294, 301
multiscaling   294

N-body simulations   113, 193, 194, 197, 214, 335
– initial conditions   193, 195, 197, 199, 200, 209
nearest neighbor
– average distance   53, 54, 115
– distribution   52, 54, 113, 114
nearest neighbor force distribution   336, 337, 340, 348

one-component-plasma 10, 88–90, 98, 193, 195–198, 246
ordinary point process 32

phase transitions 5, 39, 58, 59, 80, 83, 101
pinwheel tiling 196, 197
Poisson
– anisotropic distribution 109, 113, 114
– distribution 28, 31, 44–47, 49, 50, 52, 53, 72, 170, 171, 173, 174, 244, 402
– equation 172
– law 46
– noise 47
– power spectrum 77, 198
– substantially 46
power spectrum 73–76, 78, 83
– definition 74–76
– estimator 411
– oscillations 183–185, 187
probability density functional 32

radial density profiles 109, 111–114
random trema dust 104, 372
random walk 64, 66, 68, 103, 211
redshift definition of 221, 222
regular point process 32
renormalization group 4, 5
Renyi dimensions 144
richness-clustering relation 304, 308

sandpile model 6
scale transformations 4, 5
scale-invariance 58–60, 72, 189
– discrete 121, 123
scale-invariant power spectrum 167, 168, 171, 179, 181, 189
scaling correction to 119, 120, 123, 124
Schechter function 221, 293
self-organization 3
self-similarity 5, 58–62, 72, 102, 108, 119–121, 124, 142
Shannon entropy 55, 69, 70
shot noise 47, 253, 411, 412
shuffled lattice 10, 196, 209–211, 217

Sierpinski carpet 120, 370
Sierpinski gasket 121–126
Sloan digital sky survey 225, 227, 228, 231, 235, 251, 252, 333
southern sky redshift survey 225, 226, 228–230
spherical harmonics 180
spin glasses 3
stationary point process 28, 31–35, 41, 44–50, 54, 71, 72
stationary stochastic process 27, 29–35, 40–43, 46, 55, 71, 72, 91
statistical
– homogeneity 33, 35
– isotropy 33, 35
– stationarity 31, 34, 35, 91
– translation invariance 31
structure factor 74
sub Poisson 80, 83
substantially Poisson 46, 78, 83, 97, 169, 236, 238, 244, 319, 325, 326, 332
super Poisson 80, 83, 169
super-homogeneity 43, 46, 314
super-homogeneous
– condition 80, 83, 90, 98, 196, 314
– distribution 9, 80, 83–85, 88, 89, 98, 169, 170, 179, 189, 196, 236–238, 245, 246, 325–327, 330

turbulence 6
two-point reduced correlation function
– bump 185, 187, 189
– estimator 240, 402

uniformity scale 28

volume limited samples 220, 225–227, 250, 251, 253–255, 257–260

white noise 77, 95, 316, 317
Wick theorem 57, 94
Wiener-Khinchin theorem 75, 93, 173, 175
window function 78, 79, 81–83, 411

Zeldovich approximation 195, 200